CROP SCIENCE
AND
FOOD PRODUCTION

CROP SCIENCE
AND
FOOD PRODUCTION

Douglas D. Bishop

Department of Agricultural and Industrial Education
Montana State University
Bozeman, Montana

Lark P. Carter

School of Agriculture and Natural Resources
California Polytechnic State University
San Luis Obispo, California

Stephen R. Chapman

College of Agricultural Sciences
Clemson University
Clemson, South Carolina

William F. Bennett

College of Agricultural Sciences
Texas Tech University
Lubbock, Texas

Gregg Division
McGraw-Hill Book Company

New York   Atlanta   Dallas   St. Louis   San Francisco
Auckland   Bogotá   Guatemala   Hamburg   Lisbon
London   Madrid   Mexico   Montreal   New Delhi   Panama   Paris
San Juan   São Paulo   Singapore   Sydney   Tokyo   Toronto

Sponsoring Editor   Roberta Moore
Editing Supervisors   Patricia Nolan and Sharon E. Kaufman
Design and Art Supervisor   Caryl Valerie Spinka
Production Supervisor   Laurence Charnow
Photo Editor   Mary Ann Drury

Text Designer   Sharkey Design
Cover Photographs   Joe Azzara, The Image Bank
Richard Megna and Kip Peticolas for Fundamental Photographs

Library of Congress Cataloging in Publication Data
Main entry under title:

Crop science and food production.

   Includes index.
   1. Agriculture.   2. Food crops.   3. Food
industry and trade.   I. Bishop, Douglas D.
SB91.C78   1982   631   82-13017
ISBN 0-07-005431-2

Crop Science and Food Production

   4 5 6 7 8 9 0   DOCDOC   8 9 0 9 8

ISBN 0-07-005431-2

CONTENTS

PREFACE

Crop Science and Food Production focuses on five major instructional areas in crop production: general principles of plant science in agriculture; production and management of crops; identification and control of crop pests; harvesting and storage of agricultural crops; and processing, marketing, and utilization of plant products. The text is designed for use by students interested in acquiring skills needed for successful employment in production agriculture and agribusiness. It can also be used as a supplemental text for nonagriculture students who want to gain some familiarity with general agricultural concepts and principles.

Crop Science and Food Production provides an innovative approach to the study of crop production by integrating the practical aspects of plowing, planting, cultivating, and harvesting of crops into a study of the total utilization of agricultural products from the field to the table. This approach provides students with the technical skills and knowledge needed for production and production-related occupations while, at the same time, providing them with a broader interpretation of how this knowledge can be utilized and applied to occupations in agricultural industry and agribusiness.

The *Crop Science and Food Production* program contains the following components.

■ Textbook: The textbook is divided into twenty-one chapters, beginning with an overview of how crop production fills the worldwide need for human nutrition and then focusing on the five major instructional areas listed above. Each chapter begins with a set of goals to indicate to the student specific skills that can be developed by carefully studying the material in the chapter. At the end of each chapter there are "Questions for Review," which provide an opportunity to review the major chapter concepts.

■ Activity Guide: The student activity guide is a consumable workbook with exercises for each of the twenty-one text chapters. Each chapter begins with a brief summary of the conceptual material for the corresponding chapter in the text and a review of the chapter goals. Important conceptual questions and technical vocabulary comprise the first two sections of the activity guide. A problem-solving exercise is included in each chapter. This exercise gives students the opportunity to solve a practical problem related to the chapter content. The next section is a set of laboratory or out-of-class activities. These activities are designed so that the student can put into practice some of the concepts learned in the chapter. At the end of each chapter, an "Achievement Chart" provides a tool for rating the student's ability to perform the chapter competencies.

■ Teacher's Manual and Key: The teacher's manual and key contains general suggestions for implementing the program; answers to the text end-of-chapter questions; answers to the activity guide exercises; and suggestions for handling the problem-solving and laboratory activities.

The overall goal of the program is to provide students with a foundation of skills for successful entry and advancement in both on-farm and off-farm agricultural occupations. For those students who are planning to enter a college or university program, this text will provide the basic knowledge necessary to go on to advanced study.

ACKNOWLEDGMENTS

The authors would like to extend their appreciation to the following people who reviewed the program during its development: Dr. Robert Terry, Head, Department of Agricultural Education, Oklahoma State University; Mr. Darrell Strain, Instructor, Alton R-4 Schools, Alton, Missouri; Mr. Clark Cleveland, Instructor, Hinsdale High School, Hinsdale, Montana; Mr. Thomas Neely, Instructor, Monterey High School, Lubbock, Texas; Dr. Vernon Luft, Professor, Agricultural Education Department, North Dakota State University; Mr. Robert Cone, Instructor, Salem High School, Salem, Illinois; and Mr. Jay L. Eudy, Instructor, Plainview High School, Plainview, Texas.

Douglas D. Bishop
Lark P. Carter
Stephen R. Chapman
William F. Bennett

Chapter
1

Human Nutritional Needs

In the United States, one of the most important benefits many of us seem to take for granted is a plentiful supply of wholesome food. Many people worldwide are either underfed (do not have enough food) or are poorly fed (have the wrong kind of food) or both. America's natural resources—soil, climate, water—have been developed and managed to provide the best and most diverse diet available to humankind. This is illustrated in Figure 1-1. American agriculture is responsible for the diverse diet we enjoy as well as for the development, management, and protection of the natural resources on which our diet, as well as the diets of many of the world's people, depends.

Production of crops and livestock is the foundation of American agriculture, but there is far more to our agricultural system than crop and livestock production. American farmers and ranchers are indeed very efficient. Less than 10 percent of our population produces the raw products on which our diets depend; each farmer or rancher produces enough to feed from 50 to 70 people. The vast majority of Americans can pursue varied careers, because they are assured that food will be available to them. It is this fact that has allowed our nation to become the world's leader in such fields as science, technology, medicine, and the arts.

The growing efficiency of human power in agricultural production is illustrated in Figure 1-2. At the same time, the magnitude of our total agricultural system is expressed in the fact that over 50 percent of the jobs in the United States are in some way linked with agriculture. Although the odds that a particular person will actually operate or work on a farm or ranch may be small, it is very likely that that person's employment will in some way relate to agriculture. Thus there are good practical reasons to understand basic agricultural principles. In addition, all humans have the same basic nutritional requirements, and agriculture in some form provides the food by which these requirements are satisfied. To understand human nutrition and nutritional needs, an aware-

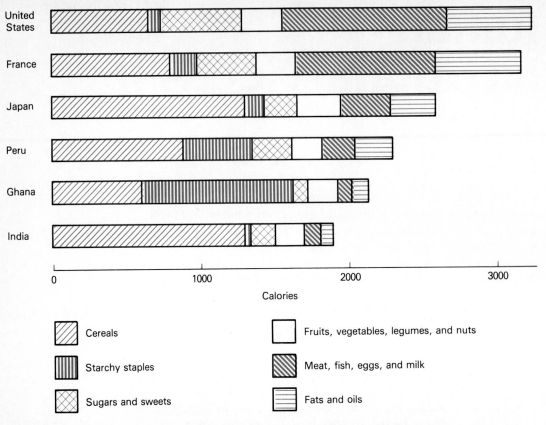

Figure 1-1. Caloric content and composition of diets in six countries. All the diets have essentially the same components, but the proportions vary. Less developed, heavily populated countries are highly dependent on cereals.

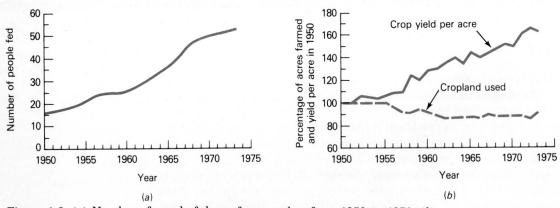

Figure 1-2. (*a*) Number of people fed per farm worker from 1950 to 1973. (*b*) Relationship between crop yield per acre and acres farmed for the same period. (*USDA data*)

ness of the fundamentals of agriculture is most beneficial, if not essential; to enjoy a career in some phase of this critical industry, such an understanding is essential.

CHAPTER GOALS

After studying this chapter you will be able to:

■ Describe human nutritional needs and explain how agriculture provides for them.
■ Name the components of a balanced diet and the specific nutritional needs they fill.
■ Understand why a balanced diet is necessary and how it can be obtained with a variety of foods.
■ Explain the crucial role plants play in providing food for humans.

PROBLEMS AND CHALLENGES

Many factors contribute to the world problems of undernutrition and malnutrition. Population growth is one. Every day the world's population increases by about 100,000 persons. Obviously, this means there are more people to be fed. In addition, with a growing population, more space is needed for housing, stores, highways, and similar features of our civilization. There is limited space on the earth, and not all of it is suitable for crop (food) production. Unfortunately, as the population grows, prime farmland is converted to housing developments and supermarkets. As a result, more food must be produced from each acre (or hectare, the metric unit of area).

Other problems confront agriculture in meeting the challenges of providing adequate food for the human race. Soil conservation, particularly the protection of soil from erosion caused by wind and water, is essential. Soil must be considered a basic resource on which crop production depends. Agriculturists have been, and must con-

tinue to be, leading conservationists. In addition to soil conservation, protection of water resources is critical, because irrigation is becoming increasingly important to increasing crop yields. Both streams and lakes, as well as underground (well) water, must be protected from pollution that might render them unfit for use. Agriculturists must also be concerned about other sources of pollution that might be caused by agricultural activities (pesticides or animal wastes) and pollution, such as smog, that can damage crops.

To meet the challenge of feeding humankind, agriculturists must also help reduce losses of food from pests and spoilage. This includes controlling or managing pests while crops are growing (insects, diseases, and weeds) and protecting the crop and food products following harvest. In some parts of the world, over 25 percent of each crop harvested is lost as a result of spoilage or contamination by insects or rodents such as rats and mice.

Looking to the immediate future, both the supply and cost of fuel will be even more serious concerns to farmers in industrialized nations such as the United States. To maintain our world-supporting agricultural productivity, mechanization is essential (Figure 1-3). Our sophisticated equipment is largely responsible for the fact that one farmer now produces enough to feed over fifty people. In addition, fertilizers will be more important as we strive to further increase yields, and fossil fuel is required to produce most nitrogen fertilizer. Farmers must be able to secure fuel both for equipment and for essential fertilizer. Remember, too, that in America, farming is a business; farmers must make a profit. All too frequently we see the profit fall to the point that the individual farmer is virtually forced to give up. The loss of the American family farm is a tragic loss to our nation.

You may feel that the future is dark indeed. To be certain, there are serious problems challenging us. Remember that, although the exact problem may have changed, the challenge of producing or securing food is as old as humankind. Our record of success is dramatic. In the United States, in addition to the development of the most sophisticated farming machinery and food-processing and packaging equipment, in the past 50 years the yields of corn and cotton have more than doubled; soybeans have changed from a questionable crop for horse feed to one of the most versatile and valuable crops we grow today; and many crop pests have been controlled or managed through a variety of discoveries—the development of cultivars of wheat that are resistant to rust diseases and of safe chemicals for the control of insects in vegetables, plus recent breakthroughs in the biological control of pests, are outstanding examples. Yes, there are real problems, but there are even greater opportunities for you and for everyone who is willing to obtain an adequate ed-

ucation in agriculture and the sciences on which it depends.

ENERGY AND FOOD

In the simplest sense, food provides energy for all the functions of the human body. From physics or other science classes, you know that *energy* is defined as the ability to do work. Energy exists in one of two states and in a multitude of forms. The two states are potential energy and kinetic energy. *Kinetic energy* is energy being used to do work, such as electric energy used to run an electric motor or the work done by the motor, such as turning a fan. *Potential energy* is like money in the bank; a source of potential energy may seem entirely without action, until certain conditions are encountered that enable the potential (inactive) energy to be converted to kinetic (active) energy. Gasoline is a good example: in the tank of an automobile it represents potential energy; and in the complex reactions in the engine, the potential energy becomes the kinetic energy that powers the car.

Both kinetic and potential energy are expressed in many common forms. Heat energy is a common form, and energy units are expressed in calories or kilocalories. A calorie (cal) is the energy required to raise the temperature of 1 gram (g) of water 1 degree Celsius (°C); a kilocalorie (kcal) (= 1000 cal) is the energy required to raise 1 kilogram (kg) of water 1°C. (In the metric system, units of energy are expressed in joules (J); this unit will gradually replace calories as our measure of energy.) Light is also a form of energy, as is motion, such as falling water or wind. In human nutrition, the most important form of energy is chemical energy; this is the energy that holds atoms together in complex molecules. When

(a)

(b)

Figure 1-3. (a) Modern combines roll through a wheat field, cutting and threshing in one operation. (*Doug Wilson/USDA*) (b) Rice harvesting being done by hand in a less industrialized society. (*United Nations*)

these molecules are broken down, the energy changes form and can be used for work by the body. Chemical energy is also expressed in the heat energy units, calories, kilocalories, or joules.

The actual utilization of food to provide energy for the body occurs in individual cells. However, before many kinds of food can be used, they must be converted to simple substances through the process of digestion, which occurs in the stomach and small intestines. From the intestines, digested food is carried throughout the body, where its potential energy is converted to kinetic energy for bodily work. Bodily work includes all types of activity, from vigorous exercise to blinking an eyelid and even thinking, as well as growth and the repair of injured tissues, from a cut finger to a broken leg.

The transformation of the energy in food from the potential to the kinetic state is extremely complex. The process has been compared to the process of burning wood, in which heat is given off as the chemical bonds in the wood are broken (as the potential energy in the chemical bonds becomes kinetic energy in the form of heat). In the cells of the body, this burning occurs in a series of complex steps, so that temperatures are closely and automatically controlled. This regulation is achieved by groups of specialized proteins called *enzymes*. In this sense, enzymes are biological catalysts. The process through which the digested food is converted to useful energy is called *respiration*. Note that the term *respiration* is also used to describe the simple act of breathing, but as used here, it is the breakdown of digested food to yield energy, carbon dioxide (CO_2), and water (H_2O). In a chemical equation, respiration would look like the equation for burning a log. As for any form of burning, oxygen (O_2) is also required.

$$\text{Food} + O_2 \rightarrow \text{energy} + H_2O + CO_2$$

HUMAN DIETARY NEEDS

When it comes to eating, there seem to be two kinds of people: some people eat to live, and other people live to eat. In the first group, you might say, are those people who "don't care what it is, so long as it's hot and plentiful"; the second group of people are the gourmets. Probably all of us fall into both categories at some time. Contrast the rushed breakfast when you are late for school to the holiday feasts of turkey and all the trimmings, or maybe a ham or a big roast. Regardless of how any of us might be classified, our nutritional needs are essentially the same, although the amount of food required to meet them may be very different and the way we satisfy these needs may also be very different.

The amount of energy your body needs depends on many factors. Three important ones are size, age, and amount of physical activity. Nutrition is an important consideration, and hospital dietitians play a major role in the successful recovery of many patients. As a guideline, the active adult requires 2000 to 3000 cal per day. A professional athlete may require over 5000, and a small person at a desk job under 2000. In order to lose weight, which seems to be a preoccupation with many Americans, the intake of calories must be less than the body's energy requirement. Thus effective dieting depends on knowing your energy needs, which in turn depend on many factors, and knowing the caloric content of the food you eat. Foods differ in the energy, or calories, a given amount contains. You must also remember that your body requires more than mere calories. Eating a balanced diet while trying to lose weight is critical. Sometimes people become ill when dieting, not because they are eating too little, but really because they have not properly planned and balanced their diets.

Respiration

Respiration occurs in individual cells. The process converts the energy in digested food to forms that can be readily used by cells for doing all kinds of work. Most commonly, the energy in the food is stored in a highly specialized chemical bond that attaches phosphorus (P) to a molecule of adenosine diphosphate (ADP), a molecule with two atoms of phosphorus. The addition of a third phosphorus atom converts the adenosine diphosphate (two phosphates) to a form with three atoms of phosphorus, adenosine triphosphate (ATP). Cells use the energy in the bond holding this third phosphorus atom to the adenosine molecule for work; the energy is readily available, although the exact way the energy is held and released is not clearly understood by scientists.

Complete respiration of food energy is much like the complete burning of a log. Food and oxygen are used, and energy (heat), water, and carbon dioxide are given off. Whether we talk about burning or respiration, the processes in a chemical sense are both oxidation, which is either the addition of oxygen to a material or the removal of hydrogen from a material. In the cell, the temperature of the burning process is regulated in respiration by very specific protein molecules. For complete oxidation of the food, two processes must occur. First, the food from digestion is oxidized through a series of chemical reactions known as *glycolysis*. Next, the products from glycolysis go through a series of steps known as the *citric acid cycle* and *terminal transport* to yield energy (ATP), carbon dioxide, and water. Oxygen is required at the last step, terminal transport, to accept the atoms of hydrogen that are removed from the food molecules to release energy. Note that two ions (charged atoms of hydrogen) combine with an atom of oxygen to form water. If the hydrogen did

not unite with the oxygen or some other substance, the cell would become too acid and be destroyed. Where oxygen is available and the end results are those we have described, the process is referred to as *aerobic respiration*.

When adequate oxygen is not available, cells do not complete the full sequence of glycolysis, the citric acid cycle, and terminal transport. Following glycolysis, cells go through an anaerobic pattern of respiration. This yields less ATP, and the food is not broken down fully to carbon dioxide and water. Instead, the products may be various types of alcohol and organic acids. The process giving alcohol is fermentation, which is very important in manufacturing many beverages and, as you will see, is critical in making silage.

There is little the farmer can do to change these basic cellular processes, but understanding them will assist you in knowing how to manage crops and crop products, from flooded fields to silage to storing seeds and/or shipping fruits and vegetables.

Types of Food

Carbohydrate is the most common form of food used by the body. There are many forms of carbohydrate, but all share several features in common. First, all carbohydrates are composed of the same three chemical elements, carbon (C), hydrogen (H), and oxygen (O); second, all carbohydrates have the same proportions of these elements—one carbon atom, two hydrogen atoms, and one oxygen atom (CH_2O). The two most common carbohydrates are sugars and starches.

Between 45 and 50 percent of the food energy consumed by the average person is in the form of carbohydrates. The government has not established a recommended dietary allowance (RDA) for carbohydrates. However, to provide energy, between 300 and 400 g

(a) (b)

Figure 1-4. Complex carbohydrates like those found in potatoes are believed to be healthier than simple sugar forms like table sugar. (*a*) Barrels of freshly harvested potatoes. (*USDA*) (*b*) Field of sugar cane from which table sugar is derived. (*Hawaii Visitors Bureau*)

of digestible carbohydrate is needed. Some nutritionists feel that a large portion of this should be in the form of complex carbohydrates, such as starch, rather than in the simpler sugar forms (Figure 1-4).

There are many different sugars; they differ in their chemical makeup and in how sweet they taste. A common simple sugar is glucose ($C_6H_{12}O_6$). This simple equation reads as follows: a molecule made up of six atoms of carbon (C_6), 12 atoms of hydrogen (H_{12}), and six atoms of oxygen (O_6). This is the typical ratio of carbon, hydrogen, and oxygen in a carbohydrate, CH_2O, or 1:2:1. (Note that a ratio of 6:12:6 is the same as a 1:2:1 ratio.) Through digestive processes, many carbohydrates are converted to (or simplified to) glucose, which is respired to yield energy. Table sugar, sucrose, comes

from both sugar beets and sugarcane. It is a disaccharide, made up of two simple sugars.

Starch is much like sucrose. Rather than being composed of only two sugar molecules, molecules of starch are composed of many molecules of simple sugars (such as glucose) and are classified as polysaccharides. They still have the common trait of being composed only of carbon, hydrogen, and oxygen in the ratio of 1:2:1. Like sugar, plants also provide virtually all the dietary starch. Cereal grains or flour and potatoes are excellent common sources.

Fats represent another high-energy portion of the diet. Fats are composed of carbon, hydrogen, and oxygen, just as carbohydrates are, but gram for gram, fats provide more energy than carbohydrates. In fact, a gram of pure fat provides about $2\frac{1}{2}$ times more energy than a gram of pure starch. Obviously, on a diet to lose weight, you could eat more starch than fat. Generally speaking, fats are provided to the body through animal products—actually fat in meat, fats in milk and milk products, and fat in some cooking materials. Some fats can be provided in the form of vegetable oils.

Proteins can serve as a source of energy, like carbohydrates and fats. In fact, in so-called fad diets the emphasis is on eating protein to the virtual exclusion of carbohydrates and fats. This is not generally recommended, because there may be dangerous buildups of truly toxic by-products of the respiration of proteins. However, the quality of the diet is determined in large part by the amount and kind of proteins it contains.

The major significance of proteins in the diet is the amino acids they supply. Amino acids are composed of nitrogen, carbon, hydrogen, oxygen, and sometimes other elements, such as sulfur. All proteins are composed of different types and arrangements of amino acids. Usually twenty different amino acids are found in humans. Of these, twelve can be made from other elements, but eight must be supplied from proteins. These eight amino acids are known as the essential amino acids. For adult humans, they are threonine, tryptophan, leucine, lysine, isoleucine, methionine, valine, and phenylalanine. Young children also require a supply of the amino acids histidine and arginine.

The body builds the proteins it needs from the amino acids—the twelve nonessential and eight essential ones listed above. The various proteins the body builds serve many functions. In many ways they regulate body activities: enzymes are proteins that regulate specific functions.

Protein quality involves two concepts. A complete protein contains all eight essential amino acids. In addition, the proportions or ratio of the amino acids in the protein is

TABLE 1-1 ANIMAL AND PLANT SOURCES OF PROTEIN AND THEIR RELATIVE VALUE IN HUMAN NUTRITION

Food	Relative Protein Score
Eggs	100
Milk	78
Casein	80
Beef	83
Fish	80
Oats	79
Rice	72
Cornmeal	42
White flour	47
Wheat germ	61
Soy flour	73
Potatoes	56
Peas	58

quite like the ratio humans require. Poor-quality protein lacks one or more of the eight essential amino acids, or the ratio of the amino acids is far different from that required by humans. The relative nutritional value of proteins from several sources is given in Table 1-1.

The deficiency of an amino acid in the diet can be very important. When the human body is manufacturing its own protein (using the amino acids from food), deficiency or lack of an amino acid required in the protein stops the building process. Moreover, the other amino acids are wasted; the body does not store them for use in the manufacture of other proteins (Figure 1-5).

Generally speaking, animal products are better sources of all amino acids for the human diet than are plant products. Meat,

Figure 1-5. The corn and bean diet of Central America is an example of complimentation. This particular combination of a legume and a cereal provides adequate amounts of all essential amino acids. (*Barry L. Runk-Grant Heilman*)

poultry, and some dairy products are excellent sources of essential amino acids. Few plant products provide all needed amino acids in a reasonable proportion and amount. For example, rice, a major food crop, has less than 3 percent protein. For an average person to obtain an adequate supply of amino acids from protein in rice, up to 50 pounds (lb) [23 kg] of rice per day would have to be consumed; this would be far too many calories.

You may wonder how vegetarians survive and are healthy. Plants differ widely in their amino acid content. Usually plants in the legume (pea) family (beans, soybeans, peanuts) are better sources of protein than cereals (wheat, corn, etc). However, certain combinations of legumes and cereals can provide all the essential amino acids in adequate amounts. For example, the corn and bean diet of Central America includes a reasonable supply of essential amino acids, as does the rice and soybean diet of many parts of Asia and the Orient. This illustrates an important dietary concept, complementation: two or more food sources combining to satisfy a particular dietary need.

Nutritional deficiencies in humans are difficult to study because experimentation may be dangerous. Certain experimental animals have provided good guidelines. Unfortunately, tragic examples of two nutritional deficiencies have become widely known in recent years. Marasmus is a disease of infants (age 1 year or less) caused jointly by undernutrition (too few calories) and inadequate protein. Kwashiorkor is a disease of children over 1 year of age. It is caused by inadequate or low-quality protein. This disease, only recently discovered, is the most common and severe nutritional problem. It is widespread in parts of South America, Africa, and Asia. The disease leads to wasting of limbs, loss of hearing, mental apathy and retardation, and death. Of course, poor nutrition opens the path for many other disease problems as well.

Vitamins. These organic substances are essential for health. They are not a source of energy, but they serve to regulate or aid many complex body functions. Vitamins are provided through foods and are required in very small amounts [1 milligram (mg)/day or less]. No one food contains all the necessary vitamins (Figure 1-6).

There are thirteen vitamins known to be essential for good health. Of these, nine are classified as water-soluble and four as fat-soluble. The water-soluble vitamins include thiamine, riboflavin, niacin, pyridoxine, cobalamin, folic acid, biotin, pantothenic acid, and ascorbic acid. The fat-soluble vitamins include vitamins A, D, E, and K.

Thiamine, or vitamin B_1, prevents beriberi, a serious disease. This vitamin is found in the outer layers of wheat and rice, and vi-

Figure 1-6. The vitamin content of food is reduced in processing and cooking. (*a*) Excess leaves, stems, and damaged portions of broccoli are trimmed away in preparation for packaging at a food processing plant. (*Larry Rana/USDA*) (*b*) Trimmed broccoli is carried through a blancher where a 4-minute spray of steam preserves its bright green color before freezing. (*Larry Rana/USDA*) (*c*) Cartons of broccoli are shown on their way to the wrapping machine where they will be wrapped, sealed, and placed on trays for freezing. (*Larry Rana/USDA*)

tamin B_1 deficiencies have been associated with increased use of refined grain products (flour) without supplemental B_1. Meat, fish, and poultry are good sources of vitamin B_1; milk also provides some.

Riboflavin, or vitamin B_2, has been known as an essential substance since 1917. Fortunately, vitamin B_2 is available in grains (although milling reduces vitamin B_2 content by about 60 percent), in many vegetables, and in animal products. This vitamin tends to be quite stable and is not lost in food preparation as much as others.

Niacin, or nicotinic acid, is a recognized antiberiberi compound. Increasing use of milled rice has led to an increase of niacin deficiency, because this essential vitamin is found in the outer layers of cereal grains. Niacin is also found in legumes (peanuts), poultry, and meat, especially liver.

Pyridoxine, or vitamin B_6, was identified as an essential vitamin in 1934. Muscle meats, liver, vegetables, and whole-grain cereals are the best sources. Freezing these foods can reduce the vitamin B_6 content by up to 25 percent.

Cobalamin, or vitamin B_{12}, is a vitamin obtained only through animal products. Folic acid, or folacin, is provided through green vegetables, as well as other plant and animal products. Biotin and pantothenic acid are other B-complex vitamins, available in many foods. Rarely are human diets deficient in these vitamins.

Ascorbic acid, or vitamin C, is known to prevent scurvy. For over 400 years, it has been known that this disease can be controlled by diet. Most plant products provide some vitamin C, and citrus fruits or juices —orange, lime, or grapefruit—especially are excellent sources. All mammals except humans, guinea pigs, and monkeys, can manufacture vitamin C.

Vitamin A is a well-known fat-soluble vitamin. It is stored in the liver, thus temporary deficiencies may not cause problems. Vitamin A, among other qualities, appears to be an anti-infection substance. Deficiencies tend to lead to respiratory ailments. Dairy products, eggs, and fruits and vegetables are good sources of vitamin A.

Vitamin D is essential for healthy bone development (the prevention of rickets) in children. Sunshine aids the body in the production of this substance. Dietary sources include milk, butter, eggs, and fish oils, such as cod-liver oil.

Vitamin E is provided through vegetable oils. It is an antisterility substance, hence important in reproduction.

Vitamin K is provided through green leafy vegetables. This vitamin plays an important role in the time it takes blood to clot, or coagulate.

Minerals. In addition to the organic compounds discussed in the preceding pages (carbohydrates, fats, proteins, and vitamins), humans also require certain specific *minerals,* or salts. There are twenty-one essential minerals. Of these, seven, needed in relatively large amounts, are called *macronutrients.* The other fourteen, required in lesser amounts, are called *micronutrients.* The macronutrients include calcium (Ca), phosphorus (P), potassium (K), sulfur (S), sodium (Na), chlorine (Cl), and magnesium (Mg). The best example of the need for these is calcium for healthy teeth and bones. The micronutrients are iron (Fe), zinc (Zn), selenium (Se), manganese (Mn), copper (Cu), iodine (I), molybdenum (Mo), cobalt (Co), chromium (Cr), fluorine (F), silicon (Si), vanadium (Va), nickel (Ni), and tin

(Sn). Although both plants and animals provide minerals, generally minerals provided by animal products are more readily absorbed by humans.

THE AMERICAN DIET

American agriculture produces large quantities of nearly all types of foods, with the exception of coffee and some spices (which in fact are not food). American farmers and ranchers produce raw products to provide a nearly endless variety of food products for a balanced diet. This is well illustrated in the fact that a large supermarket commonly has up to 10,000 different food items, from fresh fruits and vegetables to frozen pizza. In spite of this fact, the quality of the American diet is declining. This must be in part a matter of choice, or ignorance.

In 1955, the diet of 60 percent of the Americans was classified as good; that of 15 percent was classified as poor. In 1970, just 15 years later, the corresponding figures were 50 and 20 percent. Part of this decline is the result of widespread use of convenience and snack foods (Figure 1-7). Up to 15 percent of the calories in the average teenager's diet comes from snacks. Snacks are consumed at the rate of 50 lb [23 kg] per person per year! Surveys of food consumed in school lunch programs support the reality of poor diets: daily, 9.4 percent of the entree, 33 percent of the fruit, and 50 percent of the vegetables are wasted. This, unfortunately, illustrates how few people truly understand the importance of good nutrition.

Changes in the eating habits of Americans are shown in Table 1-2. We eat more meat (a 31 percent increase from 1950 to 1972)—an amount that provides more than three times the required dietary protein. The fact that our agriculture can afford to

Figure 1-7. In spite of the fact that a large supermarket commonly has up to 10,000 different food items, the quality of the American diet is declining. (*Jane Hamilton-Merritt*)

TABLE 1-2 CHANGING AMERICAN EATING HABITS: ESTIMATED AMOUNTS OF FOODS CONSUMED

Food	Pounds per Person per Year		Percent Change
	1950	1980	
Meat	145	190	+31
Dairy	466	300	−37
Chicken/turkey	25	50	+100
Fruits:			
Fresh	109	80	−29
Total processed	44	48	+11
frozen	4.3	10.6	+142
dried	4.1	1.8	−56
Vegetables:			
Fresh	115	96	−15
Frozen	3.2	10	+200
Cereals	181	150	−17

produce so much meat and animal products (e.g., poultry and dairy products) shows the wealth of resources we have as a nation. We consume more poultry products; recently this could be due in part to economic factors. Consumption of dairy products has declined. Consumption of fresh fruits and vegetables has declined in general, and far more are consumed in frozen forms, which can suffer nutritional losses.

There are areas of nutritional deficiency diseases in the United States. Fortunately, in spite of some poor eating habits, these are not widespread. Because as a nation we tend to overeat, we get enough protein, vitamins, and minerals, along with extra calories. Also, many foods that may be low in certain essential nutrients, commonly vitamins and minerals, are fortified to provide better nutrition. Milk, flour, bread, breakfast cereals, and many fruit juices are supplemented. Look at the label on a box or wrapper.

As you think about human nutritional needs, remember, too, that all animals and plants have specific nutritional requirements. Plants differ from animals in that they require inorganic salts, or minerals, and they manufacture food through the intricate processes of photosynthesis. All animals depend on plants for food, but in many ways, humans are not *directly* dependent on plants; many foods, e.g., proteins, come from animals—meat and milk. You must remember that the lower animals require plants directly as food. For your own health and well-being, understanding and meeting minimal nutritional requirements is really wise. If you produce any livestock, understanding their nutritional needs is important for economical, efficient production.

SUMMARY

Food is the source of energy for human life; it provides the calories to sustain life. However, a variety of foods are required to provide nutrition other than calories, including vitamins and minerals. Proteins are also necessary to provide amino acids, many of which the human body cannot build. Although the amount of energy a person requires varies with age, sex, and physical activities, all require the same type of balanced diet. A balanced diet provides calories from carbohydrates and fats, amino acids (and calories) from proteins, and vitamins and minerals, which are essential for many metabolic processes. Green plants can provide a complete diet for humans; and they are also used as livestock feed and are thus converted to meat, poultry, and dairy products.

QUESTIONS FOR REVIEW

1. What are the differences between undernutrition and poor nutrition? Does one exclude the other?
2. List and discuss five major problems and challenges facing American agriculture today.

3. What are the two common forms of energy? With examples, explain how they differ.
4. Define a calorie.
5. Define and describe the process of respiration.
6. Why is ATP important?
7. What happens in respiration if an adequate supply of oxygen is not available?
8. With appropriate examples, name the five major types of food and explain the role each plays in good nutrition.
9. What is an essential amino acid? Give an example.
10. Have "convenience foods" improved or harmed the nutrition of Americans? Justify your answer with examples.

Chapter

2

Food Production: Growing New Plants

If the nutritional needs of humans all over this world are to be met, it is essential that plants grow and reproduce. Humans are dependent on plants not only for food but also for shelter, fiber, and feed for domestic animals. The rapidly increasing population of the world and the loss of prime farmland to highways, airports, industrial development, and urban sprawl amplifies the need for greater productivity from every field in which plants are grown. The need to maintain a healthy place in which to live is well recognized. If the general public is informed on plant growth and reproduction, better judgments can be made relating to the preservation and protection of the environment and the management of the natural resources needed to produce plants. The understanding of plant growth and reproduction can begin with a look at the life cycle.

CHAPTER GOALS

After studying this chapter you will be able to:

- Describe the concept of the life cycle of plants.
- Identify the parts of seeds and how germination occurs.
- Determine the criteria to use in evaluating seed quality.
- Describe seed cleaning and treating and inoculation practices and procedures.
- Distinguish between sexual and asexual methods of plant reproduction.

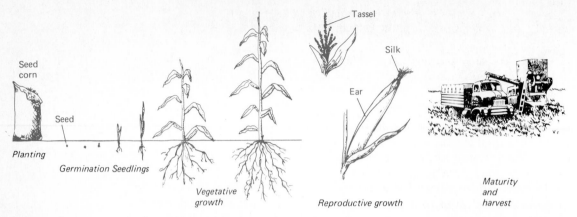

Figure 2-1. The complete life cycle of a plant.

THE LIFE CYCLE OF PLANTS

All organisms, including humans, pass through a definite life cycle. The stages of the human life cycle are easy to identify: birth, survival and growth, adulthood, reproduction, old age, and death. Plants have a similar life cycle, but the stages differ in form and length.

In general terms, the complete life cycle in plants refers to the stages that occur from the time a seed is planted until a new seed is produced (Figure 2-1). This series of events includes the planting of the seed, germination, emergence of the seedling, vegetative growth, maturity, reproduction, production of seed, and death. This general life cycle does not describe all the ways of regenerating new plants, but it gives a good basis for grasping the general concept of the stages that living plants go through and how the various species survive through time.

Modifications of the life cycle are introduced by varying characteristics of plant species. For example, plants differ in the length of time required to complete the life cycle. Annuals follow the general life cycle very closely and complete it in one year or one growing season. Winter annuals are planted in the late summer or fall and complete their life cycle the following spring or summer. Biennials are plants that generally produce only vegetative growth the first year and reproduce and complete their life cycle the second year. Perennial plants may produce seeds each year for several years and live for an indefinite period of time.

Another modification of the general life cycle can be illustrated with vegetative or asexual reproduction. *Asexual reproduction* is the propagation of plants from vegetative parts, such as leaves or stems, or from modified stems such as stolons, rhizomes, bulbs, corms, or tubers. These modifications all point to the same objective: the starting of new plants and survival of the species.

GROWING NEW PLANTS FROM SEEDS

Seeds may be described in many ways. They serve as food for humans and other animals, as well as other living organisms. They serve as the raw material needed to produce a whole spectrum of products useful to humans: everything from the vehicle or carrier for the pigment in paint, to lubricants, soap and cosmetics, and ingredients for perfume, plastics, and medicines. However, above all other purposes and uses, seeds permit the continuation and survival of their species.

Although viable seeds do not show visible signs of life in the usual sense, such as green leaves and increasing size and number of cells, they are very much alive. The formation of the seed actually begins in the parent plant as a single cell called a *zygote* that results from the union of a male and female gamete during sexual fertilization. A number of mitotic divisions occur along with growth and differentiation of cells to form the various parts of the seed. When the seed is mature, cell division and growth of cells virtually come to a standstill. Seeds, then, are living young plants existing in an arrested state of growth. The seed is made up of (1) an *embryo;* (2) a supply of stored food, which is called the *endosperm* in monocots and the *cotyledon* in dicots; and (3) a *seed coat* (Figure 2-2). Each part of the seed performs an important function related to survival of the plant species. The embryo is a tiny living plant within the seed that will germinate and grow to produce a new plant. The supply of food found in either the endosperm or the cotyledon nourishes the young plant from the time it germinates until photosynthesis begins and the plant can produce its own food. The seed coat contains and protects the embryo and the supply of food.

Seed is formed in a fruit. Sometimes the entire fruit is treated as a seed, e.g., a kernel of wheat, and sometimes the seed is removed from the fruit, e.g., a pea (seed) removed from the pod (the fruit).

Seeds respire at a very slow rate, and this respiration provides the energy for them to sustain life. They must be carefully harvested, handled, and stored to maintain their viability. They must have air. They

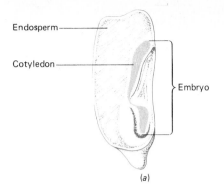

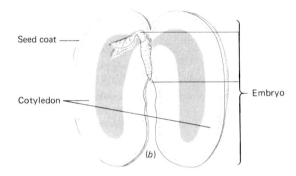

Figure 2-2. The cross sections of (a) a corn seed and (b) a bean show the differences in how food is stored. The endosperm stores food for seed of grasses including corn. Cotyledons supply food storage for nongrasses and are part of the embryo.

need to be stored in a dry place and should not be exposed to extreme temperatures. Seeds must be protected from insect pests and rodents.

Germination

The renewal of a rapid rate of cell division in the embryo and growth and differentiation of cells into the various plant tissues is called *germination.* Several conditions must be met before germination will occur. The seed must be alive. A favorable temperature is required, and oxygen must be available. Water must be available in sufficient quantity and for the period of time required for establishment of the new plant. In some species, light is required for germination.

Assuming that these conditions have been met, the germination process begins with the absorption of water and oxygen by the seed. The rate of respiration increases. During the germination process, the food stored in seeds is chemically changed. An enzyme called *diastase* is produced by the embryo. This enzyme changes starch to sugar, which is translocated to the growing points. This accounts for the sweetness of seed sprouts. Cell division and cell enlargement begin, and different types of cells develop to form the roots, stems, and leaves of the new seedling. The new plant continues to absorb water and nutrients from the soil solution. The roots become embedded in the soil, and shoots emerge from the soil. Up to this point the new plant has been dependent on the food stored in the seed for energy. As leaves form and photosynthesis begins, the plant begins to produce its own source of food for further growth and development. If one stops to think about it, this is a truly miraculous process that we often take for granted.

Evaluating Seed Quality

High-quality seed is an important factor in producing high-quality crops, therefore seed should always be purchased from a reliable dealer. High-quality seed implies seed of known ancestry. This seed will be genetically pure, and you can be confident that you are getting the variety you want and that it is free of weed seeds and of mixtures from other species.

Another important criterion for evaluating quality of seed is the percentage of live seeds that will germinate. Often bargain-priced seed will be low in germination percentage and thus of low quality.

Quality seed refers to purity and germination of the seed. Purity is expressed on the basis of percentage by weight of pure seed, without varietal mixtures, other crop seeds, weed seeds, and inert material such as straw, chaff, leaves, dirt, or small stones. Germination is expressed as the percentage of the pure seeds that are alive. This is determined by carefully controlled germination tests in seed-germination laboratories.

Labeling laws in each state require that every lot of seed that is sold commercially be labeled properly. These labeling laws differ from state to state, but in all cases the purity and germination percentages as determined by a seed-testing laboratory must be included on the label.

Seed should also be examined to assure that it is free from damage caused by insect and disease pests. Seed is sometimes damaged during harvesting and processing, and the buyer should examine for possible mechanical damage. Good quality is also indicated by plump, full-size seeds.

For most food and fiber crops, many varieties, or cultivars, are available. The term *cultivar* refers to a cultivated type or *variety* within a plant species that differs in some respect from the rest of the species. Cultivar is the new term accepted in agronomy; the term variety is still used in the field and, more generally, in horticulture. In this book the term cultivar is used.

In each state there is an agricultural experiment station with scientists who are able to evaluate the cultivars and provide information regarding their adaptability to a specific environment and their growth and quality characteristics. Often brochures are available that provide information that is useful to the producer or the agribusiness person.

Certification of Seed. The benefit of planting high-quality seed has been recognized for many years by both producers and people in the agribusiness community. To assure sources of high-quality seed, each state has established its own certifying agency. These agencies are usually called seed or crop improvement associations—e.g., the Iowa Crop Improvement Association. These associations are made up of voting members who are farmers and commercial firms that produce certified seed. The association itself does not sell seed. Other people or firms who are interested in promoting the general welfare of the association can hold associate (nonvoting) memberships. The association is generally designated by state law as the official seed-certifying agency for that state (Figure 2-3).

The reason seed certification has been established is to maintain sources of high-quality seed and plant-propagating materials of superior cultivars and to make them available to the public. The certifying agencies work with agronomists and plant breeders who develop and evaluate new and improved cultivars. They closely monitor the production and distribution of this seed, meeting rigid inspection and handling requirements.

The usually recognized classes of certified seed are:

1. *Breeders seed* This class of seed is produced under the direct supervision and control of the plant breeder and is not available for sale to the general public. It comprises a limited amount of seed coming from the breeding and selection program of the plant breeder to maintain a genetically pure source of a strain or variety. Breeders seed is increased to produce foundation seed.

2. *Foundation seed* This class of seed is the progeny of breeder seed that has been carefully produced to maintain genetic identity and purity. The certifying agency designates the breeders eligible to grow foundation seed and the conditions that must be met in the production and

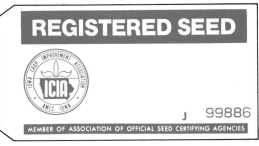

Figure 2-3. Seed certification labels for foundation, registered, and certified seed from the Iowa Crop Improvement Association.

distribution of this seed. Foundation seed is labeled with a white tag issued by the official certifying agency.

3. *Registered seed* The progeny of foundation seed is designated *registered seed*. This seed is produced and handled as designated by the certifying agency to maintain satisfactory genetic identity and purity. Registered seed bears a purple tag issued by the official certifying agency.

4. *Certified seed* This most widely available class of seed is the progeny of regis-

tered seed. Blue-tagged certified seed must also meet high standards of genetic identity and purity.

Cleaning Seed. When seed is harvested, the operator makes every effort to make adjustments in the harvesting equipment so that the seed is clean and free from mixtures, weed seeds, and inert material. However, even the best harvesting equipment, operated by a careful, experienced person, provides seed that must be processed further to have the desired purity.

The contaminants that are usually found in seed include materials that are relatively near to the size and shape of the desired seed. However, there are many differences in seed size, shape, weight, texture, and color which permit mechanical devices to be used to separate out the undesired materials.

Producers can have their seed cleaned by a commercial seed processor which will clean the seed at an established price per unit of seed. Laws in most states require that such processors operate in a way that ensures thorough cleaning without introducing mixtures from other sources.

Small-scale seed cleaners are also available for use on the farm. The fanning mill is an example of a farm-scale seed cleaner which separates seeds from foreign material and also can separate seeds of different sizes and weight. It uses screens and moving air to make these separations (Figure 2-4).

Other seed-cleaning devices are available that use differences in specific gravity, length, width, and resistance to airflow to make difficult separations. A conscientious, experienced person can clean almost any undesired material from the seed by selecting the right combination of cleaning devices.

Cleaning seed properly can be expensive, but experienced producers consider it a good investment. It costs much less to remove the

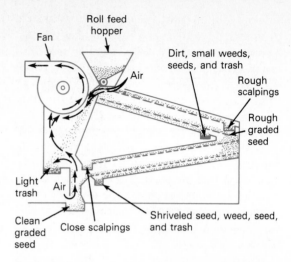

Figure 2-4. The fanning mill, an air-screen seed cleaner, is schematically illustrated to show how impurities are separated from the desired seed. (*USDA-ARS Handbook 179*)

weed seeds prior to planting than to control the weeds that result later. Just removing the weed seeds will more than pay the cost of seed cleaning in most cases. Additional benefits derived include freedom from plugging of the feed mechanism in planters and a more uniform stand as a result of even distribution of seeds. Also, crop mixtures can lower the quality and value of the crop being grown.

Disease-free Seed and Propagating Material. Plant diseases cause vast losses in crop production and income each year. A number of diseases are seedborne, with the organism either on the seed coat or inside the seed. To assure disease-free seed and propagating material, the seeds and propagating material are often produced in areas where seedborne and soilborne diseases do not occur. They are also produced from plants which have been certified to be free of disease. For example, potatoes produced for propagation are often grown in areas isolated from the large commercial potato-producing regions. These plants are carefully inspected while being grown, and leaf samples are tested for the possible presence of viral or other diseases. Fields that are found to be free of disease are certified, and the producer has assurance that this is high-quality, disease-free seed.

In some states, barley seed produced as certified seed must be tested for barley stripe mosaic virus. In these cases, buying certified seed assures the purchaser that the seed is free of this serious, yield-reducing disease.

Use of disease-free seed is a way of reducing the risk of having serious losses from certain diseases. This method of disease prevention does not cover all diseases, but disease-free seed should be used whenever it is available.

Seed Treatment. Plant *pathogens,* or organisms causing disease, may be found in the soil as well as in or on the seed. Environmental conditions in the soil that are favorable for seed germination are often favorable for the reproduction and growth of undesirable organisms as well.

As with humans and many animals, the early stages of life and infancy are the periods in the life cycle when plants are most susceptible to the detrimental effects of attack by disease organisms. If the seed and the newly emerging seedling can be protected from invasion by these organisms during similar crucial stages, the chances of having good, healthy plants are much improved.

Seed treatment refers to the coating of seed or propagating materials with specific chemicals that prevent or control the reproduction and/or growth of pathogens. Not all lots of seed and propagating materials are infected with disease organisms, nor are plant pathogens found in all soils, but producers generally consider seed treatment to be relatively inexpensive insurance against

Figure 2-5. Commercial seed treating elevators. *(Gustafson, Inc.)*

the losses that can occur from attack by these organisms (Figure 2-5).

The chemicals used for treating seed must be approved by federal agencies. Examples of chemicals that can be used to treat cereal-crop seed are hexachlorobenzene (HCB), captan, carboxin and thiran (Vitavax 200), thiobendazole (Mertect), and pentachloronitrobenzene (PCNB).

Propagating materials such as potato seed pieces may be treated to kill the pathogenic organisms on the surface. The seed pieces may be dipped in a 10% formaldehyde solution, then placed on a clean surface and covered with sacks overnight to allow time for penetration. The seed pieces should be planted the next day. It is also possible to treat seed potato pieces with a fungicide such as Mertect or captan and an antibiotic

Food Production: Growing New Plants

such as aureomycin or streptomycin prior to planting.

Most of the compounds used to treat seed are toxic to people and livestock as well as to the organisms being controlled, so proper precautions must be taken. First and foremost, one must read the directions regarding application rates, time and frequency of application, and hazards to the handler. All seeds and plant surfaces should be covered to give proper protection, and only the seed that is to be planted should be treated. Treated seed cannot be sold in the regular grain market where it could be used for human food or livestock feed.

Other Protective Practices. In some cases it may be economically feasible to fumigate the soil prior to planting to control organisms such as damping-off fungi or to kill nematodes. Fumigation can kill many weeds also. Fumigation is usually practical only with high-value crops or for plant-producing beds or in gardens.

Some plant pathogens survive from year to year on decomposing plant materials or in soils. It is possible for populations of these organisms to build up if the same plant species is grown in the same location year after year. This buildup can be discouraged by rotating the species, using unrelated crops and allowing an interval of 2 years or more to elapse before planting the same species again in the same place.

Some soilborne pathogens thrive in waterlogged soil. By good drainage, the activity of these pathogens can be curtailed.

Diseases can be transferred on plant refuse that is diseased. Proper sanitation practices, removal of diseased plant parts, and taking care not to carry diseased material from one field to another on equipment, tools, or clothing will improve a producer's chances of establishing and maintaining a good plant population in the field.

Inoculation. A unique class of bacteria are able to convert gaseous atmospheric nitrogen to a form that is available to plants. Bacteria of the genus *Rhizobium* are the best-known organisms with this capability. They are able to coexist with plants of the family Leguminosae, which includes high-protein crops such as field peas, beans, peanuts, alfalfa, and the clovers. The relationship between these bacteria and the legume plant is mutually beneficial, with the plant furnishing sugars and other compounds needed by the bacteria and the bacteria converting atmospheric nitrogen into a form available to plants. This relationship is referred to as *symbiosis*.

Effective strains of these nitrogen-fixing bacteria are not always present naturally in the soil. Cultures containing these organisms can be introduced by adding them to the seed before planting. This practice, called *inoculation,* can result in a significant increase in the amount of nitrogen available for crop growth at a low investment cost.

It is known that different strains of *Rhizobium* are required with different legume species. The different strains of bacteria are produced for sale by commercial firms. Cultures containing the strains specific to each legume species or group of species are offered for sale from most seed stores. The inoculation groups are shown in Table 2-1.

The inoculum can be applied to the seed in a mixture of culture and water called a *slurry,* by mixing the culture with peat or clay and stirring it onto the seed or by wetting the seed slightly and sprinkling the culture onto the seed and mixing it. The important thing is to get some inoculum on each seed and to plant the seed within 2 days to assure that the organisms are still alive at planting time. Seed that has been inoculated should not be treated with chemicals for disease control, since the chemical would, in

TABLE 2-1 CROSS-INOCULATION GROUPS FOR MAJOR LEGUMES

Group	Legumes Included in Group
Alfalfa	Alfalfa, sweetclover, bur clovers, spotted clover, black medic, and sourclover
Clover	Red, crimson, alsike, white, strawberry, subterranean, zigzag, and hop clovers
Pea	Peas, vetch, lentils
Bean	Garden, navy, great northern, pinto, and other beans
Soybean	All cultivars of soybeans
Cowpea	Cowpea, lespedeza, peanuts, and some beans
Lupine	Lupine
Birdsfoot trefoil	Birdsfoot trefoil
Big trefoil	Big trefoil
Sainfoin	Sainfoin
Vetch	Milk and crown vetches

most cases, kill the desired nitrogen-fixing bacteria in the inoculum.

Establishing a Stand. The number of plants per unit of area that survive and grow until harvested is referred to as *plant population* or *stand*. Maximum yields depend on the establishment of a good stand or a uniform distribution, with the desired population of plants per unit of area (e.g., plants per acre or plants per foot of row). This is determined by seeding rate (pounds of seed per acre in a crop like wheat), and by seeding rate and row spacing in crops such as corn, cotton, and soybeans. The grower should always use high-quality seed, but this alone will not establish a good stand. A suitable seedbed must be prepared that will provide an environment favorable to germination of the seeds and growth of the new seedlings.

For rapid germination, good seed-soil contact is necessary to permit absorption of moisture by the seeds. *Soil tilth* is the physical condition of the soil as it relates to texture, structure, granulation, aggregation, moisture content, aeration, drainage, and capillary capacity. A soil with good tilth provides a favorable environment for stand establishment.

Because of the high cost of energy, it is desirable to keep cultivation of the soil to a minimum. Also, each time the soil is cultivated, soil moisture is reduced as moist soil is brought to the surface, where moisture evaporates more readily. Another factor that must be considered when preparing the seedbed is weed control. A weed-free seedbed will help the crop to become well established and more competitive with weeds.

All these factors must be considered when determining the best plan for seedbed preparation. For each crop, special considerations affect the weight given to each factor when determining number of cultivations, method of cultivation, and depth of cultivation. These are discussed in subsequent chapters.

Soil in proper physical condition does not necessarily ensure a good stand. Seeds require oxygen, and although moisture is needed, they cannot survive in flooded or saturated soils for long periods of time. The temperature must be favorable for germination and growth. Finally, for seeds of some species, which require light for germination, depth of seed placement must be carefully adjusted to allow penetration of light.

Planting the Seed

Once the seedbed has been prepared, the seed can be planted. There are many me-

Food Production: Growing New Plants

Figure 2-6. (*a*) Multipurpose planting implements, such as the small two-row planter shown, can be used by the farmer to plant peanuts, corn, cotton, and soybeans. (*John Deere*) (*b*) Hydraulically folded wings on some planters facilitate safe and convenient transportation on public roads, narrow lanes, and through gates. (*International Harvester*) (*c*) Grain drills, such as the one shown, are used mainly for small grains, such as wheat, barley, and oats. (*International Harvester*) (*d*) The unique, highly maneuverable, two-drill hitch shown allows the farmer to double seeding capacity, saving time and fuel. (*International Harvester*)

chanical planting implements available (Figure 2-6). In evaluating which method of seeding to use and which seeder will do the best job, two critical factors must be considered: (1) rate of seeding and (2) depth of seed placement. When the seed is placed uniformly at the proper depth with the optimum number of seeds per unit of area, the chances of having a good stand are greatly increased.

Depth of seeding varies for different species and soil conditions. In general, small-seeded crops such as ladino clover or bluegrass are planted at shallower depths than large-seeded crops such as corn or soybeans. Larger seeds are able to emerge from greater depths because of the larger amount of stored food, and seeding deeper gives a better chance of having optimum moisture to establish the seedlings. For example, a large-seeded crop like corn can be seeded at 2 or 3 inches (in) [5 to 7.5 centimeters (cm)], while a small-seeded crop like alfalfa should be placed at about $1/2$ in [1.27 cm].

Rate of seeding refers to the number of seeds or weight of seed planted per unit of area. Recommended seeding rates for crops and cultivars in any given environment are available from state agricultural experiment stations. Researchers have determined the rates that give the best chance for optimum yields under various growing conditions, and these are probably the most reliable rates to use.

Recommended seeding rates assume that the seed is 100 percent pure live seed. Since seed available on the market seldom, if ever, meets this standard, it is necessary to adjust the seeding rate to account for the lower purity and germination of the seed being used. To adjust the seeding rate, the following formulas can be used, assuming a seed lot with 92 percent germination and 80 percent purity and a recommended seeding rate of 60 pounds per acre (lb/acre) [67.2 kilograms per hectare (kg/ha)].

Percent germination × percent purity = PLS (pure live seed)

$$92\% \times 80\% = 73.6\% \text{ PLS}$$

$$\frac{\text{Recommended seeding rate}}{\text{PLS}}$$
= adjusted seeding rate

$$\frac{60}{0.736} = 81.5 \text{ lb/acre } [91.3 \text{ kg/ha}]$$

Computing the pure live seed ratio can also be helpful in comparing different lots of seed. The cost per unit weight of pure live seed is a more reliable measure of value than the cost per unit weight of the seed.

VEGETATIVE OR ASEXUAL PROPAGATION

Reproduction of plant species that are propagated by true seeds is called *sexual reproduction*. However, some crop plants (sugarcane, for example) do not normally produce seeds, and the seeds of some seed-bearing plants, such as sweet potatoes, bulb plants, and some fruit trees, do not produce plants that conform to important characteristics of the parents such as disease resistance, size, or shape. These plants are normally propagated vegetatively, a process referred to as *asexual reproduction*. Budding, layering, and grafting are examples of asexual methods of reproducing plants. Vegetative reproduction is accomplished by "borrowing" a piece of stem or root from an existing plant and stimulating the growth of buds on the borrowed part, resulting in the production of a whole new plant like the donor parent—i.e., the rooting of a cutting.

The methods available for vegetative propagation are dependent on the structure of the plants being reproduced. Some plants have parts below the soil surface that are most effectively used, such as tubers (potatoes), rhizomes (smooth bromegrass), bulbs (onions), corms (timothy), and in some cases, root cuttings (phlox).

There are several ways of regenerating new plants by using plant parts above the soil surface (Figure 2-7). Common examples include:

1. *Cuttings* Some species can be propagated by taking stem cuttings or leaves from a plant and inserting them into a rooting medium to produce a number of new plants (Figure 2-7a). With leaf cuttings, the petiole of the leaf is inserted in the rooting medium and roots form on the petiole, as, for example, with African violets (Figure 2-7b). In another leaf-cutting method, the whole leaf is laid flat against the rooting medium and held down with small pebbles, and the main veins are cut through. New plants develop where the cuts are made. Rex begonias can be propagated this way.

2. *Layering* New plants can be started from stems while they are still attached to the parent plant (Figure 2-7c). This is

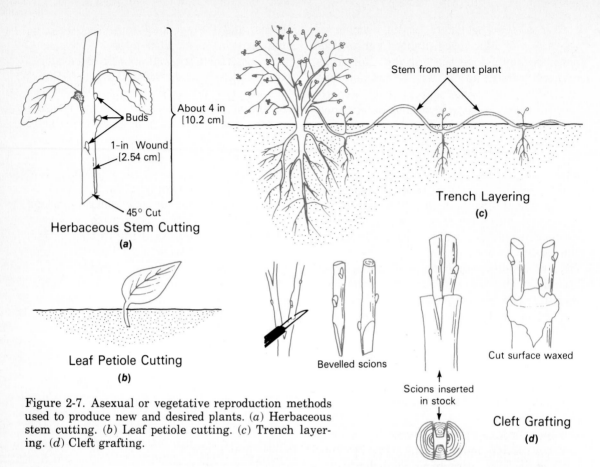

Figure 2-7. Asexual or vegetative reproduction methods used to produce new and desired plants. (*a*) Herbaceous stem cutting. (*b*) Leaf petiole cutting. (*c*) Trench layering. (*d*) Cleft grafting.

commonly done in propagating raspberry, forsythia, and many other species.

There are two methods of layering: (1) ground layering and (2) air layering. In ground layering, a stem that is growing near the ground is notched with a knife and the notched area of the stem is placed below the soil surface. After roots have formed, the new plant is cut from the parent plant and can be replanted elsewhere.

Air layering is accomplished by cutting a deep notch on a ring around the stem in the area to be layered. The wounded area is then wrapped with sphagnum moss that has been thoroughly moistened. Polyethylene plastic film is then wrapped around the moss to keep the moisture

from evaporating. Roots form in the area of the wound, and the new plant can be cut away from the parent plant and potted or placed in the garden or field.

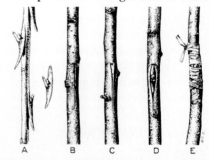

Figure 2-8. T or shield budding as illustrated here is the most common method of budding. (*a*) Bud stick. (*b*) T-shaped cut through the bark of the stock. (*c*) Bark raised to admit the bud. (*d*) Bud in place. (*e*) Bud wrapped. (*USDA*)

3. *Grafting* This method involves taking a shoot (called a *scion*) from one plant and inserting it into a prepared stock, under-stock, or rootstock of another plant in such a way that the scion will unite with the stock and grow into the desired plant (Figure 2-7d). It is essential that the cambium tissue of each be in contact with that of the other. This propagation method is used to produce desired-quality fruit trees on hardier or disease-resistant rootstocks or to produce semidwarf or dwarf trees with the same quality characteristics as the standard-size tree (see Chapter 17).

4. *Budding* This method is similar to grafting but involves the transfer of a single bud from one plant to another. *T budding,* sometimes called *shield bud-ding,* is a common budding technique in which a T-shaped incision is made into the bark of the stock plant and a single bud is inserted, with care that the cambial tissue of each part is in contact with that of the other (Figure 2-8). By this technique, a single hybrid tea rose stem containing ten buds can yield ten new hybrid tea rose plants on hardy rootstocks.

SUMMARY

In general, plants pass through a life cycle which includes planting the seed, seed germination, emergence, vegetative growth, maturity, repro-duction, production of seed, and death. Once seeds have been produced it is important to maintain their quality. Seed certification is a method of producing and providing to growers seed that has been produced and processed under carefully controlled conditions. Producers of crop plants should be aware of methods of cleaning and chemical treatment of seeds and inoculation of legume seeds.

The establishment of a good stand is essential for economical plant production. The optimum rate and depth of seeding varies for different species and soil conditions.

Plants may also be propagated asexually from vegetative parts such as buds, leaves, or stems. Methods used include budding, layering, and grafting.

QUESTIONS FOR REVIEW

1. Describe the general differences in life cycle pattern among plants classified as annuals, biennials, and perennials.
2. What is the primary function of seeds? Give some other examples of how seeds are used.
3. What are the parts of a seed, and what is the function of these parts?
4. What conditions must be met for germination to occur?
5. What criteria should be considered in evaluating the quality of seed for planting in the field?
6. What are some common seed-processing practices used to prepare seed for planting? Describe each, and show why it is important.
7. Describe the difference between asexual (vegetative) reproduction and sexual reproduction of plants. Give examples of each.

Chapter

3

Soil Resources and Crop Production

As a nation, in recent years Americans have expressed a growing concern about conserving natural resources. Conserving energy from fossil fuel has become a major concern to most of us. Conservation of fuel (or energy) resources is essential, but the need to protect and conserve our soil resources is of equal, or perhaps greater, importance.

The importance of plants to human life was noted in Chapter 1. Although plants can be grown in substances other than soil, and "farming the sea" will yield food for the world's people, in the foreseeable future, soil will remain essential for crop production, hence essential for human life (Figure 3-1). Soil must therefore be considered a prime natural resource. Although soil is formed by natural processes, the rate of formation is exceedingly slow. As a result, we cannot consider soil a truly *renewable* natural resource.

Consequently, understanding the different values of soils is important to everyone, not just agriculturists. In the future, land use, a subject closely tied to soils, may become a political issue. Wise voting must be based on facts.

CHAPTER GOALS

After studying this chapter you will be able to:

- Describe the nature of soil and some key physical and chemical properties of soil.
- Explain the critical importance of soil in crop production, including differences in the value of soil in producing crops.
- List the elements that are essential for plant growth and describe the form in which each is taken up by plants and the roles some of the more important ones play.
- Describe how plants remove elements from the soil and how these can be returned with fertilizers or by natural means.

THE SOIL

Soil is a complex, naturally occurring substance. Commonly it is defined or described as the weathered upper surface of the earth's crust capable of supporting plant growth. By this definition, there is a firm link between soils and plants. Usually the surface soil—soil from 50 to 100 cm [1.5 to 3 feet (ft)] deep—is considered as the agriculturally important part of the soil, although some soils are well over 1000 cm [33 ft] deep.

There are many different kinds of soil, which differ in suitability for crop production. All soil is composed of three basic parts:

1. The *mineral part,* which comes from weathering of the stony parent material from which the soil is formed. The mineral component determines important physical and chemical properties of the soil. The mineral part may be composed of different proportions of the three basic soil particles. The largest particles are sand, the medium-sized particles are silt, and the smallest particles are clay. Clay particles are chemically the most active, and this chemical activity affects the fertility of the soil.

2. The *organic matter,* which comes from dead plant and animal matter. When plant or animal material decays, it is called *humus.* Humus may include leaves, stems, twigs, and roots of plants; animal droppings; and even entire dead animals. Organic matter content varies greatly among different soils. The natural fertility of a soil and how the soil reacts to the application of commercial fertilizers are greatly influenced by the kind and amount of organic matter and how rapidly it decays. Organic matter also has a great effect on soil structure (discussed later in this chapter) and on how the soil adsorbs and holds water (the rate of infiltration of water and the soil's water-holding capacity); and it tends to protect soil from erosion.

3. The *biotic component,* which consists of living plants and animals, including microorganisms such as bacteria and fungi. The microorganisms play key roles in the decay of organic matter; some may cause serious diseases of plants; others help convert nitrogen in the air to forms which can be used by plants. The biotic component of the soil also includes some animals, from worms and insects to burrowing rodents, foxes, and similar animals. Earthworms are important parts of the biotic component of the soil. Their burrowing activities move large quantities of soil and help air move into the soil. This aeration of the soil is essential for plant root growth.

In addition to the mineral, organic, and biotic parts of the soil, the soil pore space (the part of the soil not occupied by solids) is of major importance. Water, which is essential for the growth of plants, is held on the surface of the soil particles. When a soil is flooded or saturated with water, the pores may be filled with water. Pores also contain the soil's supply of air; the oxygen in the soil air is essential for respiration by the roots, hence for healthy plant growth.

THE ROLE OF SOIL IN PLANT GROWTH

The soil serves three major functions for plants. First, it serves as a medium to anchor the roots and thus support the plants. Second, it serves as a reservoir to store and supply essential moisture to plants. Finally, except for carbon, hydrogen, and oxygen, all the chemical elements essential for plant growth are provided by the soil.

(a)

(b)

Figure 3-1. Soil is a fundamental resource for crop production. Although tillage is necessary, and sometimes extensive, soil must be managed to protect it. (*a*) Disk plowing a field on a farm in Texas. (*Fred S. White/USDA*) (*b*) Alternating strips of corn and small grain help protect a Maryland farm from erosion. (*Tim Mc-Cabe/USDA-Soil Conservation Service*)

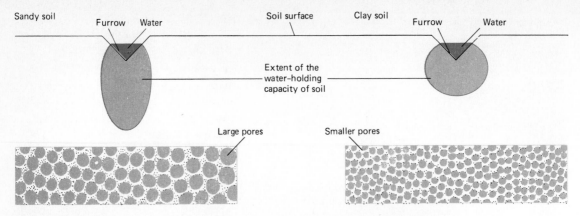

Figure 3-2. Clay soil can hold more water than sandy soil because of the number and size of its pores. Water moves differently in clay soil compared to sandy soil.

Sometimes the anchorage function of soils is not considered as important as the other two functions; for some crops, however, it is very important. For example, root and tuber crops such as sugar beets and potatoes are grown on soils from which they can be fairly easily uprooted; that is, commonly on more sandy soils, where anchorage may be less secure. The anchorage function becomes critical on soils that tend to erode badly. In such cases, farmers and other conservationists must understand both the anchorage capacity of the soil and the pattern and extent of the root growth of plants sown on the soil to help hold and protect it. The anchorage function is also affected by how readily the soil freezes and thaws and how it is moved up and down, or heaved, as a result of freezing and thawing.

Soils differ greatly in the amount of water they can hold or store. The amount of water stored is a direct function of the soil texture (discussed later in this chapter) and the amount or volume of pore space. Soils with large amounts of clay hold relatively large amounts of water—up to 4 in of water per foot of soil depth [10 cm/30 cm]. Very sandy soils hold much less water, 1 in or less per

foot of soil depth [2.5 cm/30 cm]. The concept of an inch of water, whether it is rainfall or irrigation, is simple. Generally, an inch of water is 1 in over an area of 1 acre, although it can be over any specified area. The relationship between soil texture, pore space, and soil moisture-holding capacity is shown in Figure 3-2.

Plant-soil-water relationships are very complex. To some extent they can be managed by the farmer, through irrigation, for example. An understanding of these relationships is essential for sound farm management of most crops. Four terms are commonly used to describe plant-soil-moisture relationships or conditions. These terms are *saturation, field capacity* (FC), *permanent wilting percentage* (PWP), and *available moisture.*

Saturation is the condition of a soil after an irrigation or heavy rain; all pores are filled with water. When additional water is added, it either flows through the soil or runs off the surface, frequently causing erosion. When a saturated soil drains naturally, the water held in it, as layers along the surface of soil particles, represents the water held at field capacity. The FC of a soil is the water the soil holds against the pull of gravity, the natural forces of drainage, e.g., 4 in/ft, [10 cm/30 cm] for a clay soil.

Plants remove moisture from the soil through the process of transpiration, but they cannot remove all the moisture. When plants wilt and fail to recover until water is added, the moisture remaining in the soil is its PWP. The moisture between FC and PWP is available for plant use. However, it is not uniformly available; as plants remove moisture from the soil and the PWP is approached, the plants must spend more energy to remove the water. To avoid forcing plants to expend excessive energy, at least with irrigation, it is desirable to add water when approximately half the available water has been used. Thus the farmer must understand the soil and the rate at which crops use water. This rate is affected by many factors, such as the crop, its age, temperature, and wind. Water is essential for crop growth and is a valuable and frequently expensive resource when used for irrigation. Careful management is needed to avoid economic waste as well as to conserve natural resources.

Soils provide the chemical elements, except for carbon dioxide, which are essential for plant growth. The simplified equation for photosynthesis

$$CO_2 + H_2O \xrightarrow{\text{Light}} \text{food} + \text{oxygen}$$

shows only carbon, hydrogen, and oxygen (C, H, and O) as being essential. But all plants in which this life-sustaining process occurs require a number of chemical elements for normal growth in order to carry out the process. These elements can be divided into two general categories. Those needed in relatively large amounts are the macronutrients, and those needed in lesser amounts, the micronutrients. All elements must be in specific forms, frequently oxides, before they can be taken up by plants.

ESSENTIAL ELEMENTS

Macronutrients

Twenty elements are essential for normal growth of plants. Many agricultural soils can be improved by adding several of the macronutrients as fertilizers. Micronutrients are used in smaller amounts, and although they too can be added to soil as fertilizers, this is not as common as the addition of macronutrients.

Three essential elements, carbon (C), hydrogen (H), and oxygen (O), occur plentifully in nature and are rarely added as fertilizer. Carbon and oxygen come from carbon dioxide (CO_2), which exists in the air. Normally, the atmosphere contains about 0.03% CO_2. Because CO_2 is essential for photosynthesis, the rate of photosynthesis under some conditions can be limited by this relatively small amount of CO_2. Under very special conditions, CO_2 may be added to the atmosphere of a greenhouse to increase the rate of photosynthesis. Hydrogen and oxygen as nutrients become water (H_2O) as a result of complicated biochemical reactions associated with the process of respiration. Remember that oxygen is essential for respiration; for this purpose, most oxygen is taken from the soil's atmosphere, the rhizosphere, as O_2.

Seven additional elements are considered macronutrients: nitrogen (N), phosphorus (P), potassium (K), calcium (Ca), magnesium (Mg), sulfur (S), and sodium (Na). The first three, N, P, and K, are most frequently applied in the form of commercial fertilizer in the United States. These three elements are given a special name, the *primary plant foods*.

Nitrogen is the element most frequently deficient in soils, hence the element that most frequently limits plant growth and crop yield. Nitrogen is essential for the healthy green color of plants and is part of

Figure 3-3. Spreading lime on soil helps prevent calcium deficiency. (*Grant Heilman*)

all amino acids; amino acids are the building blocks of proteins. Usually nitrogen is taken up by plants in the ammonium form (NH_4^+) or the nitrate form (NO_3^-). Air contains about 78% nitrogen in the gaseous (N_2) state, but this form cannot be used directly by plants.

Phosphorus plays a key role in key energy exchanges at the cellular level during photosynthesis and respiration and is needed for strong root development. It is usually taken up in the phosphate form, expressed as HPO_4^{2-} or $H_2PO_4^-$.

Potassium performs several major functions in plants. It is involved in carbohydrate metabolism and nitrogen metabolism and plays a role in the water balance of the plant. It is taken up by plants as K^+ which is the ionic form. Sulfur is a component of two essential amino acids. It is taken up in the ion form, SO_4^{2-}. Sulfur is not commonly deficient in agricultural soils as are nitrogen and phosphorus, although it is applied as a fertilizer where appropriate soil tests indicate that adequate amounts are not available for normal plant growth.

Calcium is needed for normal cell wall formation. Tips of shoots and roots and buds fail to develop and grow normally without adequate calcium for the plant. In addition to its role as an essential element, calcium in the form of lime is frequently applied to soils to increase pH (or reduce acidity). This practice is common in humid areas (areas with 30 to 40 in [75 to 100 cm] or more annual precipitation), where basic elements are leached from the soil. Over 2 tons of lime per acre [4.5 metric tons (t)/ha] may be applied every 2 to 3 years. Limestone is relatively insoluble; its effect on the soil pH is influenced by how finely the material is ground before it is applied (Figure 3-3).

Magnesium is part of the chlorophyll molecule. It is essential for chlorophyll development, hence for the normal, healthy green color of plants. The symptoms of magnesium deficiency displayed by a plant are similar to those of nitrogen deficiency—commonly light-green to pale-yellow foliage and, of course, ultimately reduced yields. However, there are detailed differences. Magnesium deficiencies are noted first on lower leaves and appear as light and darker stripes; nitrogen deficiency is expressed first as a light green or yellowing down the midrib of the leaf.

The role of sodium is not clearly understood by scientists. It may substitute in part for certain functions of potassium. Rarely, if ever, is sodium applied as a fertilizer, and it may be inappropriate to consider it as essential.

Micronutrients

Ten other elements constitute the micronutrient group. These are iron (Fe), zinc (Zn), copper (Cu), boron (B), molybdenum (Mo), manganese (Mn), chlorine (Cl), cobalt (Co), vanadium (V), and silicon (Si). Of these, zinc and boron are most frequently applied as fertilizers, but the amounts applied are far smaller than the amounts of primary plant foods (N, P, and K) used or other macronutrients. Although Co, V, and Si are considered essential elements, their roles in the growth and development of crop plants are not well understood.

SOURCES OF ESSENTIAL ELEMENTS

When a crop is harvested, essential elements it has used are removed from the field. The amount of various elements removed depends on the crop and the yield; the greater the yield, the greater the amount of essential minerals removed from the soil. The supply of essential elements in the soil is not inexhaustible. To retain the productive capacity of the soil, these elements must be replaced, either through decomposition of organic matter or, more frequently, by the application of commercial fertilizers, or by weathering processes that break down nitrogen.

The removal of essential elements from the soil is well illustrated by wheat production in Montana. Each bushel of wheat harvested contains, hence removes from the soil, about 1.5 lb [about 0.7 kg] of nitrogen.

A moderate yield of 40 bushels per acre (bu/acre) [3.48 cubic meters per hectare (m^3/ha)] would remove from each acre about 60 lb [about 27 kg] of nitrogen ($1.5 \times 40 = 60$). Merely replacing this amount of nitrogen would require the application of nitrogen fertilizer that contained 60 lb of nitrogen. Ammonium nitrate fertilizer is a common source of nitrogen. It contains about 33% N. Thus to replace the 60 lb of nitrogen removed in the grain would require the application of 180 lb [82 kg] of ammonium nitrate. There are many forms of fertilizer, all differing in the amounts of essential elements they supply. It is beyond the scope of this text to discuss these differences. However, from this example you should realize that (1) crops remove large amounts of essential elements and (2) that even larger amounts of fertilizers are required merely to maintain the fertility of the soil. Other examples of mineral removal are given in Table 3-1.

Commercial Fertilizer

You do not have to understand the entire chemistry of fertilizers to understand what amounts of different fertilizers are comparable to one another. However, some background in chemistry ultimately is most helpful in managing the fertility of the soil and in purchasing and applying fertilizers. Commercial fertilizers are labeled in a standard way throughout the United States. To understand what elements are applied in what amounts, you must understand this labeling. This will also allow you to determine the price per pound (or ton) of the essential element in any fertilizer (remember, fertilizers are not pure essential elements). A 100-lb [45-kg] bag of ammonium nitrate fertilizer, NH_4NO_3, contains 33 lb [15 kg] of nitrogen (N); the other 67 lb [30 kg] is oxygen, hydrogen, and inert material.

TABLE 3-1 APPROXIMATE AMOUNTS OF NUTRIENTS CONTAINED IN VARIOUS CROPS

Crop	Yield per Acre	lb/acre							
		N	P_2O_5	P	K_2O	K	Ca	Mg	S
Corn:									
Grain	180 bu	170	69	30	48	40	4	16	14
Stover	4 tons	70	36	18	192	160	35	34	16
Soybean:									
Grain	60 bu	250	48	21	8	72	17	17	12
Straw		85	16	7	58	48	10	10	13
Wheat:									
Grain	80 bu	140	46	20	46	22	2	12	5
Straw	3 tons	40	9	4	132	110	4	12	15
Cotton:									
Seed and lint	3750 lb	95	39	17	42	35	4	11	7
Stalks, leaves, and burs		85	25	11	82	68	30	24	23
Tobacco:									
Leaves	3000 lb	145	14	6	156	130	80	15	12
Stalks	3600 lb	95	16	7	114	95	15	9	7
Peanuts:									
Nuts	4000 lb	140	23	10	35	29	2	5	10
Vines	5000 lb	100	16	7	150	125		20	11
Rice:									
Grain	7000 lb	77	46	20	28	23	5	8	5
Straw	7000 lb	35	14	6	139	116	15	6	7
Alfalfa	8 tons	450	80	35	480	400	200	40	40
Orchard grass	6 tons	300	101	44	360	300	50	25	35
Bermudagrass	10 tons	500	137	60	420	350	100	45	45
Potatoes	400 bu	150	80	35	264	220	6	12	12
Sugar beets	30 tons	125	16	7	240	200	35	27	10
Sugarcane	100 tons	160	92	40	330	275	45	40	50
Sorghum:									
Grain	60 bu	50	25	11	15	12	4	5	5
Stover	3 tons	65	20	9	95	79	29	18	12

Source: *Our Land and Its Care*, p. 42.

Any bag of commercial fertilizer will have on its label a line showing three (or more) figures (Figure 3-4). Consider only three figures for the moment, with a bag labeled 10-5-5. Reading from left to right as usual, these figures express in *percentages* the

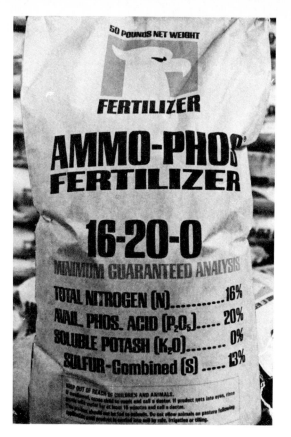

Figure 3-4. A 50-lb [23-kg] bag of fertilizer showing percentages of nitrogen, phosphorous pentoxide, potassium oxide, and sulfur contained in the fertilizer. (*Jane Hamilton-Merritt*).

amounts of N (10 percent), P_2O_5 (5 percent), and K_2O (5 percent). Note that the value is given for elemental nitrogen (N) and for the oxides of phosphorus and potassium. None of these elements actually exists in these forms in the fertilizer, nor are these forms of these elements taken up by plants. Nonetheless, all forms of these elements are normally *expressed* in terms of the element or the oxide. This allows comparison of the costs of various fertilizers. For example, in buying phosphorus fertilizer, a ton of 0-10-0 contains half as much P_2O_5 (or its equivalent) as a ton of 0-20-0 and thus should cost half as much. If a label has other figures, these

will be identified in terms of the elements they represent. The key point to remember is that the figures represent percentages, and the first three (including zeros), in order, are N, P_2O_5, and K_2O.

Alternatives to Commercial Fertilizers

Nitrogen Fixation.   Fertilizers are a major expense in the production of many crops. Nitrogen fertilizers are the most frequently required type of fertilizer. In addition to the expense of buying the fertilizer, the manufacture of nitrogen fertilizer requires the use of considerable amounts of fossil fuels, which are expensive and becoming limited in supply. Thus there are two good reasons to seek alternatives to the use of commercial types of nitrogen fertilizer.

Nitrogen in the air (a major source of plant-available N) is converted to a form usable by plants through the action of specific bacteria living in the roots of leguminous plants, such as alfalfa, soybeans, and peanuts. The bacteria belong to the genus *Rhizobium*. The process of converting N_2 to NH_4^+, which in chemistry is a reduction process, is referred to as *nitrogen fixation:* the N_2 is fixed in a form useful to plants. These microorganisms cause the formation of nodules, or bumps, on the roots of plants. Atmospheric nitrogen, N_2, is transformed to ammonium (NH_4^+) by the action of the bacteria in these nodules. Nitrogen in this form can be utilized by the plant. In addition, when the plant dies, the nitrogen fixed as NH_4^+ is available for use by other plants. More nitrogen is generally fixed than the legume plant will use. For these reasons, such plants serve as soil builders or soil improvers and thus are very important in crop rotations. Depending on the legume, the nitrogen available for plant use in the soil, and climatic factors (temperature and moisture), legumes can add from 50 to 100 lb [about

Soil Resources and Crop Production

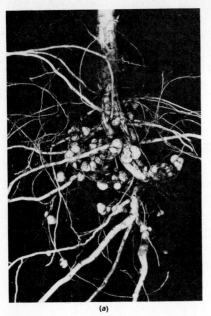

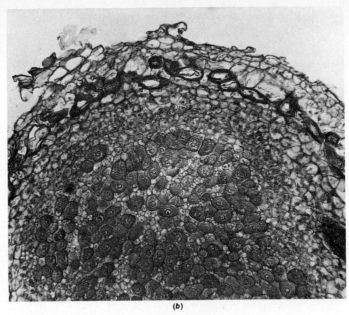

(a)　　　　　　　　　　　　　　　　　　(b)

Figure 3-5. (a) Nodulation on soybean roots. (USDA) (b) Cross section of soybean root nodule. This nodule contains nitrogen-fixing bacteria of the genus *Rhizobium* sp. (magnified 200×). (*Runk/Schoenberger-Grant Heilman*)

23 to 45 kg] of plant-available nitrogen to the soil annually.

Different species of legumes require different strains of *Rhizobium* bacteria. Some strains of *Rhizobium* can inoculate several legume species effectively, while others are effective in only one species. Not all species of legumes have been successfully inoculated with *Rhizobium*. Successful inoculation is indicated by formation of the nodules in which the bacteria actively fix nitrogen, as shown in Figure 3-5a. Generally, a crop is inoculated by treating the seeds to be planted with a dust containing the appropriate bacterial strain (see Table 2-1). The relationship between the bacteria and the legume plant is commonly considered to be a symbiotic one: two organisms living together for their mutual benefit. The legume provides carbohydrates (food) and water to the bacteria. In turn, the bacteria fix essential nitrogen. Crop producers must

understand this symbiotic relationship in order to take advantage of this valuable source of an essential element. They must know how to manage legume crops so as to encourage optimal nitrogen fixation. This includes knowing what strains of *Rhizobium* to apply, how to apply the inoculum, and how to manage the plants so as to encourage nitrogen fixation. Frequently management includes irrigation and the use of other fertilizers to ensure that legumes are not weakened as a result of deficiencies of elements other than nitrogen.

At the opposite extreme from nitrogen fixation by *Rhizobium* in association with legume plants is the actual temporary tie-up of nitrogen by bacteria that cause decay of organic matter and ultimately release nitrogen. Many of the bacteria that cause decay of organic matter and ultimately release nitrogen for plant growth require nitrogen in the same form as crop plants and compete effectively for this nitrogen. Certain types of crop residue or stubble are very high in sugars, or relatively simple carbohydrates. This material is utilized as a source of energy by bacteria, which flourish if adequate nitrogen is available to them. The bacteria temporarily tie up nitrogen through their metabolic activities. As a result, the crop growing in the field may suffer from a nitrogen deficiency. When the stubble is decayed, the bacteria die and the nitrogen is again available for plants. Stubble from crops such as sorghum, wheat, and corn commonly leads to this problem. Sometimes nitrogen fertilizer must be applied to fields to support seedlings of a crop following these crops in the rotation. The crop manager must be aware of the phenomenon of nitrogen tie-up as organic matter decays and must know how this can be overcome.

Manures. At one time, various animal manures were commonly used as fertilizers. In recent years they have been generally replaced by synthetic fertilizers, which are easier to transport and apply and which are higher and more uniform in their content of essential elements. Today there is renewed interest in manure as a means of providing essential elements. First, with marked increases in feedlots and similar intensive, large-scale livestock operations, disposal of manure has become a problem, and spreading it on the soil represents a possible solution. Second, manure may represent a valuable by-product, since plant-available nitrogen and other essential elements are found in manure, and it is produced without the use of fossil fuel.

In evaluating the potential value of manure, several factors must be considered. The cost of transporting and applying the manure is important. Wet manure cannot be moved long distances without significant costs in energy as well as money, and drying animal wastes before moving them is usually too expensive. Besides these costs, the actual value of the manure as a fertilizer must be considered in relation to the total cost of other forms of fertilizers. In terms of providing essential elements for plants, there are marked differences among various types of manure; some of these are shown in Table 3-2. On a wet basis, waste from feedlot beef is a comparatively good source of the primary plant foods (N, P, and K) and is generally superior to waste from dairy cows. Poultry represents a potentially more valuable source of manure, but total supplies are smaller. Whenever any product is applied to the soil, there is a danger of its leaching into groundwater and ultimately polluting water sources. This can be a serious problem in using animal wastes as fertilizers; but of course, it is also a serious problem in other ways of disposing of such wastes.

TABLE 3-2 PERCENTAGES OF AVAILABLE PLANT NUTRIENTS FROM VARIOUS ANIMAL WASTES

Animal Manures	N	P_2O_5	K_2O	CaO	MgO	S
Cattle	0.6	0.3	0.6	0.3	0.2	0.7
Hog	0.5	0.3	0.4	0.8	0.1	1.4
Horse	0.7	0.2	0.7	1.1	0.2	0.7
Chicken	3.4	3.4	2.8	2.7	0.8	0.7

Source: *Our Land and Its Care,* p. 61.

SOIL TESTS

When a soil fails to provide a plant with an adequate amount of specific essential elements, the plant develops characteristic deficiency symptoms. The symptoms are quite similar for nearly all crop plants, and many factors cause plants to display similar symptoms. In the case of deficiencies of essential elements, recognition of the fact that the plant does not appear to be normal is necessary for good economic management and highest crop yields. It is not as critical to recognize the *cause* of the abnormal condition, because you should seek expert help when an abnormal condition is recognized. This help will frequently include a scientific test of the soil to determine the availability of essential elements as well as other aspects of the soil that might affect crop growth, development, and yield. In addition to the soil test, sometimes plant tissues are tested to determine whether the plant is receiving adequate amounts of essential elements. If a deficiency is detected early enough in the crop's growing season or life cycle, it can often be corrected by applying appropriate fertilizer, without serious yield reductions. Plants require a balanced supply of essential elements and in some cases regulation of the pH of the soil. Soil tests and plant tissue analysis provide the information to supply a crop with the elements it needs in appropriate amounts. To manage crop production effectively, you should learn not only to recognize abnormal symptoms in crop plants but also the types of tests that can determine the cause of the problem and who can do these tests. (Frequently your county extension agent, vocational agriculture teacher, or private firms can perform the tests or tell you who can.) Of course, you also must learn how to interpret test results, with the help of the expert who conducted the test. Finally, when the tests have been conducted and in-terpreted, as a farm manager and/or owner, you must decide what action you will take. A typical report from a soil test is shown in Figure 3-6.

PROPERTIES OF SOIL

Fertility

The natural fertility of a soil is one of its very important properties. This property reflects the capacity of the soil to provide plants with the elements they require for normal growth and development. Two aspects of this soil property should be considered. The first is the natural supply of elements in the soil. Many elements came from the organic matter in the soil. When a soil is farmed, the natural cycle of organic matter typical of a soil in a forest, meadow, or rangeland may be disrupted. Other elements came from the rocky parent material from which the soil is formed. The rate at which these elements become available to plants may be slow, but the basic supply is almost limitless compared with nitrogen which comes from organic matter. The other property to be considered is how a soil holds essential minerals. This is a basic chemical property of every soil and is related directly to electrical charges of atoms and molecules in the soil, usually associated with clay particles in the soil and with the soil organic matter. This property is expressed as the soil's exchange capacity (or its base or cation exchange capacity). *Exchange capacity* reflects the number of charges a unit of soil has that will hold a positively charged atom or molecule (e.g., Ca^{2+} or NH_4^+). Clay soils have a higher exchange capacity than sandy soils. As a result, larger amounts of nutrients can be held by clay soils for use by crop plants than by sandy soils. This is important

MONTANA STATE UNIVERSITY

SOIL TESTING LABORATORY

DATE RECEIVED: [illegible]
DATE REPORTED: 07/[illegible]

PLANT AND SOIL SCIENCE DEPARTMENT
Montana State University
Bozeman, Montana 59717

GROWER:
ALLEN PERRY
C/O DICK THOMAS
BOX 126 BELGRADE, MT 59714

EXTRA COPY:
None requested

LAB NUMBER 2691

FIELD : 1

SOIL ANALYSIS RESULTS:

DEPTH in.	pH	SALT HAZARD	ORGANIC MATTER %	NITRATE NITROGEN ppm	#/A	OLSEN PHOSPHORUS ppm	POTASSIUM ppm	TEXTURE
0 TO 6	7.8	0.9	3.2 L			7	228	CL

IRRIGATED FEED BARLEY FOLLOWING ALFALFA
THIS RECOMMENDATION IS FOR AN EXPECTED YIELD OF 80 BUSHELS PER ACRE.

THIS IS YOUR FERTILIZER RECOMMENDATION:
SOURCE: Montana Fertilizer Guides

****

NITROGEN	PHOSPHORUS(P205) BANDED	OR BROADCAST	POTASSIUM(K20)
POUNDS/ACRE			
105	35	60	25

SPECIAL COMMENTS:

****

If you band fertilizer with the seed, do not exceed 10 to 20 pounds of
N plus K20 with a 12 inch row spacing and average soil moisture.
Additional N or K may be broadcast in the fall or spring.
A soil nitrate value of 30 pounds per acre was assumed

Figure 3-6. Typical soil test report which a farmer receives after soil sample has been analyzed.

in managing fertilizer application to avoid waste.

Soil pH and Lime

Soils differ greatly in their relative acidity or alkalinity. The pH scale ranges from 0 to 14, with pH 7 being neutral; lower values indicate acidity, and higher, alkalinity. The pH scale is logarithmic; a pH of 5 is 10 times more acidic than a pH of 6. Generally crops are grown at a pH of 4 to 10; however, crops are most productive at a pH of 5.5 to 8.2. Although crops differ with respect to their tol-

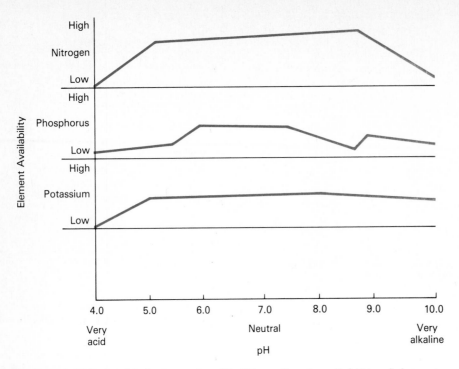

Figure 3-7. Relationship between the pH of the soil and availability of elements that are essential for plant growth.

erance to very high or very low pH conditions, the effects of pH are commonly associated with changing availability of essential elements. This phenomenon is illustrated in Figure 3-7. Note the differences between the elements nitrogen, phosphorus, and potassium.

As a rule, rye is the most tolerant of acid soil conditions. This crop also tolerates cold, damp conditions better than other crops. Sugar beets favor high (alkaline) pH conditions, and barley seems to thrive on somewhat alkaline soil. Corn will tolerate acid conditions, but like most crops, it will grow better on soils that are near neutral (pH 7). Examples of how different crops tolerate pH conditions are given in Table 3-3.

Acid soils are formed under conditions of high rainfall, typical of the eastern and southeastern United States. Alkaline soils are more common to the West and Southwest. One point should be explained: an alkaline soil is merely one with a pH above 7 (Figure 3-8). In addition, more commonly under alkaline than acid conditions, soils may be saline; that is, they have natural accumulations of salts that inhibit plant growth. Sometimes misuse of fertilizers and other cropping practices contribute to salt buildup. Excessive use of sodium nitrate as a source of NO_3 has led to damaging accumulations of sodium; in parts of Montana, the Dakotas, and Canada, peculiar soil conditions and an alternating crop-fallow rotation common to cereal production have led to increases in saline seepage areas following periods of above-average rainfall.

TABLE 3-3 PLANT RESPONSE TO SOIL pH. NOTE MANY FACTORS OTHER THAN pH AFFECT PLANT GROWTH.

Very Acid pH 4–5	Strong to Medium Acidity pH 5–6	Slightly Acid pH 6–7	Slightly Base (Alkaline) pH 7–8
Lespedeza	Corn	Alfalfa	Lettuce
Redtop	Wheat	Red clover	Onions
Millet	Oats	Sweetclover	Barley
Buckwheat	Potatoes	Sugar beets	Sugar beets
Rye	Field beans		Beans
Soybeans	Tobacco		Peas

Rarely are efforts made to decrease soil pH. However, it is quite common to apply lime to increase the pH (Figure 3-9). Proper liming has brought large areas of acid soils back to productivity. The effects and persis-tence of the benefits depend on the type of lime applied and how finely ground it is. Finely ground material has a more immedi-ate effect but is not as persistent as coarser materials. Applying lime does not raise the

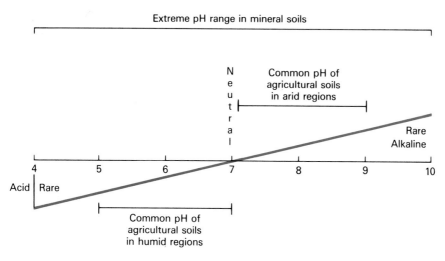

Figure 3-8. Range of soil pH for most agricultural land and the common pH ranges of agricultural soils found in arid regions and humid regions.

Figure 3-9. Soil scientist testing for lime or pH. (*Hugo Bryan/USDA*)

pH permanently; lime must be applied every 1 to 3 years on cropland to regulate soil pH in many areas with high rainfall.

Soil Texture

The concept of soil texture was mentioned earlier in the discussion of soil moisture-holding capacity. Soil texture plays an important role in crop production. There are a number of textural types or classes of soil. These are determined by the relative proportion of the three soil particle groups—sand, silt, and clay—in the surface layer of the soil. The textural classes are illustrated by the soil triangle in Figure 3-10. (Each side of the triangle represents one of the soil particle groups.)

In addition to moisture-holding capacity, texture affects other important soil characteristics. Light-textured (coarse) soils—soils that are relatively high in sand—tend to heat and cool more rapidly than heavy-textured (fine) soils—soils that contain relatively large amounts of clay. *Note:* Light soils are also referred to as coarse soils and heavy soils as fine soils. Under comparable conditions of slope and ground cover, erosion hazard is less with coarse-textured soils than with fine soils. In some areas, crops grown for roots, such as sugar beets, are grown on coarse soils for ease of harvesting. However, these crops can be grown successfully on heavier soils, although more powerful equipment may be required for various farming operations.

Soil Structure

Soil structure relates to how fundamental soil particles group together, or form aggregates. The infiltration rate of water (rain or irrigation) is higher in soils with good structure. With poor structure, water tends to run off, thereby causing erosion. Thus soil structure has a great bearing on the amount of water retained by certain soils.

Organic matter tends to improve soil structure. Thus manures or cover crops are beneficial for more than the essential elements they may add to the soil. Soil structure can be damaged. Excessive cultivation, excessive traffic over a field, and working the soil when it is wet all tend to damage the soil structure. Flooding can also be damaging. Structure can be restored naturally, but long periods of time are required. Adding organic matter helps, but preventive measures such as not overworking the soil and not working it when it is wet are better.

Other Soil Factors

Many other aspects of a soil affect its ultimate suitability and potential for crop

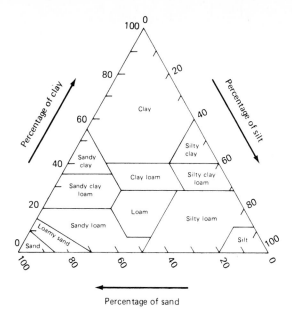

Figure 3-10. Soil triangle showing relationship between silt, sand, and clay proportions and soil texture.

production. Color (light vs. dark) influences how rapidly a soil heats and cools. Dark soils tend to absorb heat more than light-colored ones, but coarse-textured, sandy soils exchange heat most rapidly. The depth of a soil may be important, whether this is the depth to the rocky parent material, to a hard layer or hardpan, or to standing water. The depth of soil influences the area in which roots may grow, thus influencing the potential soil water reservoir and available elements. The slope and exposure of a given land area influence soil temperature (south slopes generally warm more quickly in the spring than north slopes) and potential water runoff or erosion hazards. Because plants are static and must complete their life where the seed germinates and grows, successful farmers and all those working with and for farmers must strive to understand how to protect and foster this critical resource, soil.

AVAILABILITY OF LAND FOR CROP PRODUCTION

About 75 percent of the best cropland in the United States is currently being used to produce crops. This shows a good use of a valuable resource. It also shows that a critical resource, land for crop production, is not limitless. Unfortunately, thousands of acres of prime cropland are lost every year to urban uses—homes, roads, shopping centers, etc. Many of the factors that make land suitable for crop production make it suitable, but *not* essential, for other human activities. In some states zoning laws have been proposed to protect the critical land resources.

There is a common joke that since the surface of the earth is seven-eighths water and only one-eighth land, people are expected to fish 7 days and work only 1! To feed the human race, obviously farmers must work more than 1 day in 8. Unfortunately, not all the land area of the earth's surface is fully suitable for crop production. In the United States, the Soil Conservation Service (a branch of the U.S. Department of Agriculture) recognizes eight classes of land. The descriptions of these classes are given in Table 3-4. The first three of these classes are generally suitable for cautious mechanized crop production, with safeguards against erosion. The fourth class has greater erosion hazards and poor moisture penetration and is shallow and of less value for cropping. The last four classes, for a variety of reasons, are not suitable for crop production. Many can be grazed lightly and/or used for recreational purposes. In virtually every instance, lands in these classes may be subject to severe erosion.

In establishing appropriate land-use criteria, it must be remembered that the land is an essential natural resource that must be protected. Land that can best be used to

Soil Resources and Crop Production

TABLE 3-4 LAND-USE CLASSES ESTABLISHED BY THE U.S. DEPARTMENT OF AGRICULTURE

Land Capability Class	Land Use	Conservation Treatment (%)	
		Adequate	Needed
I–III	Cropland	38	62
V–VII	Cropland	31	69
I–IV	Pastureland	30	70
V–VIII	Pastureland	20	80
I–IV	Rangeland	32	68
V–VIII	Rangeland	28	72
	Forestland	41	59

Source: "U.S. Department of Agriculture Statistical Bulletin 461," 1967.

produce food should be preserved for that purpose; housing and other needs should be satisfied on non-prime-agricultural lands. Questions of land use will become more pressing in the future. To help make sound decisions for future generations, you must understand what makes land valuable for various purposes.

Soil Requirements for Crop Plants

Although some type of plant life will seemingly flourish on nearly any soil from desert sands to the shallow, rocky soils of mountains, for economical crop production, a soil must have at least the following six attributes or qualities:

1. In this day, it must be suitable for the use of farm implements. Excessively steep or rocky areas cannot be cultivated, although they may be suitable for grazing or similar uses.
2. A soil must have minimal losses from erosion under normal farming operations. Obviously, farming operations can

be adapted to meet this requirement in many instances—from farming on the contour (plowing around hills rather than up and down them) to modern techniques of minimum tillage which leave the soil surface protected with plant material, to leveling or terracing the land and installation of drainage ditches and pipelines.

3. Soils should also have suitable moisture-holding capacity. Thus extremely sandy soils typical of the western deserts and the perpetually wet, boggy swamp soils in certain southern areas may require special treatment to become productive; frequent irrigation of sandy soils and drainage of swampy soils may be needed. Soil moisture-holding capacity can be increased in sandy soils by the addition of large quantities of organic matter over long periods of time.
4. A soil must have good aeration to be productive. Aeration is a function of soil texture, of the activities of organisms such as earthworms, and of soil structure. Soil structure is influenced by the

amount of organic matter in the soil, and it can be destroyed by overcultivation of the soil or by cultivation or traffic on the soil when it is wet. Thus this requirement can be managed to a fair extent by farmers. Good conservation practices tend to favor a suitable moisture-holding capacity and good soil structure.

5. For successful farming, a soil must provide essential elements to plants. Soil fertility can be lowered by continually cropping. Over long periods of time, as land lies idle, fertility may be restored naturally, but in practice, soil fertility is managed and maintained by the use of appropriate fertilizers. Manure and other organic wastes also may provide essential elements.

6. The soil must be free from toxic substances, including natural accumulations of salts, such as the chlorides and sulfates of sodium and magnesium. Excessively acid or alkaline conditions affect the availability of essential minerals more than they affect the plants directly. In some soils, toxic levels of specific elements may occur. Aluminum toxicity in acid soils in humid southern areas is not uncommon. Careful management of fertilizers and of the soil can minimize problems of fertility and toxic substances.

The modern crop producer or manager must be aware of potential problems so that preventive or corrective measures can be taken before the problem becomes insurmountable.

SOIL CLASSIFICATION AND NOMENCLATURE

Part of any science includes classifying and naming the objects under study in some organized, systematic manner. Examples range from the periodic table of chemical elements to the science of plant and animal taxonomy. Soils also have been the subject of both classification and organized naming. The system currently in use was developed by U.S. Soil Conservation Service scientists in the period from the late 1950s to the 1970s. Additions and modifications make this system a constantly growing project; it has been recognized as the Comprehensive System of Soil Classification and Nomenclature. This system includes two crucial elements: the classifying and the naming of soils. Thus it serves to group different soils by their similarities and to provide a common basis for classification and nomenclature. These features allow the method to apply on an international basis and foster exchanges of research findings among scientists of different nations.

A given soil is ultimately recognized by soil series and type. The soil *series* is specified by characteristics of the soil profile, a vertical view of the soil from the surface downward to a base layer or horizon, generally considered to be parent material. Soil profiles differ markedly in the color of the soil, the depth and differentiation of horizons, and the presence of distinct patterns or structures. Soil *type* is defined as the texture of the surface layer of a soil. Thus a soil identified as a Yolo fine sandy loam has the profile characteristics of the Yolo series, a soil common to parts of the Sacramento Valley near Davis, California; the surface soil has a fine, sandy loam texture. An example of a soil profile is shown in Figure 3-11.

About 100,000 different soils representing more than 7000 series have been described in the United States. Obviously if these are treated as unrelated units, the mass of unrelated data will be of little value. The comprehensive system of soil classification first groups a soil into one of ten soil orders, based on specific measurable characteristics,

Figure 3-11. Closeup profile of Lofton silty clay loam showing layers of horizons. Profile development results from materials being leached from upper layers of the soil and accumulated in lower layers. (*L. L. Jacquot/USDA-Soil Conservation Service*)

which commonly include consideration of the general climatic conditions under which the soil order is found. To a large extent, this very general classification identifies major factors which limit agricultural potential, e.g., arid or desert conditions or extreme cold. Each order is divided into two or more suborders. Suborders are divided into great soil groups, which are in turn subdivided into subgroups. Each subdivision gives more specific characteristics and more precise identification of the soil. The classifications are based on measurable properties of the soil and climate rather than on description of a profile, so that scientists can compare soils at widely different locations. The subgroups are divided into families, and families into the ultimate classification, the soil series.

An integral part of the system of classification is the system of nomenclature associated with it. Although not described here, the nomenclature is valuable to soil scientists and those who work with soils. Crop producers can better understand the soil if they study soil surveys. Obviously, soils differ in their potential for crop production. Understanding the various potentials and the factors that affect potential is critical for business people who make decisions concerning farm credit and real estate appraisals. Finally, as we move to the reality of agricultural zoning and land-use decisions, an understanding of the soils involved makes it possible for laws to be based not on emotion but on the best possible evaluation of an area's potential for crop production.

SUMMARY

Crop production is essential for human life. We must be able to produce more on every acre. Soil is a fundamental resource for crop production, and prime land is diminishing. Plants require specific elements for healthy growth, and these can be provided by synthetic fertilizers, but energy is required to make most fertilizers. Crop rotation affects soil fertility; including a legume in the rotation can increase available nitrogen as a result of symbiotic nitrogen fixation. Manures may also provide essential elements, but they are not concentrated sources. Soil classification is important to many people other than soil scientists. The Comprehensive System of Soil Classification and Nomenclature includes a specific naming procedure that describes many attributes of soils.

The soil is the fundamental resource for crop production. Soils have many different physical and chemical properties, including fertility, pH, texture, and structure. Soils differ in the amount of moisture they can retain and provide for crop growth. They differ in their value and their suitability for crop growth. When crops are harvested, essential elements are removed from the soil. All crops require the same essential elements. The macronutrients are elements needed in large amounts; the micronutrients are elements needed in relatively small amounts.

QUESTIONS FOR REVIEW

1. What are the three basic parts of the soil? Why is each important to agriculture?
2. List the macronutrients.
3. List the micronutrients.
4. Could you with certainty identify a nutrient deficiency from field symptoms? Why or why not?
5. A 50-lb [23-kg] bag of fertilizer is labeled 16-0-5. How many pounds of N, P, and K does it contain? What do the three numbers (16, 0, and 5) stand for?
6. What is nitrogen fixation? Why is it important? With what organisms do you associate it?
7. Discuss the purpose of a soil test.
8. What is the purpose of liming the soil? Where is liming most common? Why?
9. Is available land a serious problem in crop production?
10. What are the major factors that set the potential of land for crop production?

Chapter

4

Managing the Crop and Crop Environment

The *environment* can be defined as the sum of all external factors affecting an organism. Organisms differ with respect to how they react to environmental factors, and the relative importance of different factors may change as an organism grows and develops. How the same environmental factor affects organisms differently can be illustrated simply in several ways. Consider heat, or temperature, which is an important environmental factor, and its effects on humans. If you have a baby brother or sister, more than likely your home is kept somewhat warmer than a home with only teenagers and adults. You may have noticed that your grandparents also tend to keep their home warmer than the temperature that is most comfortable for you. Recall, too, that when you are sick, doctors frequently suggest staying warmer than usual. These familiar examples show three things. Temperature (or heat) is an environmental factor. People respond differently to this factor depending on their age and health, and, within some limits, we can modify this factor for our comfort and well-being.

Like other organisms, crop plants are exposed to a changing environment. Various environmental factors affect crop growth and ultimate crop yield and quality. It is important for agriculturalists to understand what factors affect crop yield and quality, how they affect yield and quality, and how these factors can be modified or otherwise managed. This understanding is important in all steps of the food chain from the farmer producing the crop to the processing of the crop products to the ultimate preparation of a meal in the family kitchen, because at each step environmental factors may affect the nutritional value of the crop product or food. A major worldwide goal is to provide better nutrition to the human race. To do this, we must produce more and higher-quality crops and process these crops to minimize nutritional losses. This requires an understanding of the environmental factors that affect crops and crop products.

CHAPTER GOALS

After studying this chapter you will be able to:

■ **Identify major physical and biological factors that affect crops.**
■ **Describe how crops are affected by these factors.**
■ **Explain how a farmer may either change a factor or manage a factor or crop to increase yields.**

WEATHER AND CLIMATE

Weather and climate are both important aspects of the environment. It is important to realize that weather and climate are not the same and that the environment is composed of more than weather and climatic elements. *Climate* is a general concept that describes the long-term weather (or meteorological) conditions. We consider climates in terms, for example, of hot summers, wet winters, cold dry winters, cold snowy winters, etc. General climatic conditions significantly affect what crops are grown and how they are grown. For example, cotton requires a long growing season; thus it is not grown in areas with a frost-free period (an important climatic factor) of much less than 180 days. In areas with a dry climate, like much of the southwestern United States, successful crop production depends on irrigation to overcome limitations on crop growth imposed by low or inadequate rainfall. In the United States, some of our major areas of crop production are determined by climatic conditions. This is illustrated in Figure 4-1.

Weather describes current or short-term conditions; thus it is more variable. Both climate and weather affect crop yield and quality. Climate may dictate the regions within which a crop can be grown (e.g., cotton) and how it must be grown or managed in the field (e.g., irrigation). Weather, because of its short-term effects, is more difficult to manage but may have disastrous effects on crops. Hail storms as cereal crops approach maturity may cause complete crop failures, and unexpected frosts may severely reduce fruit yields of many orchards, unless the farmer takes decisive action in time. We have improved our ability to forecast weather dramatically in the past 20 years. To a large extent, this has reduced certain risks in crop production, as long as growers understand the forecast and its implications for management of their crops. At the same time, scientists have discovered more about how weather affects crops and how to minimize damage from unfavorable conditions. In some areas, actually changing the weather, mostly through rain-making, may become a practical reality.

ENVIRONMENTAL FACTORS

Naming or listing parts of the total environment is relatively simple. In doing this you must remember that the effects of these factors change, depending on the organism you are considering and the stage of growth or development of the organism. To further complicate the situation, you must also realize that these factors are frequently not independent; two or more factors may combine to affect an organism in a manner different from the separate effects of the factors involved. For example, the effects of high temperatures are greatly intensified by dry, windy conditions.

For convenience, environmental factors can be classified as occurring below ground

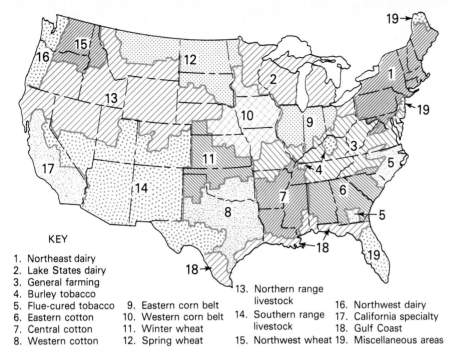

KEY
1. Northeast dairy
2. Lake States dairy
3. General farming
4. Burley tobacco
5. Flue-cured tobacco
6. Eastern cotton
7. Central cotton
8. Western cotton
9. Eastern corn belt
10. Western corn belt
11. Winter wheat
12. Spring wheat
13. Northern range livestock
14. Southern range livestock
15. Northwest wheat
16. Northwest dairy
17. California specialty
18. Gulf Coast
19. Miscellaneous areas

Figure 4-1. Major farming areas in the United States, including pasture (dairy) and rangelands.

or above ground. Keep in mind that some factors, for example, temperature, have a role both above and below the surface of the soil.

Below-Ground Factors

Minerals. The below-ground portion of the environment includes a number of components. The mineral portion of the soil must be considered. This component determines many of the physical and chemical properties of the soil, including some aspects of soil fertility and soil texture. The soil atmosphere is the soil's air, found in the pore space of the soil. (Remember that the pore space is determined in part by the soil's texture.) Remember also that roots must have oxygen for respiration: flooding and poor cultural practices can disrupt the soil's structure, hence the soil-plant-air relationship in the root zone, or *rhizosphere*.

Biotic and Organic Matter. The soil also contains a multitude of living organisms, from moderately large animals—squirrels and other rodents—to worms, insects, and microorganisms including fungi, bacteria, nematodes, and viruses. These organisms vary in their impact on plants: some affect plants indirectly by disturbing the soil, others directly by actually eating the plant (e.g., some rodents and insects). Certain insects and microorganisms are directly beneficial to plants, particularly the bacteria and fungi that are largely responsible for the breakdown and decay of organic matter and subsequent release of essential minerals for use by other plants.

Managing the Crop and Crop Environment

Soil organic matter is the product of plants and animals living in or on the soil. It may range from the raw leaf litter of the forest's floor to the fully decomposed materials which provide essential minerals for plants. Organic matter affects the soil, hence plant growth, in a multitude of ways. The organic matter is a source of essential minerals. It also affects the physical character of the soil, notably soil texture, which greatly affects the rate of infiltration of water into the soil, the rate of temperature change, and the general tilth of the soil.

Above-Ground Factors

Moisture. The most commonly mentioned environmental factor is moisture. The effects of moisture on crop growth can be attributed to at least three of its aspects. The first is the amount. Crop species differ drastically in how much water they require for satisfactory yields. Generally, rice is considered to require the most (up to 60 in/acre, or about 150 cm/ha). Some species of range grass require the least, but among cultivated crops, proso millet, common millet, and grain sorghum require relatively little moisture. Because it is grown under cold conditions, winter wheat in Montana may produce an economic yield with from 8 to 10 in [20 to 25 cm] annual precipitation, but generally there is an additional 3 to 4 in [7.5 to 10 cm] of moisture stored in the soil.

The distribution of moisture—that is, the season of the year when it falls—can be nearly as important as the amount of precipitation. The soil can store some moisture; the amount depends on soil texture. When the soil is saturated with moisture, additional precipitation does not benefit plants and may cause damage by flooding. The most effective moisture occurs when plants are growing actively and removing moisture

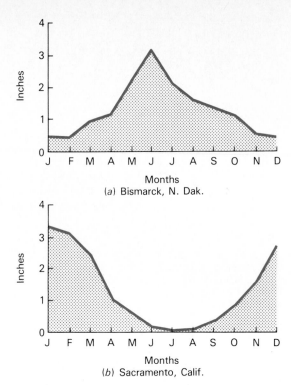

Figure 4-2. Differences in seasonal distribution of rainfall. (a) Bismarck, North Dakota. (b) Sacramento Valley, California.

from the soil. Generally this is in the spring and summer months. In some areas, such as the southeastern United States, total precipitation is a misleading figure. The total amount, commonly over 35 in [about 90 cm], suggests that moisture never limits crop growth, yet periodically there is a dry period of 2 weeks or more in midsummer that drastically reduces the yields of crops such as corn and soybeans. As a result, irrigation is becoming more common. In semiarid regions, such as Montana, winter wheat requires late summer or early fall rains for good stand establishment. Drought conditions in the fall, although not common, are a continuous concern to farmers in such areas. Figure 4-2 shows the difference in the distribution of rainfall in the Great Plains and on the West Coast.

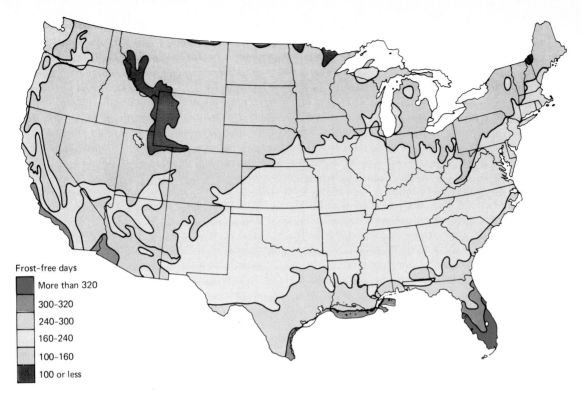

Frost-free days

More than 320
300–320
240–300
160–240
100–160
100 or less

Figure 4-3. Average number of frost-free days for various parts of the United States.

Finally, the type of precipitation is important. Many people think only of rainfall as precipitation which affects crops directly. This is not entirely true. Snow provides moisture for crop use and is a major source of water supply for rivers, lakes, and reservoirs for irrigation. In addition, snow serves as a critical layer of insulation to protect crops from freezing damage. The intensity of storms must also be considered. Remember, if moisture does not reach the roots of the plant it is of little value. Although heavy downpours typical of thunderstorms may contribute a great deal to the total annual precipitation, they are of little benefit to crops, because the rate of precipitation is far greater than the infiltration rate of the soil (the rate at which water enters the soil). Thus, the rain runs off and is

lost. Of greater consequence, this runoff is a major contributor to the nation's most serious conservation problem, soil erosion.

Temperature. As noted above, temperature is related to crop water use. Remember that both air and soil temperature are important in considering the effects of temperature. Two aspects of temperature are of prime concern: temperature extremes and temperature duration. Temperature extremes refer to the highest and lowest temperatures. Duration refers to the time a given temperature condition persists. The frost-free period is an important aspect of temperature in crop production. It is defined as the number of days from the last killing frost in the spring until the first killing frost in the fall. Figure 4-3 shows the average number of

Managing the Crop and Crop Environment

frost-free days for various parts of the continental United States. Crops differ remarkably in the minimum frost-free period they require. Sugarcane and cotton have the longest—from 150 to 200 days or more for sugarcane—and certain cultivars of winter wheat the shortest, around 90 days. Of course, winter wheat takes advantage of frost-free days in the fall to germinate and establish seedlings prior to winter dormancy; allowing it to complete its life cycle rapidly in the following spring and summer.

Soil temperature is just as important as the temperature of the air. It is affected by soil texture, color, and the amount of water in the soil. Generally, sandy-textured soils heat and cool more rapidly than clay-textured ones. With all other factors equal, a dark soil absorbs more heat than a light one, and moist soil changes temperature more slowly than dry soil. Because seeds of different species germinate at different temperatures, understanding soil temperature is important. In addition, certain soilborne plant pathogens become less harmful as temperatures change; the diseases they cause are managed in part by planting according to the soil temperature rather than according to date.

Light. Light is a critical factor. Crops are grown because they convert solar energy to food through the process of photosynthesis. Obviously this requires light. Three aspects of light must be considered: duration, quantity or amount, and quality.

Duration of light refers to the length of days. Days vary in length in two ways—by the season (long summer days and short winter days) and by geography (near the equator there is much less difference between the longest and shortest days than there is at the north pole). Duration of light is important in terms of the photoperiodic response of plants, which is the flowering response to relative length of light and dark periods as the season progresses. Some crop species require increasing day length to initiate flowers, while others require decreasing day length. Some seem not to react to the relative length of day and night. In some crops, such as soybeans, cultivars differ in their reaction to the photoperiod. Seed producers and greenhouse operators must understand photoperiodicity and how to manage it.

Light quantity is frequently measured in terms of intensity or brightness, although light energy is the important force to measure. Modern scientific instruments have allowed us to measure actual light energy, rather than merely brightness, although from a practical point of view they are closely related. Light energy received by a crop can be affected by many things. Shading or competition from weeds or as a result of an excessively heavy seeding rate are examples. Sometimes the amount of light must be reduced. This is important in the production of ornamental plants and some types of tobacco, in which artificial shade is employed. Light energy is related to temperature, and scientists have developed equations that predict how rapidly crops will grow, when they will mature, and their water needs on the basis of these so-called photothermal units. A *photothermal unit* is a measure of both heat and light energy. The ability to predict when a crop will mature is most helpful in scheduling plantings and harvests of crops that require some type of immediate processing, such as peas or corn for freezing.

Light quality in the simplest sense is the color of the light. Sunlight, which appears to be white, is composed of light of different colors. This is easily seen in the colors of a rainbow. The color of light is determined by its wavelength, which also determines the energy of the light. Red light has a rela-

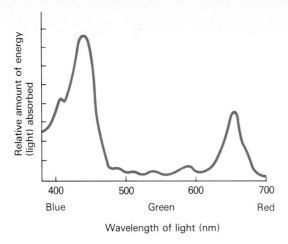

Figure 4-4. Light used in photosynthesis, or absorbed by leaves.

tively long wavelength—about 750 nanometers (nm)—and is lower in energy than blue light, with a wavelength of about 425 nm.

Not all light is visible to the human eye, and not all the light our eyes detect is used by green plants for photosynthesis. However, green plants use only light within the range visible to human (about 425 to 775 nm) for photosynthesis. Figure 4-4 illustrates plants' use of more red and blue light in photosynthesis and transmission of green light.

Biotic Factors. Biotic factors make up an important part of the plant's environment. Humans can have the greatest impact; they can intentionally alter the plant's environment, to the extent of controlling it completely in some instances. This is done only on a limited scale, by research scientists using sophisticated, extremely expensive equipment. On a larger scale, however, partial modification occurs in many farming enterprises. Producing crops in greenhouses or glasshouses is a good example of human management of certain elements of the plant's environment. Unfortunately, some human activities modify the plant's environment in an unfavorable manner. The best examples of this appear to be the impact of air pollution (smog) from cars in parts of southern California. Careful studies have revealed specific damage to citrus trees resulting from smog. Damage could reach a level at which production in some citrus groves would no longer be economic. Frequently agriculture is charged with pollution arising from the use of agricultural chemicals, from fertilizers to weed killers, insecticides, and similar chemicals. The criticisms directed against farmers appear to be far more severe than evidence justifies. Nonetheless, anyone using an agricultural chemical should use it with extreme caution and follow all directions carefully.

Grazing animals (both domestic livestock and wild animals) are important biotic components of the plant's environment. In the absence of human interference, a natural balance seems to be maintained for the wild animals. However, the introduction of large numbers of domestic livestock to parts of the Great Plains and western rangelands has led to long-term, if not permanent, changes in the native plant communities of these regions. Livestock, like humans, are selective eaters. They eat first what they like best, but when the forage is scarce or there are too many animals, they will eat nearly anything and nearly denude an area. This changes the competitive relationships among plant species, and when ground is left bare, opens the door for serious erosion problems. Range scientists monitor areas for overgrazing. One of the first signs of impending trouble is a change in the proportion of species making up a given range area. Frequently plant species that are not preferred by livestock increase while the proportion of desired species decreases. A good illustration of well-managed rangeland is shown in Figure 4-5.

Figure 4-5. In some areas, range is the best use of land, ensuring a thriving livestock industry. (*Frank Roadman/USDA-Soil Conservation Service*)

Insects and various microorganisms, in addition to other larger animals, are part of the above-ground environment. In some instances these are beneficial, if not essential. to the plant. For example, certain plant species depend on particular insects to serve as pollinators, without which reproduction, hence seed and fruit production, will not occur. Other insects are serious pests. The grazing or feeding of insects such as grasshoppers can devastate crops and rangeland. Other insects, such as aphids, serve as vectors by carrying disease-causing organisms such as viruses between infected and healthy plants. A major task confronting agricultural scientists today is determining what types of damage can be tolerated and how to minimize significant insect-related damage without disturbing or polluting the environment. This field is now recognized as *integrated pest management* (IPM); it will become increasingly important in the face of a continuously growing population, with growing demand to produce more per acre and decreasing tolerance for losses of crop yield from pests.

THE GOAL OF CROP PRODUCTION

Crop plants directly or indirectly provide us with necessities of life: food, feed, and fiber. Many crop products are or can be consumed more or less directly as food. These include common foods such as fruits and many vegetables. Other crops must be processed before they are consumed. The conversion of wheat to flour and of flour to a multitude of baked goods is an excellent example. The conversion of sugar beets or sugarcane to a bag of refined sugar represents a more complex level of processing.

In addition to providing food more or less directly to humans, plants serve as feed for livestock, which in turn produce food for humans. To some extent, the livestock process the feed so that it becomes an acceptable form of food for humans. In many instances livestock, because of their peculiar digestive systems, utilize plant products that could not be utilized by humans.

Finally, plants are the source of important fibers. Cotton is no doubt the best known, but linen from flax and rope from hemp are other products. Fibers also come from livestock which has fed on plants; wool is the most common example. As the shortage of fossil fuel becomes more intense, natural (plant) fibers will again become more important and popular, because petroleum products are used in the production of most common synthetic fibers.

The fact that humans depend on plants for food, feed, and fiber cannot be denied. Green plants provide these varied products using solar energy, water, and simple chemical elements. The key process is photosynthesis. The ultimate goal of crop production is to obtain the maximum amount of photosynthesis per acre (or other unit of area). This goal may be modified to the maximum amount of useful products of photosynthesis. We now also realize that while attaining maximum photosynthesis is a desirable goal, provisions must be made for the protection of essential natural resources such as the soil. A practical goal is to attain the maximum yield of desired crop products (products of photosynthesis) without permanently harming or losing essential natural resources. To achieve this goal requires the understanding and managing of those factors that affect crop growth and yield.

PHOTOSYNTHESIS

Photosynthesis is one of the most important biological phenomena known to humans. For all practical purposes, the sun is seen as an inexhaustible source of energy, but this energy can be used directly only by green plants and a few highly specialized bacteria to create nutritious food from nonnutritious, simple chemical elements. The process of photosynthesis captures the energy from the sun and stores it in food. The simple chemical equation

$$CO_2 + H_2O \xrightarrow{\text{Light}} \text{food} + O_2$$

does not describe the complexities but shows why photosynthesis is so important. Carbon dioxide and water are in some way combined, using energy from the sun, light, to produce food (the direct product is a carbohydrate food composed of only carbon, hydrogen, and oxygen in the ratio of 1C:2H:1O, or CH_2O, such as sugar). Oxygen is given off as a by-product. Plants and animals, including humans, use the oxygen for respiration.

The process of photosynthesis occurs in relatively few cells of the leaves and stems of green plants. These cells contain special structures called *chloroplasts,* which contain the pigment on which photosynthesis depends, *chlorophyll.* This is the pigment that causes a plant to appear green, because red and blue light are absorbed by it and used in the photosynthetic process, the green light is transmitted to be seen.

The light energy excites or energizes electrons in the chlorophyll molecule. These electrons then provide the energy for two critical events. First, there is energy to split a water molecule, H_2O, into its components, hydrogen and oxygen. This requires several steps. Next, energy is used to join the hydrogen (H) with carbon dioxide (CO_2) to yield carbohydrate (in shorthand, CH_2O). The oxygen is given off. Little can be done to change the process of photosynthesis, but because of its life-sustaining significance this detailed discussion should be of interest to you.

Figure 4-6. Controlled environment chambers are used by agricultural researchers to simulate "real-life" conditions. (*USDA*)

ENVIRONMENTAL MANAGEMENT

Under certain specific conditions, virtually all major environmental factors affecting crop growth can be managed directly. Researchers have developed highly sophisticated, controlled-environment chambers in which the effects of various environmental factors on plant growth and yield can be studied (Figure 4-6). These facilities, however, in spite of their scientific importance, are far too small to be important in the actual production of food, feed, or fiber crops.

In recent times, very large greenhouse facilities have been built for the production of several specialty crops. Normally these are used in the production of ornamentals, mainly fresh flowers. In addition, such facilities are used to produce valuable vegetable crops, such as tomatoes. Even the largest of

these facilities, which may extend over 100 acres [40 ha], contribute little if anything to solving the world's food problems, but they illustrate how environmental factors can be controlled, particularly temperature, moisture, and certain aspects of light. The development of large greenhouse facilities is becoming increasingly common.

Before considering how the crop environment is managed under field conditions, two concepts must be clearly understood. Management of the environment involves two distinctly different ideas. First, management may involve directly changing an environmental factor, as by using irrigation to overcome moisture conditions which limit crop growth. Second, management may involve handling a crop in such a way as to minimize the harmful effects of an environmental factor that cannot be changed: this is the case when a particular soybean cultivar is planted to utilize a given photoperiodic condition at a particular latitude.

Moisture

Arid Conditions. Of all the major factors that affect crop growth, available moisture is the one most frequently actually changed, through irrigation. In recent years, the use of irrigation has spread from semiarid regions of the western United States, where it is essential for economic production of many crops, to the more humid (wetter) areas in the East and Southeast, where it is a form of insurance against seasonal dry periods.

Irrigation has become a precise science. No longer is it a matter of merely passing water over the surface of the soil. The goal of irrigation is to get water to the rooting zone of crops in an amount and at a time when it can be used and is needed by plants. More and more, water is recognized as a valuable resource that should not be wasted. Efficient irrigation is important also because of the actual cost of pumping water and, in some instances, of buying water from a water company or governmental agency.

The goal of irrigation is to supply the right amount of water by an acceptable method at the right time and place. First, the grower must understand the pattern of root growth of the crop to know from what part of the soil the plants are removing moisture. This varies with different crops and as a crop grows and matures. The grower must also know how much water a crop uses in a given time period (commonly per day or week), plus total seasonal use, and how the use will be accelerated with unusually dry, hot, windy weather conditions. The grower must also know how much water the soil can store in the plants' rooting zone. Added to this,

knowledge of the soil's infiltration rate is important. Growers must know approximately how rapidly water will penetrate the soil. This determines how fast water can be applied to the soil's surface and may dictate the irrigation method to be used. The topography or geography of fields also dictate irrigation methods: on rolling or uneven terrain, various types of sprinklers have solved many problems. A modern sprinkler system is shown in Figure 4-7.

On the level valley lands of California and Arizona, for example, surface irrigation is practiced, using siphon tubes or special irrigation pipes to direct water down furrows between rows of crops such as many vegetables, corn, and sugar beets. Alfalfa for hay, pastures, and some orchards are irrigated, using low levees to divide fields or orchards into level borders or strips and basins

Figure 4-7. Modern sprinkler systems can cover large areas of uneven land with a minimum amount of labor. (*Valmont Industries, Inc.*)

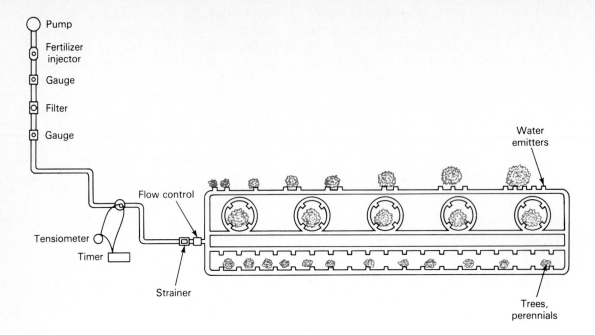

Figure 4-8. One example of a piping pattern used in drip irrigation. In orchards, where the trees are small and close together, the pipes can be laid parallel to the rows of trees with numerous orifices or water emitters. With large trees, the pipe is looped around each tree with several emitters. Tensiometers measure the moisture in the soil so that the delivery of water can be scheduled efficiently.

which are filled with water to irrigate crops.

Trickle irrigation, supplying a more or less continuous, light supply of moisture to individual plants, such as orchard trees or vines in a grape vineyard, is gaining popularity. It is the most effective method of conserving moisture but requires heavy investments in control equipment to regulate flow. An example of the special equipment used in this type of irrigation is shown in Figure 4-8.

As energy problems become more serious, irrigation will play an increased role in the application of various chemicals. When methods give precise control over the

amount of water applied—commonly sprinkling or trickle methods—fertilizers and some pesticides can be applied through the irrigation water. This procedure ensures movement of the chemical directly to the root zone; however, there is significant danger of causing pollution, since excess water and chemicals may drain into groundwater and ultimately contaminate streams, ponds, and even wells. This problem must be another concern of crop producers.

Irrigation is a satisfactory method of overcoming problems of inadequate precipitation if a source of water is available: rivers, lakes (including artificial lakes in appropriate watershed areas), or wells. As the use of irrigation spreads, laws governing water rights and uses will become more important and crop producers will require expertise also in this complex field.

Excessive Moisture. At the opposite extreme from arid conditions are conditions of excessive moisture. Most commonly these conditions are associated with known weather patterns, e.g., runoff from melting snow or periodic flooding from heavy seasonal rains. In some instances flooding dangers have been reduced by the development of lakes and reservoir systems, which provide water for irrigation later in the growing season. Generally such facilities are developed with state and federal assistance. On a single-farm basis, the installation of drainage ditches and sometimes subterranean drain lines may be required. This is expensive and, because lines may ultimately clog, may not be a permanent solution. Land leveling may also be effective in reducing flooding. The wise use of cover crops and protection of watershed areas are also important preventive measures. Finally, the selection of crop species or variety may be dictated by moisture conditions that cannot be altered effectively. Under drought conditions, for example, grain sorghum is better adapted than corn, and in areas where spring flooding occurs, forage grasses are useful because, in general, they tolerate flooded conditions better than legume species.

Temperature

Most people tend to think that little if anything can be done to change temperature, but this is not entirely true. It is true that changing temperature is not as simple as adding moisture through irrigation, and the extent of the potential change is not too great, yet actually changing the tempera-ture is possible through either heating or cooling.

Heating and Cooling. Commonly, temperature change is considered in terms of increasing temperatures. For most major field crops, e.g., corn, wheat, soybeans, little direct change is feasible. For crops that are expensive to produce and bring relatively high profits, mainly orchards and vineyards, heaters are used in the field (Figure 4-9a). The use of orchard heaters is shown in Figure 4-9b. The growth pattern of the plants, the canopy of leaves and branches, tends to hold heat (or warmed air) in the orchard. Effective use of heaters depends on numerous factors. As the cost of fuel continues to rise, these factors will become increasingly important. Orchard heaters are usually used in the spring as trees initiate growth following winter dormancy. The effectiveness of these heaters depends on how low the temperatures fall; usually heaters do not offer good frost protection if temperatures fall below about 27°F [−2°C]. The placement of heaters in the orchard is important so that warm air will not be blown out of the orchard. Wind speed and direction obviously are also important factors, as is the duration of the cold period. The size, shape, and topography of the orchard are important considerations, because exposed hilltops may chill rapidly and warm air from heaters may be lost due to wind. Growers who manage orchards and vineyards pay strict attention to weather forecasts and may rise in the middle of the night to check temperatures. Needless lighting of heaters is an expensive waste of fuel; failure to light heaters which might eliminate frost injury can result in damage to trees or vines, leading to severe yield reductions if not to crop failure.

In some foothill areas, large fans are erected in orchards and vineyards. If tem-

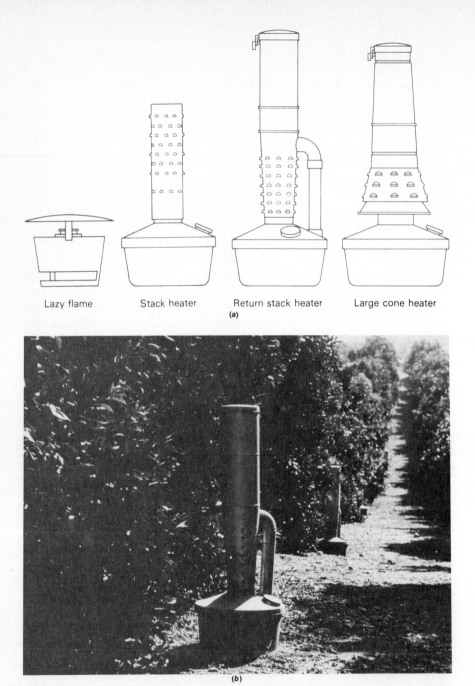

Lazy flame   Stack heater   Return stack heater   Large cone heater

(a)

(b)

Figure 4-9. (a) There are several types of orchard heaters: lazy flame, stack, return stack, and large cone. (b) Placement of heaters in an orchard depends on many factors: the type of crop, the type of heater, and the topography. (*Alan Pitcairn-Grant Heilman*) peratures fall slightly below freezing (29 to 30°F, or −1 to −1.5°C), operating these fans may minimize frost by stirring warm air in the orchard, as shown in Figure 4-10. The artificial wind also inhibits frost formation. Because water is a good heat exchanger, irrigation can also serve to warm an area. As temperatures approach freezing, operating sprinklers may protect a crop from frost. Various methods of surface irrigation are less effective than sprinkling.

Irrigation can also be used to cool a crop. In vineyards, excessively high temperatures as the fruit is maturing can be very harmful. Both sprinkler and surface irrigation methods may lower the temperature, but sprinklers are more effective, lowering temperatures 5°F or more.

Farming or production practices are more important in considering temperature management than direct methods to change the temperature. For example, seeding methods for winter wheat that place seed in a deep furrow protected by a ridge of soil tend to trap snow. A layer of snow 4 to 6 in [10 to 15 cm] deep over the seedlings tends to protect them, much as an igloo stays warm even in subfreezing weather. The rows should be at right angles to prevailing winds to minimize drifting of the snow and loss of this protective blanket.

Figure 4-10. Thirty-ft [9.1-m] high wind machines, like the one shown, pull higher warmer air down into the orange trees during cold winter nights. (*USDA*)

Crop Species. Two of the major management elements related to temperature are choice of crop species and of the cultivar. Species vary markedly in tolerance of cold. Tolerance also varies with growth or developmental stage of the plants. Among the major food crops of the world, corn, rice, and sorghum are warm-season crops and do not tolerate cold weather well. Sunflowers are also warm-season plants, but seedlings of sunflowers tolerate frost far better than corn seedlings. Thus, in areas where there is a danger of late spring frosts, growing sun-flowers may be preferable to growing corn. Wheat, oats, barley, and rye, however, are cool-season crops. Rye tolerates wet, cold conditions better than any of the major food crops. Although yields of winter wheat may surpass those of spring wheat, in some areas, such as northern Montana and North Dakota, winters are so severe that even the most cold-tolerant cultivars will not survive, thus spring wheat must be grown.

Dormancy is a characteristic of many cold-tolerant cultivars. There are marked differences in cold tolerance and dormancy.

Managing the Crop and Crop Environment

Among cultivars of alfalfa, *Medicago sativa,* two major groups—dormant and nondormant—are recognized. The nondormant types are grown in areas in which winters are mild and there is little danger of freezing damage. In regions with cold, severe winters, dormant types must be grown or stands are lost from freezing. The biology of dormancy of plants is an important topic. The trait is inherited and is known to be very complex. Biochemical changes at the cellular level have been associated with the onset of dormancy; plant hormones also play a major role in the onset of and return from dormancy.

Planting Date.   Part of the management for temperature centers on field operations, particularly planting date. Obviously, if seedlings of a crop are sensitive to frost, the planting date should be delayed until the danger of a late spring frost has passed. This means the grower must be familiar with weather and climate records and data. In some instances, the planting date is delayed until the soil temperature falls to or below a specific point at which harmful pests are not active. This is an accepted method to reduce damage from fungi that attack the seedlings of winter wheat, for example.

Soil-Temperature Relationships.   A number of factors affect the soil temperature. Obviously, the air temperature and changes in it are critical. Rarely will changes in air temperature affect the soil immediately to a depth of more than 6 in [15 cm]. Of the soil factors that influence how the soil temperature changes, texture is quite important: sandy or coarse-textured soils change temperature more rapidly than clay or fine-textured soils. The color of the soil may also be important: dark soils tend to absorb heat more rapidly than light soils. Soil organic matter is directly related to soil color: high

organic matter content is tied to dark color. The moisture status of the soil also influences how rapidly the temperature changes: soils that are at or above field capacity change temperature more slowly than drier soils.

The use of plastic foams and mulches to increase soil temperature is becoming more common in the production of some vegetable or truck crops. Functionally, this practice converts an entire field to a cold frame for seedlings. Machinery has been developed to lay down strips of plastic and insert seeds through small openings. In addition to increasing the soil temperature and thereby fostering rapid germination and seedling growth, the plastic tends to smother weeds and reduce water loss from evaporation from the soil's surface (Figure 4-11).

Shading.   For some crops, artificial shade is used to reduce temperatures and to protect plants from intense full sunlight. In some instances, plantings are made on one side of a building or fence either to take advantage of added warmth or to avoid excessive heat. It is well known that southern exposures are warmer than northern. In early spring this phenomenon can be clearly observed as the first growth of lawns (and weeds) appears adjacent to buildings.

Light

Light is the energy source on which life depends. Unfortunately, among the factors that affect plant growth, light is one of the most difficult to change directly; thus management is limited compared, for example, to management for moisture through irrigation.

The maximum quantity of light energy available—far more than plants can use—is fixed by the sun. Management of plants to ensure that they receive optimal amounts of

Figure 4-11. Black plastic mulch is being placed down in this field to help foster watermelon production. (*Charles V. Hall/Iowa State University*)

light includes weed control to reduce shading and competition for light by weeds and the use of proper seeding rates and row spacings of minimize shading of crop plants by one another. Species and cultivar selection may also be important. For example, certain corn hybrids tolerate dense planting rates which lead to self-shading better than others. Obviously, if a dense rate is used, the proper hybrid should be planted. (Re-member, too, that if seeding rates are increased, higher levels of fertilizer and more water may be needed.)

Natural sunlight is the ideal source of energy for photosynthesis. Anything that alters the composition of sunlight may lower light quality, hence reduce total photosynthesis and ultimately reduce crop yields. Smoke, haze, dust, and smog all act as filters and can change light quality. You have seen the sun through smoke or haze appear more as a red-orange disc than its normal brilliant white image. Managing light quality, then, must include reducing air pollution. In greenhouses or in buildings, light quality is affected by the type of covering on the building (some plastics filter sunlight excessively) and by the type of indoor lights used. Nursery operators and others who work with indoor plants know that the standard light bulb produces too much red light and not enough blue. Special lights may be needed for producing plants indoors.

Finally, the natural length of the day cannot be changed. Management of the response of plants to the duration of light is achieved in two ways. First, crop species and varieties differ with respect to their photoperiodic response. Soybeans are an excellent example. They are very sensitive; some cultivars are better adapted to the long summer days in more northern regions than to southern conditions. There are also differences in corn hybrids. Second, assuming that other conditions are favorable, frequently the time of planting can be adjusted to ensure that seedlings will be exposed to appropriate conditions of light duration.

Air Movement

Winds are another facet of the plant's climate that must be considered, although

Managing the Crop and Crop Environment

Figure 4-12. Tall trees are often used to protect crop plants from wind damage.

little can be done to change them. In some areas from the Great Plains to the Salinas Valley of California, various types of windbreaks have been planted to protect crop plants from wind damage. Of course, these are effective only where winds virtually always come from one direction. The windbreaks may be large trees or one or more rows of relatively tall-growing plants, such as sunflowers, grown in fields planted to beans, tomatoes, lettuce or other high-value crops (Figure 4-12). Sometimes snow fences and other artificial barriers are used.

Damage from wind effects takes various forms. In the most severe cases, plants are uprooted or stems are broken and leaves damaged. In other instances, leaf tissue is seemingly burned or cut by blowing sand and silt. This can be a serious problem with leafy crops such as lettuce and cabbage. Usually windy conditions, except during storms, are associated with reduced humidity, which in turn means a higher plant transpiration rate, or higher rate of water use. Thus wind effects are sometimes expressed as drought damage. Windy conditions may greatly accelerate irrigation schedules, and farmers must closely monitor soil water availability and, where possible, provide irrigation as needed. Wind, of course, also affects irrigation efficiency: high winds increase evaporation from ditches, streams, and fields and can distort the precipitation pattern of sprinkler irrigation systems. This distortion becomes even more serious if chemicals are being applied through the sprinkler system.

Biotic Elements

The plant's environment is affected by a multitude of living organisms. Humans have the greatest impact, unfortunately at times a negative one. Insects are also important considerations. Both larvae and adult forms may damage crops in the field (or in storage in some instances), and the egg-laying activities of some insects (e.g., wheat stem sawfly) can be most damaging. See Chapter 7 for a discussion of insect damage to crops.

Not all insects are harmful. Some seem to be neutral, while others are directly beneficial, such as the bees needed to pollinate alfalfa and many orchard crops. Sometimes insects play a major role in the spread of a disease; in this sense their damage is secondary. An excellent example is the transmission by aphids of the virus that causes yellow dwarf disease in cereals (barley, oats, and wheat). Management requires selection of resistant cultivars where possible, insect control where feasible, and elimination of the weedy species in which the virus apparently survives over winter periods.

In addition to insects, multitudes of microorganisms affect plant growth. Viruses, bacteria, and fungi all may cause serious plant diseases. It is essential to be able to recognize disease symptoms and take appropriate remedial steps. Of greater importance is the ability to recognize the conditions that favor disease development and prevent the problem. The best single bit of insurance is the planting of certified seed of a crop cultivar that is resistant to the most serious local diseases. Plant breeders and seed producers strive to keep a continuing supply of adapted, high-yielding cultivars available to meet changing disease problems.

In addition to bacteria, fungi, and viruses, nematodes of various types cause major crop losses. These soil inhabitants must be identified by an expert. Control is by the growing of nonsusceptible crop species and resistant cultivars, prolonged rotation away from sus-

ceptible crops, and actual fumigation of entire fields.

Not all microorganisms are bad. Bacteria of the genus *Rhizobium* associate with particular species of legumes and through a symbiotic relationship convert atmospheric nitrogen, which cannot be used by most plants, to a usable form. This explains why legumes are often considered to enrich the soil. In addition, many bacteria and some fungi are critical in the breakdown of plant and animal tissue in the soil and the subsequent release of the minerals in these tissues for use by plants.

Many other organisms are part of the plant's environment. Worms are significant in the amount of soil they move and their impact on soil aeration and water infiltration. Burrowing animals also stir the soil and add organic matter, and of course many animals graze directly on a wide range of plant species. Overgrazing has become a significant concern of stock growers, range scientists, and conservationists. Overgrazing reduces the nutritional value of rangelands, but even worse, it may leave the land bare and subject to erosion. Also, overgrazing destroys, or at least disrupts, the natural balance with wildlife (Figure 4-13). Understanding the environment and plant growth as well as grazing and wildlife needs leads to range management plans that make livestock production compatible with the best range and wildlife conservation practices and also allow vast areas not suited for the production of food crops to be used in meat production.

Soil

The soil, discussed in detail in Chapter 3, is also a major and fragile part of the plant's environment. Farming activities must be regulated to (1) minimize erosion—still our

Managing the Crop and Crop Environment

Figure 4-13. Overgrazing reduces the nutritional value of rangelands, subjects the land to severe erosion, and disrupts the natural balance. (*USDA—Soil Conservation Service*)

nation's number one conservation problem; (2) ensure sustained productivity through proper fertility management and pH regulation through liming; and (3) protect productivity by maintaining optimal levels of soil organic matter. Remember that, other than human resources, soil is the most important natural resource for crop production.

SUMMARY

Crop environment relationships are extremely complex. They must be understood if the joint goals of maximum photosynthetic productivity and resource conservation are to be realized. Meeting these goals will require an increasingly higher level of scientific knowledge for farmers and men and women working in professions serving farmers. No one person will be able to keep up in all the complicated fields of study.

With increasing technical knowledge, we will be prepared to recognize problems, even those we may currently be unaware of, and to develop solutions. Knowledge will lead to better overall decision making in management.

In addition to the direct importance of technical knowledge to farmers and other agricultural professionals, this knowledge is important to all citizens, since all of us depend on agriculture and need to appreciate the problems of the farmer and the complexities of crop and livestock production. As voters, this knowledge should help us make wiser decisions as we become more involved in decisions on, for example, land use, zoning, and pesticide regulations.

QUESTIONS FOR REVIEW

1. Define *environment*. Is the environment for a crop constant? Why or why not? Give examples.
2. Discuss moisture as a factor in the crop environment.
3. Discuss temperature as a factor in the crop environment.
4. What aspects of light affect crop growth, health, and yield?
5. What biotic factors affect crop growth? Give examples.
6. What are the goals of crop production?
7. With regard to the process of photosynthesis, explain why the sun is considered to be the source of all energy.
8. Describe the basic chemical process of photosynthesis.
9. With examples, explain how temperature, light, and moisture can be managed by farmers.
10. Describe three methods of irrigation and give their benefits and examples of where each can be used.
11. Give examples of how fields can be heated or cooled to manage temperature.
12. Name six biotic elements that affect crop growth, quality, and yield.
13. What is the most important conservation problem facing the United States?

Chapter
5

Weeds

Weeds are probably the most damaging pest agriculturists must control. Losses in crop yields alone amount to billions of dollars each year. It has been estimated that wild oats, one of the world's worst weeds, causes 300 to 500 million dollars in crop loss in the United States in a single year. The weed also causes approximately 500 million dollars loss in crop yields in Canada per year. The dollar loss in crop yields accounts for only a part of the total cost of weeds in this country. If the cost of all efforts to prevent, eradicate, and control weeds were added to the loss in yield, the dollar figure would be even more staggering.

The reduction in crop yield is an obvious loss to the producer, but how might such a loss affect the consumer? Research in North Dakota has shown that 100 wild oats plants per square meter [about 11 ft^2] can reduce the yield of fertilized wheat by approximately 30 to 35 percent. What would such a loss mean in loaves of white bread? Assume that a field of wheat without a wild oats infestation would yield 50 bu/acre [4.4 m^3/ha]. One bushel of wheat will make enough flour to bake 73 one-pound [$^1/_2$-kg] loaves of bread. Assume that the wild oats reduced the yield by 30 percent, or 15 bu/acre [1.3 m^3/ha]. The 15 bu of wheat lost because of wild oats would make enough flour to bake 1095 one-pound loaves of white bread. Other cereal crops also are affected seriously by wild oats.

Our entire food supply is affected by weeds. At some point all field crops, vegetables, horticulture crops, and root crops are subject to some damage from weeds. The productivity of marine life and livestock used for food is also affected. As the demand for food continues to increase, it becomes increasingly necessary to reduce the adverse effects of weeds on food production.

CHAPTER GOALS

After studying this chapter you will be able to:

- Describe how weeds cause crop losses.
- Describe how weeds are classified.
- Identify the ways in which weeds spread.
- Describe how weeds can be controlled on both agricultural and nonagricultural land.
- Describe the growth characteristics of weeds.

IMPORTANCE OF A KNOWLEDGE OF WEEDS

Weeds are here to stay. Losses caused by weeds occur daily on the nation's farms and ranches. Weeds are very familiar plants in the environment, even to most people outside of agriculture. Weed-infested lawns, roadsides, gardens, forests, and parks are familiar sights. Even that favorite fishing hole may be full of aquatic weeds. Whether you live in the city or in the country, you are not immune to the adverse effects of weed infestations. Consumers as well as producers of food products are affected by the weeds found in our environment. The reduction of our food supply, the effects of using chemicals to control weeds, and the added costs incurred by the producer in controlling weeds are problems of concern.

A general knowledge of weeds and the effects of weeds on everyday life will help to develop an appreciation for the enormous problem that weeds create. A general knowledge of weeds will lead to more effective control, as each citizen seeks to prevent the introduction and spread of detrimental weed species in agricultural and nonagricultural land.

It is most unlikely that weeds will ever be eradicated. Control has been and will continue to be the first line of defense in the battle against this crop pest. Weed control cannot be achieved by agricultural producers alone. Weeds can thrive in a city park, backyard, suburban lawn, and community golf course as well as they can in the fields and ranges of our nation's crop- and feed-producing areas (Figure 5-1).

WHAT IS A WEED?

Everyone will agree that some plants should be classified as weeds. These are the plants that cause problems for everyone. Other

Figure 5-1. Unused areas around buildings are commonly infested with many species of weeds. (*Northcutt*)

plants are weeds to some people, but to others they are beneficial plants because of environmental or topographical conditions. Sweetclover to the beekeeper is most beneficial, but to many producers it is a plant of little value. The controversy continues, but no one seems able to find a definition that is satisfactory to everyone. Generally speaking, a weed is a plant that is growing where it is not wanted and doing more harm than good. This definition will be satisfactory for the discussion in this chapter.

Although this concept is generally accepted, at a given time in a given location and under a particular set of circumstances, it may be difficult to categorize a plant as a weed using this definition. Some plants normally considered weeds may reduce soil erosion on steep slopes. Other weeds serve as food for domestic livestock, wildlife, and birds. Still other weeds have such colorful flowers that they add natural beauty to an area. A knowledge of weeds allows us to weigh the good qualities of a plant in a particular location against the more undesirable or "weedy" tendencies of the plant.

In some cases one member of a plant family is a detrimental weed and another plant in the same family is a valuable crop. An example is the oil-rich safflower, which is in the same family (Compositae) as the Canada thistle. Other uses are being discovered for other plants which have in the past been considered weeds. Some weeds have been found to contain hydrocarbons from which rubber and plastic can be produced. Geneticists are experimenting with the use of selected strains of wild oats to improve the productivity of domestic oat varieties.

CLASSIFICATION OF WEEDS

Weeds, like crop plants, have individual characteristics. Knowing these character-istics, one is more likely to appreciate why weeds must be controlled. Once weeds are recognized, knowing the weed's growth habits will permit selection of an appropriate control method. In addition to specific plant characteristics such as shape of leaf, type and color of flower, root system, and type of growth, weeds are generally grouped on the basis of their life cycle as annuals, biennials, or perennials.

Annuals

Annual weeds are those that come up from seed, flower, produce seeds, and die in a year or less (Figure 5-2a). When small, most annual weeds are easy to control. The prime concern is to prevent these weeds from going to seed.

Annual weeds are further classified as summer annuals and winter annuals. The *summer annuals* (Figure 5-2b) germinate early in the growing season, grow and produce their seed during the summer months, mature, and die. In other words, their life cycle is completed during the period from spring to fall. Lamb's-quarters, pigweed, crabgrass, ragweed, and foxtail are examples of common summer annuals.

The *winter annuals* complete their life cycle during the period from fall to spring. They germinate in the fall and winter, live during the winter period, produce seed in the spring or early summer, and die (Figure 5-2c). Downy brome, field pennycress, and tansy mustard are some of the more troublesome winter annuals.

Biennials

Biennial weeds complete their life cycle in 2 years. During their first year of growth, biennials produce an extensive root system below the ground and a cluster of leaves, or rosette, above the ground. Weeds in this classification are dormant through the

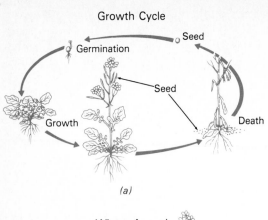

Growth Cycle

(a)

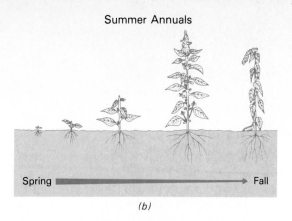

Summer Annuals

Spring ➝ Fall

(b)

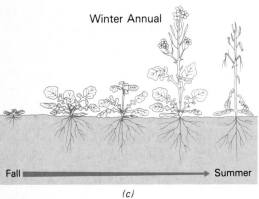

Winter Annual

Fall ➝ Summer

(c)

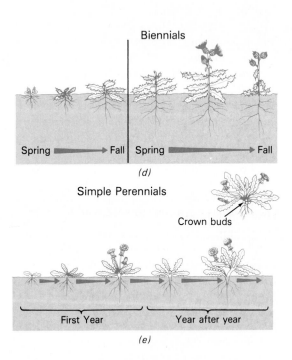

Biennials

Spring ➝ Fall   Spring ➝ Fall

(d)

Simple Perennials

Crown buds

First Year   Year after year

(e)

Figure 5-2. (a) Annuals complete their life cycle from seed in less than one year. (b) Summer annuals germinate in the spring, make growth, set seed, and die before fall. (c) Winter annuals germinate in the fall, over winter, mature, set seed, and die in the spring or early summer. (d) Biennials complete their life cycle within 2 years. (e) Simple perennials spread by seed, crown buds, and cut root segments. (*Montana State University Bulletin 669*)

winter months. When the spring growing season arrives they send up flowering stalks, set seed, mature, and die (Figure 5-2d). Biennials such as sweetclover, burdock, bull thistle, and wild parsnip are less common than the annual weeds. Control is most effective when the plants are in the rosette stage of their first year of growth.

Perennials

Perennial weeds live for 3 or more years. During the winter months the above-ground portion of the plant may die back, but early in each growing season the above-ground growth develops from the underground parts of the plant (Figure 5-2e). These weeds produce seeds plus an extensive root system, which may include rhizomes, tubers, or bulbs. Perennials not only reproduce by seeds but may also spread by asexual means, making control more difficult. Common perennials such as field bindweed, leafy spurge, Canada thistle, johnsongrass, and quackgrass are common throughout much of the United States.

Primary and Secondary Noxious Weeds

Some state weed laws classify weeds as primary noxious and secondary noxious weeds. Other states classify them as prohibited and restricted noxious weeds. The primary noxious weeds spread by both sexual and asexual means. Secondary noxious weeds are usually less difficult to control and somewhat less persistent. The weed law in the individual states will identify those weeds that fit into each category. The weeds included in these classifications vary from state to state, depending on the seriousness of the weed in the crops grown. Strict regulations control the amount of such weed seed that can be contained in seed sold for planting.

HOW WEEDS SPREAD

Weeds spread by both natural and artificial means, but most often the movement of weeds from one distant locale to another is associated with the movement of people. Clearing of land, plowing, fertilizing, and generally disturbing the soil through cultivation causes weeds to grow and become established more easily. Weed seeds cannot move across mountain ranges and span oceans without human help. Many of the more common weeds in the United States today were introduced into this country by the early pioneers.

Weeds spread quickly from one place to another when their reproductive parts are carried to a distant location and placed in contact with an acceptable growing environment. The annual and biennial weeds can spread only if their seeds are transported from one place to another. Perennials may be spread by seeds or by vegetative means (Figure 5-3).

Natural Dissemination

Natural dissemination occurs when the different reproductive parts of the plant are carried from place to place by wind, water, and animals, excluding humans. Many weed seeds are especially equipped to be moved easily by the wind. Wind is one of the most common means of natural dissemination of weed seeds. The dandelion seed has a feathery pappus that catches the wind and moves easily from one place to another. Usually weed seed adapted for wind dissemination is small and light in order to be carried by the wind. However, some plants are tumbled over the ground by the wind, dropping their seeds as they roll. Russian thistle and tumbling mustard are good examples. Other weed seeds are blown along the ground or over the snow by the action of a strong wind.

Flowing water moves millions of weed seeds from place to place. The amount of seed and the type of seed moved by water depends on the volume and velocity of the stream and the size and weight of the weed seed. The water in irrigation ditches often carries weed seeds. One producer's weed seed may be spread to another producer's fields as the water is used for irrigation. Leafy spurge has spread to infest new areas by this method. The erosion of the soil through runoff following heavy rains and subsequent flooding spreads weed seeds far beyond the banks of the stream. As with the

Figure 5-3. Weeds are spread in many ways. (*a*) Cultivation disturbs soil causing weeds to grow more easily. (*International Harvester*) (*b*) Transporting crops can spread weed seeds. (*Louisiana Tourist Development Commission*) (*c*) Water flowing through irrigation ditches can carry weed seeds. (*J. H. Pogue USDA-Soil Conservation Service*) (*d*) Birds can distribute weed seeds across vast distances. (*Vernon Ekedahl/US Fish and Wildlife Service*) (*e*) Animals grazing in fields contribute to the spread of weeds. (*Joseph Zaryl/USDA-Soil Conservation Service*) (*f*) Hikers can spread weeds when exploring backwood trails. (*Jane Hamilton-Merritt*)

wind, the smaller, lighter seeds are carried farther by flowing water.

The animals that graze the fields, pastures, and ranges are often carriers of weed seed. The barbed, prickly, and sometimes sticky seeds attach themselves to the hair or wool on the bodies of passing animals. These seeds may be carried great distances before they are shaken or brushed off the animal's body onto the ground. Weed seeds may be carried in the mud clinging to animals' legs and hair. Many weed seeds are carried in the digestive tract and are still viable when excreted in the feces of both wild and domestic animals. Birds are common carriers of weed seed, spreading the seeds over great distances as they excrete them in their droppings.

Artificial Dissemination

People and their activities provide the other means by which weed seed and vegetative parts of weed plants can travel from place to place and start new weed infestations. Weed seeds may cling to the trouser legs of a hiker in the forest and remain there for miles before being shaken loose to fall to the ground in a new location. The railroad cars and trucks that haul grain and produce from where it is produced to where it will be processed or consumed also distribute weed seed along the way. Even the tourist taking a new plant home from a distant state may carry a reproductive part of a perennial weed in the potting soil in which the plant is being grown. People moving animals have significantly spread many seeds, both in the animals' feed and in the feces.

Although weeds travel from place to place in a variety of ways, by far the most serious introduction of weed seeds into croplands is through planting crop seed contaminated with weed seed. Every year, farmers use their modern planting equipment to distribute weed seed of various types evenly across their fields.

WEEDS ARE COSTLY

The loss from weeds can mean the difference between profit and loss for the producer and higher food prices for the consumer. During any given year, farmers in the United States lose billions of dollars because of weeds. But the losses in crop yields do not tell the entire story. Higher production costs are incurred through additional tillage operations, the use of costly chemicals to fight weed pests, and dockage loss when weedy products are sold to the consumer or processor. Livestock producers suffer an estimated 3 to 5 percent death loss in cattle, sheep, and horses from poisonous weeds found on the range and pastures and in the hay and forage used for livestock feed.

The food-processing industry sustains huge losses each year in extra cleaning and processing costs. It has been reported that in the decade from 1969 to 1979, 16 million tons [14.5 million t] of grain were shipped to terminals at Thunder Bay on the Pacific coast of Canada. The grain contained an estimated 487,000 tons of wild oats [about 442,000 t]. At that time, excess transportation costs to ship the weed seed amounted to 2 million dollars, and the additional cleaning costs amounted to 8 million dollars (Figure 5-4).

What does the loss from weeds really mean to producers across the country? The soybean producer could lose from 10 to 20 percent of the yield. Assume that a soybean farmer's yield would have been 30 bu/acre [2.6 m³/ha], but 10 percent of the yield was lost to weeds. The farmer who harvested 300 acres [about 120 ha] would lose 900 bu [31.3 m³], which at 8 dollars per bushel would amount to 7200 dollars. Corn produc-

Figure 5-4. Wild oats can cause severe losses in small grain crop yields and increase a farmer's production costs. (*Northcutt*)

Figure 5-5. Millions of dollars are spent yearly controlling Russian thistle along freeways. (*USDA*)

ers may lose 15 to 25 percent of their yield because of giant foxtail and redroot pigweed within the rows of corn. Such a loss, in the approximately 65 million acres [26 million ha] of corn grown in the United States, amounts to a tremendous total. Researchers have found that just two Canada thistle plants per square yard [0.836 m²] caused an average field loss of 15 percent in wheat yield. The livestock producer can also suffer losses. In 1971, some 1200 sheep died in one night in Utah after grazing on a stand of halogeton, a poisonous weed.

Nonagricultural Losses

It was estimated that in 1978, the control of weeds in some 200,000 acres [81,000 ha] of weed-infested waters in Florida was costing the taxpayers of that state between 12 and

14 million dollars annually. A large part of this annual cost went toward weed control, while the remainder was used for research to find new ways to control aquatic weeds.

Millions of dollars are spent each year cutting, spraying, and burning weeds along our roadways and in our public parks, forests, and recreational areas (Figure 5-5). If the weed pests in these areas are not controlled, they become seed producing areas which supply weed seed to continue the invasion of the surrounding farms while reducing the beauty of our recreational areas.

HOW WEEDS CAUSE LOSSES

The most obvious way that weeds affect crop production is through their interference with the growth of the crops themselves.

Weeds reduce crop yield by competing with the crop plants for nutrients, moisture, and light. But this is only one of several adverse effects that weeds have on the producer and on our society.

Interference with Crop Growth

Weeds, like crop plants, require nutrients, light, moisture, heat energy, and carbon dioxide to grow successfully. Weeds are aggressive and perpetuate themselves at the expense of crop plants. Like crop plants, they thrive when associated with good soil and high fertility. Ultimately, such competition results in lower crop yields and reduced profits for the producer.

Often the weeds are more aggressive than the crop being grown. Thus the stronger competitors are able to absorb a larger percentage of the available nutrients and moisture. In some cases the weeds emerge and use up valuable nutrients before the crop plants are large enough to compete effec-

tively. In all cases there is some reduction in crop yield. The severity of the loss is related directly to the aggressiveness of the weed and the duration of its association with the crop. The more dense the weed infestation, the greater the loss in crop yields.

Weeds need water to grow. In fact, some weeds require more water for maximum growth than the crop in which the weeds are growing. For example, it has been estimated that one Russian thistle growing in a field of grain sorghum will use as much water as three sorghum plants. A mustard plant may use as much water as three oat plants, and a large sunflower in a corn field may use as much water as two and one-half corn plants. Weeds compete with crop plants for water, and the result of this competition depends on the environmental conditions, the root structure of the weed, and the crop being grown (Figure 5-6).

Plants must have light to grow. Weeds, particularly the tall broadleaf species, tend to shade crop plants. Low light intensity

Figure 5-6. A single weed in an irrigated field can reduce the yield of the crop in the immediate area of the weed.

slows down and limits the growth of the crop plant. Broadleaf spreading-type weeds growing with grass or grasslike crops are particularly adapted to compete for the sun's light. Experiments have shown that kochia, a tall plant often called fireweed, will shade irrigated sugar beets during the latter part of the growing season to such an extent that the weed will seriously lower the tonnage of beets harvested.

Some weeds live as parasites on other plants. These weeds attach themselves to the host plant and take their food directly from the host plant by sending rootlike projections into the stem or root of the plant. Field dodder is an annual generally found attached to clover, alfalfa, and lespedeza. Witchweed is a parasitic weed that attacks corn, sorghum, and sugarcane and can reduce corn yields by as much as 50 percent; it attacks the roots of the crop soon after germination, so the damage occurs before the weed emerges from the ground. Mistletoe parasitizes trees and shrubs in areas with an environment favorable for this species.

Many weeds act as host plants for diseases and insects. In most cases the plant serves as a host during one stage in the life cycle of the insect or disease. The common barberry is a host for the fungus that causes black stem rust in wheat in the northern wheat region. Johnsongrass is a host for maize dwarf mosaic virus, which attacks corn during the growing season, and the curly-top virus of sugar beets is hosted by weeds. Chinch bugs and European corn borers are often found overwintering in the stems of dead weeds. Horse nettle is a host for the Colorado potato beetle.

Weeds do more than compete for nutrients, water, and light. Some weeds secrete chemical substances which can inhibit plant growth. When these substances are released into the environment, the germination rate of some crop seeds is reduced and the growth rate of other plants is retarded. Quackgrass has been shown to reduce corn yields even though ample nutrients were available. Water extracts of spotted knapweed reduce both germination and growth of many seeds. In other cases these chemicals cause the formation of abnormal crop seedlings and keep the roots from elongating properly.

Effects on Agricultural Products

At the same time that weeds are reducing crop yields, they may also cause an undesirable effect on the crop or on other agricultural products produced from the crop. The palatability of hay may be seriously affected by the presence of weeds. Studies have shown that the protein content of weed-infested alfalfa hay may be lower than that of weed-free alfalfa. The price received for crop seed that is contaminated with weed seed is much lower than that for clean, weed-free seed. The presence of weeds in field crops in storage may cause enough spoilage, odor, or dockage to make the crop unsalable for human consumption.

The returns of the dairy farmer are often affected by weeds in the hay fed and in the pastures on which the cows graze. When eaten by dairy animals, many weeds such as wild onion, wild garlic, field pennycress, and ragweed affect undesirably the odor and taste of milk and milk products. Such milk cannot be sold for human consumption but must be sold at a much lower price for animal feed or simply thrown away.

Some weeds are particularly objectionable for the sheep producer. The burrs of cocklebur, sandbur, and burdock become so entangled in the wool of the sheep that the value of the fleece is greatly reduced.

Ranchers can also sustain losses from poisonous weeds growing on the range or in

Figure 5-7. Hemlock (*a*) and lupine (*b*) are common poisonous weeds that can cause extensive livestock damage. (*Northcutt*)

the forages they feed. The best way to prevent loss is to follow good grazing management practices (Figure 5-7).

Effects on Humans

Millions of Americans sneeze and cough their way through the hay-fever season each year. The pollen of ragweed and various other plants affects the respiratory system of susceptible persons. Poison ivy and poison sumac are examples of weed plants that cause various forms of skin irritation. In rare cases, children die as the result of accidentally eating poisonous plants. Other

people die on the roads and highways when tall, thick-growing weeds block the view of oncoming traffic at intersections and crossroads. Often, dry, dense weed growth creates a potential fire hazard.

Effects on Production Costs

Weeds increase production costs in many different ways. The major reason for cultivation is to destroy weed growth. Each time the producer has to cultivate a crop, the profits from growing that crop are reduced.

Machinery and equipment costs are higher because of weeds. Cultivation, planting, and harvesting equipment must be more complicated in order to function effectively in weedy field conditions. Labor and maintenance costs are higher as the weeds slow down harvesting and increase the wear on equipment.

Many physical characteristics of weeds help reduce profit. Certain weeds such as morning glory become entangled in the harvesting equipment, causing loss of some of the harvested crop. In other cases, the shape and size of the weed seed makes it difficult to separate from the harvested crop, resulting in a high percentage of dockage at the processing plant.

As weeds compete with crop plants for nutrients, fertilizer costs increase. Under proper farm management, higher crop yields mean lower production costs. Actively growing weeds that are competing with the crop will lower the producer's return while reducing the amount of food available for consumers.

WHY WEED CONTROL IS DIFFICULT

All professional athletes have certain unique characteristics that make them competitive in the position they hold on a team. Likewise, weeds have certain characteristics that make them more competitive when competing for the same raw materials as cultivated crops. These unusual characteristics have helped weeds spread and become established.

Annual weeds are the most numerous of the classes of weeds. Fortunately, they are often easier to keep under control. Most annual weeds produce a large number of seeds. The seeds are often very small in size and light in weight. This characteristic enables the seed to be moved quite easily by the wind. Although the annuals produce large numbers of seeds, many of the seeds are not viable, so the germination percentage is often low.

Most species of weeds possess the ability to produce some seed under adverse conditions. If moisture is a limiting factor, the weed produces a shorter, smaller seed stalk and fewer seeds, but usually enough seed to ensure the survival of the species.

Most crop plants are quite specific in the amount of time they require to germinate, mature, and produce viable seed. Under adverse conditions some weeds, however, can produce seed at an earlier stage in their growth cycle. Also, if conditions permit, some weeds will begin producing seed early and continue to produce seed over the entire growing season.

The seeds of many of the common weeds can be dormant for a long period of time and still germinate when conditions are favorable. The seed of some weeds will not germinate under any conditions for a period of time. Other weeds produce seed that will remain dormant when high concentrations of CO_2 are present. When brought near the surface of the soil, where the CO_2 level is low, the seed will germinate. Some weed seed needs light to germinate. Cultivation often brings the weed seed into the upper soil layers, inducing germination. Other annuals mature ahead of the crop, thus the seed shatters prior to harvest, allowing the seed to germinate quickly on the surface of the soil.

Many weeds seem to germinate and grow rapidly. This characteristic enables the weed seedling to come up with or ahead of many crops, resulting in very early competition with the crop.

Other weeds have the ability to reproduce themselves both by seed and vegetatively. As these plants grow, the infestation appears in larger and larger clusters out from the parent plants. Reserve food is stored in the underground structures to sustain the plant and enable it to produce regrowth after cultivation and to get an early start during the growing season in relation to the crop plant starting from seed. Much local spreading occurs as tillage equipment cuts and drags the vegetative parts of the plant to new parts of the field (Figure 5-8).

Weeds, as a group of plants, are highly competitive. Fortunately, few weeds possess all the special characteristics described; most, however, have enough of these features so that they are able to compete favorably with most crop plants for nutrients, light, and moisture. It is the combination of physical characteristics and growth habits of the weeds that makes them so troublesome for the producer.

PREVENTING THE SPREAD OF WEEDS

Many measures are being taken to prevent the spread of weeds. In 1974, President Ford signed the first Federal Noxious Weed Act, which was funded in 1979. Until this law was enacted, the importation of weeds was not regulated closely by federal law. More importantly, the weed act authorized the inspection of crop seed at ports of entry, gave authority for weed surveys in the United States, and allowed appropriate agencies to quarantine or hold seed in order to prevent the spread of noxious weeds.

State and regional laws regulate the movement of crop seed across state lines. These laws require that crop seed for commercial sale be properly labeled to alert the purchaser to any weed contamination. Local and regional weed districts are being established which require property owners to take measures to control noxious and secondary noxious weeds to help prevent further spreading. Weed prevention becomes a matter of not introducing weeds, destroying

Figure 5-8. Certain areas in the West are becoming seriously infested with leafy spurge. This perennial can produce thousands of seeds. (*Northcutt*)

their reproductive capacity, and preventing them from spreading from one area to another.

These laws are only as effective as producers and seed processors permit them to be. In many areas of the country, relatively little of the crop seed used is labeled as the law requires.

WEED ERADICATION

Weed eradication might be appropriately described as "mission impossible." The term *eradication* implies that a weed species and all its reproductive parts have been removed from a given area. Further infestation will

Figure 5-9. Canada thistle is a seriously dangerous weed on the range as well as in cultivated fields. (*Northcutt*)

be the result of the reintroduction of the weed into the area, either naturally or artifically.

From an economic standpoint, total weed eradication is practical only in confined areas such as greenhouses, small plant containers, or perhaps small geographic areas. Sometimes the high cost of eradication may be justified for small initial infestations of a highly noxious weed; and individual producers may eradicate persistent weeds such as Canada thistle from their land through constant management (Figure 5-9).

WEED CONTROL

Effective weed control must consider the physical characteristics of the weeds, the effects they have on crop production, the effect of various control measures on the environment, and the economic feasibility of using control measures. In most cases, effective weed control involves one or more control measures which can be classified as managerial, mechanical, chemical, or biological. Chemical and biological pest control are discussed in Chapters 8 and 9.

Managerial Control

Good farming practices, which include the proper use of land and water, are very important in the battle against weeds. Reducing weed infestations through good farming practices increases profits for the producer and reduces the need for chemical control of weeds.

Selecting and planting high-quality, weed-free seed is the crop producer's most effective weapon against weed infestation. The official inspection tag on the seed bag describes the kind and amount of weed seed that is contained in the crop seed. The producer should make sure that there are no noxious weeds and only a minimal number of the less damaging weed species in seed to be planted. Clean, certified seed will cost more, but the payoff comes in the form of higher yields. The producer must remember that the benefit of using clean seed is greatly reduced when it is planted in fields which have undesirable weeds in the seedbed (Figure 5-10).

Following good crop-rotation practices helps reduce the population of weeds that thrive under conditions similar to those required for the crop being grown. For example, raising small grain year after year in the same location gives rise to weeds that

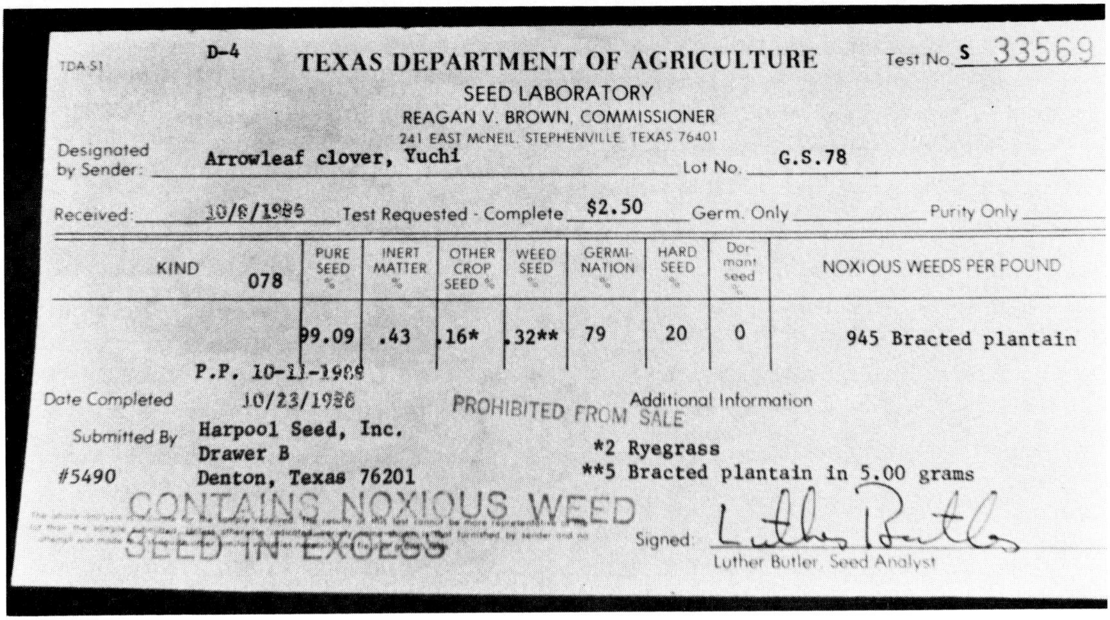

TDA-S1 D-4 **TEXAS DEPARTMENT OF AGRICULTURE** Test No. S 33569

SEED LABORATORY

REAGAN V. BROWN, COMMISSIONER

241 EAST McNEIL, STEPHENVILLE, TEXAS 76401

Designated by Sender: Arrowleaf clover, Yuchi Lot No. G.S.78

Received: 10/8/1985 Test Requested - Complete $2.50 Germ. Only ______ Purity Only ______

KIND 078	PURE SEED %	INERT MATTER %	OTHER CROP SEED %	WEED SEED %	GERMINATION %	HARD SEED %	Dormant seed %	NOXIOUS WEEDS PER POUND
	99.09	.43	.16*	.32**	79	20	0	945 Bracted plantain

P.P. 10-11-1985

Date Completed 10/23/1985

PROHIBITED FROM SALE

Additional Information

Submitted By Harpool Seed, Inc.
Drawer B
Denton, Texas 76201

#5490

*2 Ryegrass
**5 Bracted plantain in 5.00 grams

CONTAINS NOXIOUS WEED SEED IN EXCESS

Signed: Luther Butler

Luther Butler, Seed Analyst

Figure 5-10. The official inspection tag on seed bags ensures that producers know the quality of seed before they buy it. (*Jane Hamilton-Merritt*)

grow well with small grain. Good crop rotation in which a row crop is rotated with small grain helps break the weed cycle.

In the Great Plains grain belt, summer fallow has been used to aid in controlling annual weeds. The land is left idle for one growing season, and early weed growth is destroyed through various tillage operations. The key to good production in the semiarid and arid regions is to store soil moisture for subsequent crops. Weeds must be controlled before they mature and certainly before producing seed if moisture is to be conserved.

Highly competitive crops are also used to reduce the competition of weeds for light, nutrients and soil moisture. Rapidly growing crops can shade weeds and greatly reduce their growth. Well-managed legumes such as alfalfa and clover and small grains are often used where there is a heavy weed infestation danger. Delayed seeding, fall tillage, and use of forage crops as silage or green chop are other management practices used to help control weeds.

Mechanical Control

Mechanical control of weeds involves a variety of tillage operations, each intended to prevent weeds from germinating and growing or to keep them from maturing and producing viable seed. Mechanical control measures have the disadvantage that while the tillage operation may kill growing weeds, it will probably bring dormant weed seed nearer the surface, giving the seed an opportunity to germinate and grow.

Deep tillage, which usually consists of plowing and disking, is practiced in early spring and fall. These operations bring the lower soil layers to the surface while covering the growing weeds that were on the surface. Deep tillage brings many underground roots to the surface, where they may be killed as they are dried out by the sun or frozen by the winter's cold. But like cultivation, deep tillage brings weed seed to the surface, giving rise to new crops of weeds.

Cultivation or secondary tillage is used throughout the growing season to destroy many annual weeds and slow the growth of perennials. Cultivation forces plants to use stored energy for regrowth. This use may leave the plants to face stress conditions, e.g., winter's cold, with inadequate available energy. Shallow tillage operations (disking and harrowing) keep the weeds cut off just below the surface, preventing them from completing their life cycle. Regular tillage operations to cut off all shoots 3 to 4 in [7.5 to 10 cm] below the surface at definite time intervals are important for perennial weed control. This practice will help to deplete the root reserves of the weeds.

CAREERS IN WEED SCIENCE

The control and prevention of weed growth provide many job opportunities in a wide variety of working conditions. The effects of weeds are now being realized. Control of this agricultural pest will require the cooperative effort of a variety of specially trained agriculturists.

Researchers are needed to deal with the discovery and development of new products and ways and means of controlling weeds. Persons are needed to educate our society in the ways as well as the benefits of controlling weeds. Other workers are needed who are trained to market and handle the products produced by research scientists to control weed infestations. Finally, qualified persons are needed to administer, regulate, and evaluate programs designed to bring about more effective weed control.

SUMMARY

The first plant to occupy an area has a competitive advantage in that area. Naturally, this plant should be a crop or ornamental plant that serves a useful purpose. Introducing weeds into our cropland damages the crop not only the first year but for years to come. In fact, each year the weed infestation becomes more difficult to control and may damage the crop without producing any additional seed. Dormant weed seeds that may lie in the soil for years before germinating and vegetative parts of weeds that have the ability to produce new plants are a constant source of infestation. Not only producers but every citizen must remember that weeds cause more problems than all other pests and must be controlled.

QUESTIONS FOR REVIEW

1. What is a weed?
2. What are the differences among annual, biennial, and perennial weeds?
3. How do weeds spread?

4. In what ways do weeds cost both the producer and the consumer?
5. How do weeds interfere with crop growth?
6. What effects do weeds have on agricultural products?
7. Why are weeds so difficult to control?
8. How can the producer prevent the spread of weeds?
9. Why is it so difficult to eradicate weeds?
10. What careers are available to those trained in weed science?

Chapter
6

Parasitic Plant Diseases

Agricultural production is subject to enormous risks. At times these risks endanger the prosperity and subsequent long-term success of the farm or ranch. Plant diseases are one of the major risks facing agricultural producers today.

Plant diseases are capable of wiping out entire crops in a given area. In other cases, specific crops are no longer grown in an area because of the destructive potential of some diseases.

It is not possible to determine the exact amount of loss from crop diseases. In the United States alone, it is estimated that plant diseases account for over 3 billion dollars in crop losses each year. There are losses that result from the actual death of some plants, but in many cases the loss is the result of reduced yields and lower quality of the crop produced.

Plant pathology is a branch of science in which scientists, called *plant pathologists,* study plant diseases. It is an applied field of science which seeks answers to the problems of plant diseases. The field of plant pathology offers a promising career for those trained to deal with health of crop plants and with increasing food and fiber production (Figure 6-1). Those persons working in the area of plant protection have a need for practical knowledge and basic understanding of technical agriculture. Plant pathologists must have a knowledge and understanding of diverse fields such as crop science, soil science, microbiology, biochemistry, botany, and genetics. An understanding of plant diseases is one of the major areas of concern for plant pathologists.

CHAPTER GOALS

After studying this chapter you will be able to:

- Explain how plant diseases affect crop production.
- Identify the parasitic agents that cause plant diseases.
- Explain how plant diseases are spread.
- Identify ways in which plant diseases may be controlled.
- Identify some of the common symptoms of parasitic plant diseases caused by bacteria, fungi, viruses, and nematodes.

Figure 6-1. Spring wheat cultivars being grown for use in disease research. (*USDA*)

WHAT IS A DISEASE?

Plant diseases are caused by microorganisms that live in and on host plants. These tiny organisms are called *pathogens*. Plant pathologists are not always in agreement on the definition of a plant disease, but most experts will agree that a plant disease is a condition which may disrupt the normal functions of a plant at any of the plant's growth or reproductive stages. The disease may cause a change in the normal cell division of the plant, the transportation of food and water, the storage of food material, or the photosynthetic processes.

Certain diseases will interrupt the plant's activities at one or more stages in the plant's life cycle. Some disease activity involves only minor changes in the plant's normal function, and the plant is able to complete its life cycle. In fact, the damage caused by the disease may be so slight that it goes unnoticed except to the trained eye. On the other hand, a plant disease may interfere with so many of the plant's life processes that the plant dies prematurely.

It is difficult to describe what a normal plant looks like. Variations within plant populations, environmental conditions, and damage caused by grazing may cause a plant to appear diseased. From a practical standpoint, the producer is most interested in those changes in the plant's life cycle caused by disease that reduce the value of the crop and thus become an economic consideration.

SYMPTOMS OF DISEASES IN PLANTS

When a person who is sick goes to a doctor, the doctor immediately looks for known characteristics of common diseases or signs of a malfunction of the human body. These signs or characteristics that are usually associated with diseases are called *symptoms*. They are visual indicators that show the reactions of a person to a pathogen.

Plant pathologists must study symptoms to diagnose plant diseases, just as a doctor studies symptoms in order to diagnose human diseases. The plant pathologist looks for specific symptoms that commonly occur under a given set of conditions with known plant diseases. Often there is a combination of symptoms that provides the clue needed to identify a specific disease (Figure 6-2).

Years of research and observation have established that certain diseases affect one or more of the plant's life-giving functions. For example, if soil moisture is available and a plant shows signs of permanent wilting, one would suspect that the plant is suffering from a disease that affects its water uptake and the movement of water through it. A gall or cancerous growth appearing on the root or stem of the plant suggests that something is interfering with cell division. The presence of leaf rusts may result in yellowing or chlorosis, indicating that the plant's photosynthetic process has been altered.

Known symptoms are useful in identifying a plant disease, but it is important to remember that one may not be able to identify specific diseases by visual inspection. The disruption of a plant's root system or lack of the proper plant nutrients may also affect the upper part of a plant.

Considering all the diseases of plants, there are few that cause the plant to die. More often, the disease impairs the

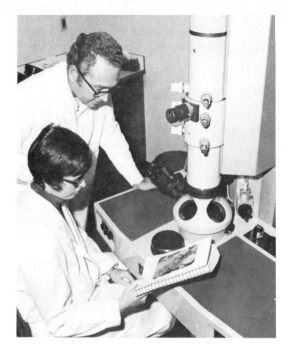

Figure 6-2. Electron microscopes are used by plant pathologists to diagnose plant diseases. (*MSU*)

plant's ability to grow and produce. Generally, the yield of the plant is reduced or the quality of the harvested crop is lowered. The stage of the plant's life when the disease attacks and its symptoms appear has much to do with its severity. In general, the younger the plant the more serious the damage resulting from the disease.

CAUSES OF PLANT DISEASE

The agents that cause disease can be classified in two groups: parasitic and nonparasitic. Every parasitic plant disease is caused by some type of agent. The parasitic agents include bacteria, fungi, viruses, and nematodes. The more common nonparasitic causes include such things as nutrient imbalances or deficiencies, air pollutants, environmental factors, and physical injuries. Only the parasitic agents causing disease are discussed in this chapter.

Parasitic Agents

A parasite is a living organism that must obtain part or all of its nutrients from other living organisms. The bacteria, fungi, and viruses are parasitic plants, while the nematodes are considered to be animals. In any case, the parasite structurally damages the plant through direct competition for the nutrients available to the plant without providing any benefit to it.

Bacteria Plant-pathogenic *bacteria* are single-celled microorganisms. They usually are rod-shaped and microscopic in size, and they reproduce themselves by individual cell division. Under the most favorable conditions, a single bacterium may divide every 20 minutes. It is fortunate that the vast majority of bacteria are beneficial to plant life.

Bacteria are unable to penetrate the plant by physical or chemical means; they must enter through the stomata or other natural openings or through a wound caused by one of the nonparasitic agents. A heavy hailstorm, insect infestation, pruning, or grafting may result in physical injury which will allow bacteria to enter the plant. Once inside the plant, the bacteria can be carried through the vascular system. When harmful bacteria attack a plant, the plant responds in different ways. Galls may appear, the plant may begin to wilt, growth may be slowed, imperfect fruit or ears may form, or there may be a distortion of the leaves.

Pathogenic bacteria must be moved from plant to plant in order to spread disease. Wind or irrigation water may move diseased parts of plants from place to place, or insects (vectors) and rain may carry bacteria from one plant to another. In other cases producers themselves spread the disease-causing bacteria through field operations such as mowing or plant propagation. Often however, bacterial diseases are spread through infected seed or vegetative material.

Diseases caused by bacteria can be controlled or prevented in several ways. Soil treatment is sometimes used but it usually too expensive for large-scale use. Plowing under infected crop residue will prevent the spread of bacteria by wind and encourage rapid breakdown of the crop residue which harbors the pathogens. It is very important, when working with plants that are vegetatively propagated, that all cutting knives and grafting equipment be sterilized between uses on successive plants. Prevention of plant disease is the most economical and effective method of controlling it. Plant breeders continue to develop varieties that are immune to disease-causing bacteria.

Fungi These pathogens are tiny, threadlike microscopic plants. However, there are some large and conspicuous fungi such as the fleshy mushrooms. Molds are among the most familiar fungi. The fungi are living plants, but they are unable to manufacture their own food because they lack chlorophyll. A large number of the fungi reproduce asexually by spores. A single pathogenic plant may support millions of tiny spores, each of which is capable of spreading the disease.

Upon germination, those fungi which are harmful to plants produce infecting structures which grow into the plant. All fungi are dependent on the plant to varying degrees. Lack of chlorophyll means that fungi must depend on the plant for all nutrients and water. After entering the plant, the fungus may cause some softening of the plant tissue. The fungal growth remains attached to the plant. The infected plants may show only slight effects or severe stunting and chlorosis (Figure 6-3). Losses in yield will occur, and in a severe attack by a fungus, the plant may die.

To spread disease, the tiny spores of fungi must move from plant to plant. Fungi cannot move on their own and must depend on outside forces such as wind, water movement, insects, birds, animals, and humans for transport. Because of their small size, the spores can be carried great distances. In a few cases, such as plant rusts, specific plants must serve as intermediate hosts on which the fungus can undergo complicated changes to enable the spores to reinfect. Smut spores in cereal crops may be transmitted on the seed; therefore, it is important to treat all seed before planting.

Viruses   The smallest of the disease-causing pathogens are the *viruses*. These rod-shaped or spherical particles are so small that they can be seen only through an electron microscope. The virus is an obligate parasite and therefore cannot grow in nonliving tissue. Viruses are not generally considered to be living organisms, but they can multiply inside living cells.

The disease-causing viruses may enter the growing plant in several ways. Some enter the plant through damaged cells, passing through the cuticle and cell wall through small breaks caused by damage from wind, animals, or other agents. Such mechanical damage to plants should therefore be avoided.

Other types of virus are carried from plant to plant by vectors. In plant pathology, a *vector* is an agent that transmits a disease. Some of the more common insect vectors that transmit disease while they are feeding are aphids and leafhoppers. When these insects feed on infected plants, their mouthparts become contaminated, and as they feed on new plants, the virus is spread.

Figure 6-3. This corn plant shows severe stunting and chlorosis caused by fungus. (*Grant Heilman*)

Some insects transmit a specific virus, while others spread many types of viruses. Whiteflies, thrips, and mealybugs also contribute to the spread of pathogenic viruses.

Many pathogenic viruses are transmitted through the seed. The virus may be found in the seed coat, endosperm, or embryo. Although some seeds carrying a virus may be somewhat lighter and smaller in size, infected seeds are hard to detect. Once an infected seed is planted, infection may spread throughout the entire field, leading to early and widespread disease. In horticulture and in vegetable and fruit crops, the virus may be spread through grafting, budding, or other means of vegetative propagation.

When a plant has been infected by a disease-causing virus, a combination of many symptoms or a single symptom may appear. The disease symptoms may be so slight that they go unnoticed or so severe that they destroy the plant. Some plants may carry a virus and show no outward visible symptoms. Only the general, more common symptoms are discussed here.

The ornamental horticultural industry is particularly concerned about the effect of viruses on some flowers. Flowering may be premature or delayed on the diseased plants. Some flowers will show breaking or variegation of color. The plant leaves may be chlorotic or yellowed.

Some viruses cause the infected leaves to be thickened or dwarfed, and rolling or curling of the leaves may occur (Figure 6-4). Some viruses cause the leaves to become stiff and brittle. In extreme cases the leaves may be completely chlorotic.

Fruit production on infected trees may be reduced through premature dropping of fruit. If the fruit is produced, it may be small or misshapen. Often blisters, scars, or cracks lower the sale value of the fruit.

The inside tissue of a virus-infected plant may be corky or woody in appearance. Because of the unthrifty condition of the plant, the root structure may be somewhat reduced. These symptoms are quite similar to those caused by the deficiency of some of the essential plant nutrients.

Nematodes These animal pathogens are slender, threadlike roundworms. They have a muscular system, a digestive system, and organs for feeding. They also have a nervous system and are well adapted to repro-

Figure 6-4. A virus has caused the leaves of this tobacco plant to thicken and curl. (*Grant Heilman*)

ducing themselves. They are considered to be animals that feed on cell sap in the plant tissue. The nematode, sometimes called the *eelworm,* is totally different from the earthworm, flatworm, and cutworm in form and habitat. Some nematodes are found in soil, others in fresh water or salt water. Many nematodes are beneficial, but a sizable number live on plants as parasites and can cause a variety of plant diseases. Researchers believe that all crop and ornamental plants can be attacked by parasitic nematodes.

The survival of the nematode depends on its reaching a plant on which it can feed. Some nematodes are hatched in the plant, but others must travel through the soil to reach the plant. As they travel through the soil, many are destroyed by predators.

Nematodes cause much of their damage while feeding on the plant tissue. The nematode has a special feeding organ called a *stylet,* which is a spearlike structure that acts like a hypodermic needle. The stylet is hollow, and when it is pushed into the plant tissue or cell, the nematode can suck out the cell contents. Usually the nematode will inject a digestive secretion into the plant cell before the cell sap is ingested. The nematode may enter the plant to feed or may feed from the plant's outside surface.

Different species of nematodes may cause different types of damage. Sometimes the plants are killed, but more often they are weakened, growth is reduced, and the yield is lowered. When the plant roots are attacked, necrotic lesions result from the salivary secretions injected during feeding. Often giant cells begin to form, causing root galls (Figure 6-5). Other nematode species feed on the root tips or growing points, thus supressing cell division and bringing about shorter roots. The nematodes that feed on the upper portion of the plant cause different symptoms. Discoloration of the green

Figure 6-5. Galls on eggplant roots caused by root-knot nematode. (*USDA*)

tissue is common. Necrosis, blotches, spots, and galls on the stems and leaves are often evident. As with root attacks, the nematode attack on growing points may cause abnormal or enlarged growth. It is believed that the abnormal growth is caused by the growth substance secreted by the nematode during feeding. Plants infested with nematodes seem to be more susceptible to other pathogens than noninfested plants.

HOW PATHOGENS SPREAD

There are several means by which pathogens are dispersed from plant to plant and field to field. Each pathogen has some unique features in its movement patterns, but in general, pathogens move through the aerial environment or through the soil, are carried by vectors, or are moved by humans. Some pathogens move great distances, while others move only a very short distance.

Air and Water

Many of the pathogens that cause plant disease make use of conditions in the aerial environment to spread from plant to plant. The small spores of fungi are released into the surrounding environment, where they may be picked up by air currents, raindrops, or moving water. If the spores are picked up by raindrops, they may be splashed onto a plant surface a short distance away. On the other hand, if the spores reach the upper atmosphere, they may be carried for thousands of miles by winds.

Bacteria can also be disseminated through the environment of the host plant. When a diseased plant sheds leaves, they may be picked up by wind or moving water and brought into contact with healthy plants. Raindrops running over diseased plant tissue fall onto the soil or may splash onto healthy plants, thus spreading a disease-causing pathogen.

Under most conditions, disease-causing viruses and nematodes are not spread by the natural movement of wind and water.

Soil

The soil often acts as a barrier to the movement of pathogens, except for very short distances. Those plant pathogens which form spores which can remain viable for long periods of time have the best chance for dispersion through the soil. Some pathogens act as saprophytes as well as parasites —that is, they can live and grow in the soil away from the host plant and thus spread more widely throughout the soil.

Some spreading could occur through the soil left on root crops, such as potatoes and sugar beets, as they are transported from the field to the collection site.

Vectors

With reference to the spread of plant diseases, the term *vector* is applied to an insect or organism that transmits or disperses the disease-causing bacteria, fungi, viruses, or nematodes from one plant to another. The vectors will vary depending on the type of pathogen and, in some cases, the strains within its different species. Some vectors are highly specific to one pathogen. In these cases, the pathogen may multiply within its vector and is usually spread as the vector feeds. In other cases, the pathogen multiplies outside the vector. The vector then becomes contaminated by the pathogen, and dispersal takes place as the vector moves from plant to plant. A vector may even become contaminated by chance and thus spread the disease-causing pathogen. Some insects, such as aphids and leafhoppers, may extract sap from diseased plants and deposit the disease organisms in other plants on which they feed. Biting insects such as grasshoppers, beetles, and earwigs are capable of transmitting disease organisms as they move from one plant to another to feed.

Beneficial insects or animals may become vectors that help spread many of the pathogens. The insects that pollinate crops may, in a mechanical fashion, be responsible for carrying disease organisms from plant to plant. Animals sometimes consume spores, which pass through the gut and are then deposited as the animals feed.

Humans and Machines

The seed that is planted can serve as a carrier of disease organisms. It is easy to see how a disease can spread rapidly if the organism is attached to the seed and the seed is planted. Disease organisms may also be attached to vegetative parts of plants that are used to establish new plants.

Indirectly, producers and consumers have done much to speed the spread of plant diseases. The consumer demands more food, so the farmer plants more seed. The demand for greater quantities of seed means that the seed used for planting must be transported for greater distances. Both these conditions mean that producers themselves can exaggerate the spread of disease by one of the most natural mechanisms, the seed itself. As a precaution, producers should always plant certified seed.

The increased use of vegetative propagation has also helped spread disease. Diseases can be spread readily if workers fail to properly disinfect tools and equipment when grafting or budding or when preparing cuttings and tubers for planting. Often the workers carry disease organisms on their clothing and shoes. Larger farms mean that workers travel over larger areas, sometimes carrying pathogens with them. The machines used to produce crops can also transport disease organisms over a wide area; and the shipment of food products from state to state and around the world contributes to the spread of plant diseases.

CONTROLLING DISEASE ORGANISMS

Plant diseases, along with other pests, can determine the success or failure of a crop in a given area. Diseases alone are costing the American producer and consumer millions of dollars each year. The producer's loss is represented by reduced yield, and ultimately the consumer's loss is represented by higher food costs. The profit of the processor is also reduced through increased cleaning and sorting costs and a lower-quality product to sell to the retailer.

In extreme cases, disease may prevent a crop from being grown even though it is adapted to an area. In other cases, although a disease may not destroy the crop, it makes the crop uneconomical to raise. Faced with the demand for more food, researchers, scientists, and producers must work together to develop and use many methods of controlling plant diseases.

Cultural Control

Some of the earliest methods of disease control included cultural practices. Today, cultural practices that help reduce losses from plant diseases include many of the modern farming methods essential to good crop production. Unfortunately, the intensive farming methods used to increase food production often favor the growth and development of disease organisms. Some of the more important cultural control practices include:

1. *Planting disease-free material.* Vegetatively propagated plants such as sugarcane and potatoes should be free of virus. Crop seeds such as corn and wheat should be treated to prevent the spread of fungal and bacterial diseases. A rapidly growing agribusiness in many communities is the production and sale of disease-free planting material and certified seed.

2. *Planting crops properly.* Placing planting material and seeds at the proper depth in an undamaged condition is critical. Breaks in the seed coat or a cut in a bulb or tuber may allow disease organisms to enter the plant and begin to grow. Planting too deep will delay germination and emergence, giving disease organisms a better chance to infect the plant before it can begin to grow rapidly. Planting time

is critical, in that pathogens are more apt to infect plants under certain environmental conditions. For example, high soil temperature favors the development of many root-rotting fungi. Planting winter wheat when soil temperature is too high may result in increased damage from root-rotting fungi.

3. *Fertilizing properly.* Excessive nitrogen keeps plants in a succulent condition, which often makes them more susceptible to disease. Rusts and mildews seem to develop more rapidly in the presence of high levels of nitrogen compounds. The skillful formulation and proper application of fertilizer are critical to vigorous plant growth.

4. *Rotating crops.* Crop rotation will rarely eliminate a pathogen, but it may break the pathogen's life cycle. Crop rotation will thus often reduce the pathogen level in the soil to the extent that the disease will cause only minor damage to crops. To be effective, the rotation must remove all the pathogen's alternate hosts from the area (Figure 6-6).

5. *Preventing plant injury.* Many pathogens enter a plant through cuts and bruises caused by careless use of machinery and equipment. In addition to mechanical damage, insects, animals, wind, and hail can cause plant damage and create an entry point for a disease organism. Careful handling of the harvested crop is also important to prevent development of disease during storage.

Chemical Control

Many are opposed to the use of any chemical to control crop pests. But used properly, approved wide-spectrum pesticides can ef-

Figure 6-6. Potato scab is a soilborne bacterial disease. Rotating the crop will reduce the incidence of the disease. (*MSU*)

fectively control some pathogens. It has been estimated that for every dollar invested in pesticide treatment, there may be a 5-dollar increase in returns for corn, soybeans, and wheat.

Most of the chemicals used in the control of pathogens are classified as fungicides. A smaller number of bactericides and nematocides are available, but to date scientists have been unable to develop effective chemical treatments for the viruses. New chemical fungicides are being developed. Some are approved for use, but others are not allowed to be sold because they are either socially or biologically unacceptable.

Sulfur and copper were among the first chemicals used to control plant diseases. These chemicals supplemented the plant's own defenses against plant diseases by helping form a barrier against disease organisms. Many of these chemicals are still in use today and are sometimes classified as protectant chemicals.

Such chemical compounds, which help protect the plant, are applied to its surface. They must be applied before the pathogen arrives. To use such chemicals effectively and economically, the producer must have some indication of a pending attack by the pathogen.

The protectant type of chemical compound has a number of disadvantages. Because it is applied to the surface of the plant, it is subject to all the elements of nature. Rain can wash the material off the plant, and heat and light may reduce its effectiveness. Also, as the growing season progresses and as new leaves, flowers, and fruits emerge, more of the chemical must be applied to protect the plant.

Systemic-type chemicals enter the host plant and help provide protection from plant diseases from within the plant. The chemical becomes a part of the plant and acts as an internal addition to the plant's own protective mechanism. The systemic chemicals act more like modern medicines used by humans. They can be used to treat diseases which are already present within the plant.

It may seem that the systemic compounds are the answer to the control and prevention of plant disease, because so many pathogens are easily controlled by these compounds. However, many pathogens are able to develop a high tolerance for systemic fungicides. Thus, in time, some of these compounds lose their effectiveness completely against certain strains of pathogens.

Physical Control

Heat treatment can sometimes be used to destroy disease organisms on planting material and in growing media. Oftentimes, fungicides do not reach those organisms that are deep in the tissue of the plant or seed. Treatment with hot water and alternate applications of hot and cold water are used to disinfect plant material against some nematodes and some species of mites. Hot-air treatment can rid some planting stock of virus pathogens.

Although heat can be an effective method of disinfecting planting material, extreme care must be exercised to prevent damage to the planting material. The margin of safety between destroying the pathogen and damaging the planting material is small. The use of heat, and particularly heat and water, often requires expensive equipment. The moisture content of treated plant material can create storage problems.

Soilborne pathogens are a serious problem of many agricultural crops. The soil and other material used for planting media in greenhouses is often sterilized with heat to destroy disease organisms. Although this is an expensive program, it is often economically feasible because of the intensive nature of the operation and the relatively small amount of material involved. The high-profit crop justifies the high cost of disinfecting the growing media.

High losses also occur in field crops, but the less profitable crops do not justify expensive soil treatments. Experimentation with the use of solar heat to control diseases caused by soilborne pathogens is under way. Under experimental conditions, sterilization of the soil using solar energy has increased both yield and quality of potatoes, peanuts, cotton, and many vegetables. In the future, solarization may become one more method of disease control used by producers.

Biological Control

Plant pathologists are beginning to recognize the potential of natural antagonists or pathogens. Although not a new concept, biological control of the plant diseases has

lagged behind the development of other control methods.

The development of resistant cultivars has reduced and will continue to reduce the severity of many plant diseases (Figure 6-7). Many cereal crops have cultivars which are resistant to rusts, smuts, and mildews. But because new strains of pathogens are constantly appearing, there is a constant need for the development of new disease-resistant plants. Chapter 8 gives a detailed description of biological pest control.

COMMON CROP DISEASES

The plant-disease picture is always changing. Even with the most up-to-date information, there are always new problems to face in individual situations. New diseases may appear for the first time in an area. A crop variety totally resistant to one strain of a pathogen may be quite susceptible to an-

Figure 6-7. Research greenhouse are used to develop disease-resistant cultivars. (*MSU*)

other strain of the same pathogen. Environmental conditions in an area may result in a pathogen's causing much more serious damage there than in an area with a different set of environmental conditions. In many cases, diseases are simply taken for granted until the damage is serious.

The following discussion describes the general types of disease caused by bacteria, fungi, viruses, and nematodes. Only a few specific diseases among the various types are discussed.

Diseases Caused by Bacteria

Losses caused by bacterial diseases vary from year to year. It is difficult to obtain precise figures on the total economic loss caused by the various forms of bacterial disease.

Some bacteria cause galls, which appear as swollen disfigurations on the plant. Bacterially produced galls are quite common on fruit trees, grapes, blackberries, and roses (Figure 6-8).

Blight diseases are commonly caused by bacteria. The blight causes rapid discoloration and ultimate death of certain portions of the plant. In extreme cases, the blight may cause the death of the affected plant. The real damage is caused by a reduction in the green leaf surface the plants need to produce carbohydrates. Fire blight in apples and pears is common.

The leaf-spot diseases caused by bacteria produce lesions on the leaves, stems, or fruit. Local spots occur commonly on the leaves. A common symptom of leaf-spot disease is the yellowing of the area immediately around the dead spots where the bacteria have invaded the tissue. Angular leaf spot of cotton, yellow leaf spot of wheat, and bacterial spot of peach are examples of leaf-spot diseases.

Vegetables and fruits are often affected by bacteria that cause soft rots. Evidence of

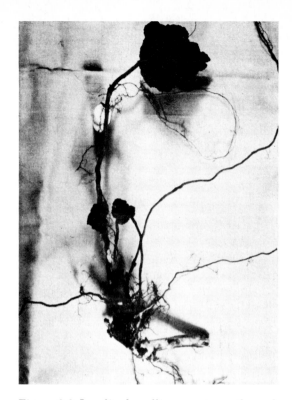

Figure 6-8. Localized swelling or outgrowths such as those caused by crown gall contain large, swollen cells caused by invading bacteria. (*MSU*)

soft-rot disease can be observed on the produce-store shelf when the stock ceases to be fresh. The soft rots develop in the fleshy tissues. The bacteria produce an enzyme that dissolves a portion of the cell wall. The bacterial action often causes a slimy foul-smelling mass. Soft rot is often followed by an invasion of some other pathogen.

The bacterial wilt diseases affect the vascular system of a plant. The bacteria multiply rapidly, producing a slimy substance that plugs the xylem of the plant. All parts of the plant can be affected, and the entire plant may wilt and die. Bacterial wilt is a serious alfalfa disease. Maize wilt, gumming disease of sugarcane, and cucumber wilt are other examples.

Parasitic Plant Diseases

Diseases Caused by Fungi

A large number of crops are affected by various species of *Puccinia* which cause rust diseases as well as the species of *Fusarium* and *Verticillium* that cause wilt diseases. The fungi that cause plants to wilt are often found in the soil. When susceptible crops are grown in the same soil for several years, the reproductive spores of many fungi are always present in the soil to infect young seedlings. The fungi penetrate the roots of the plants, begin to grow, and plug the vascular cells. The vascular wilt diseases are often caused by a host-specific pathogen. Cucumber wilt, maize wilt, wheat stripe, oak wilt, and Dutch elm disease are typical of the vascular wilt diseases. The U.S. Department of Agriculture (USDA) estimates that verticillium wilt in alfalfa will cause a loss of over 1 million dollars in reduced yields over a 3-year period in the Pacific Northwest alone, cutting the yield in half.

Very young plants may be attacked by damping-off fungi shortly after germination. The attack may occur before or shortly after the plant emerges. The soft, immature tissue of the seedling is destroyed. Plants affected have lesions or dead spots that are visible near the soil line. Some plants affected may be severely stunted, while others die. The pathogen causing damping-off usually infects only young plant roots. Cotton and sugar beets are frequently affected by this disease.

Stem anthracnose and septoria brown spot are fungal diseases that affect the stems, pods, and leaves of the soybean plant. Pod-stem blight is a fungal disease of soybeans prevalent in the eastern states. Alternaria leaf spot is common in potatoes in Colorado and is known to occur in several other states.

Figure 6-9. Ergot is a common fungal disease of barley, rye, durum wheat, and some varieties of hard spring wheat. *(MSU)*

Hundreds of strains of pathogenic fungi cause root rots. The pathogens attack the extravascular or outer layer of the roots. The affected plants then wilt, and if enough roots are affected, the plants may die. Alfalfa, cotton, and ornamentals are among the many plant species affected by root rots. Root rot in spring wheat and barley can cause millions of dollars in losses annually.

Wheat, oats, barley, and rye may be attacked by distinct species or subspecies of rust fungi (Figure 6-9). Stem, leaf, and stripe rust may be found in the cereal crops. The water requirement for plants affected by rust is higher than that for healthy plants. The straw of the affected plants

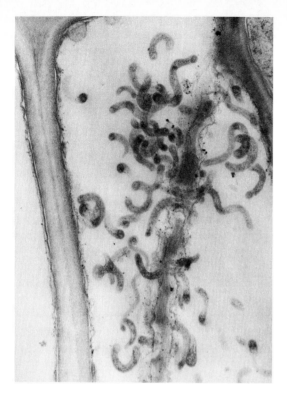

Figure 6-10. Worm-like spiroplasms—magnified 25,000 times—in corn smut infected tissue.

Figure 6-11. This oat plant is severely stunted by blue dwarf virus disease. (*MSU*)

turns brown, becomes dry and brittle, and may break off. Also, the yield is reduced, and the test weight and milling quality are lowered.

Another group of fungal diseases important in crop production are the smuts. Each year, thousands of dollars worth of wheat, oats, and barley is lost through reduced yields. When the grain is marketed, the producer receives a lower price because of the dark color and offensive odor of smut spores that are attached to the grain kernels. Corn and sorghum are also plagued by this pathogen (Figure 6-10). Loose and covered smut are commonly found in cereal grains and sorghum. Corn is susceptible to common smut and head smut.

Diseases Caused by Viruses

Almost every living organism can be attacked by some strain of virus. In humans, smallpox, mumps, and scarlet fever are examples of virus diseases. Virus diseases of plants have probably plagued producers for centuries. The virus destroys the chloroplasts, thus reducing the photosynthetic capacity of the plant. The pathogens causing virus diseases are often carried from plant to plant by insect vectors which feed on plants.

Dwarfing caused by a virus affects a number of cereal crops. The plants become stunted in the early stages of growth (Figure 6-11). The leaves turn yellow. Crop yields are greatly reduced. The chlorotic symptoms resulting from the virus are similar to symptoms found in plants with an excess of water, drought, a shortage of nitrogen, or low-temperature injury. Plants become infected at all stages; young plants are frequently killed. Barley yellow dwarf commonly infects wheat, oats, and barley.

Several mosaics caused by viruses attack the cereal grains. These viruses tend to destroy the chlorophyll in the leaf, thus slowing photosynthesis. The chlorosis, or yellowing, is not uniform but appears in patches, spots, or streaks on the leaves. Winter wheat is most seriously affected by the mosaics, but several of the viruses attack other cereals.

Diseases Caused by Nematodes

Nematodes, or eelworms, can cause plant diseases that are of economic importance.

The parasitic nematodes are generally specialized; that is, certain nematodes attack only certain plants. Some plants seem to have some immunity to certain nematode species. Plant breeders are continuing to work on the development of nematode-resistant crop varieties.

Root-knot nematodes attack the underground roots of the plant, causing the development of knots or galls. The galls most often appear on the younger plants and somewhat resemble a string of beads. When the roots are seriously affected, they begin to rot. The gall formation blocks the vascular system so that the plants appear stunted and wilted and the yields are greatly reduced. Nematode attack seldom kills the plant (Figure 6-12).

The root-knot nematode is quite common in cotton. The young larvae infect the cotton

Figure 6-12. On the left, a healthy tomato plant root system; on the right, a root system infected by root-knot nematodes. (*Runk/Schoenberger-Grant Heilman*)

plant by entering through the soft root tips. Although not considered too serious, the root-knot nematode is sometimes found in alfalfa in areas where the parasite is abundant in other crops. A number of species of root-knot nematode will attack and reduce yields of soybeans. The root-knot nematode may also be a problem wherever sugar beets are grown.

The stem nematode attacks the upper parts of the plant, usually feeding on or near the growing point of the plant. Such feeding results in shorter stems and, in the legume hay crops, may leave the crop too short to be harvested effectively. The stem nematode occurs naturally on alfalfa, sweetclover, and white clover but does not seem to adapt itself to many other host plants.

SUMMARY

Many plants that produce our food and those which provide beauty in our surroundings are subject to disease. Viruses, bacteria, fungi, and nematodes are the four groups of plant pathogens. More than 160 bacteria, 250 viruses, and 8000 fungi are known to cause plant diseases. The direct control of disease-causing pathogens involves manipulation of the pathogens themselves, removal of the host plant, modification or changing of the surroundings of the disease-causing organisms, or development of disease-resistant cultivars. In addition to these control measures, chemical and biological controls may be necessary in some cases to protect plants from disease-causing organisms.

QUESTIONS FOR REVIEW

1. What conditions affect the spread of disease?
2. How do plant diseases spread from plant to plant?
3. How do plant disease organisms enter the plant?
4. How can the spread of plant diseases be slowed or stopped?
5. What causes plant diseases?
6. What is the difference between a parasitic and a nonparasitic disease?
7. How can physical injury to a plant help spread disease?
8. What is a vector, and how does it help spread disease?
9. What are some common symptoms of plant diseases caused by parasitic agents?

Chapter
7

Insects

Insects follow weeds as the second most damaging pest agriculturalists must control. They consume and destroy significant amounts of the world's food supply, thereby contributing to the undernourishment or malnourishment of millions of people.

Insects may be aquatic or terrestrial. Insects are adaptable and can be found on the hot desert sand or living in rain-soaked jungles. Still others survive on high mountain peaks where temperatures drop well below freezing. Insects are not limited to fields where crops are grown or to the feedlots where livestock are produced. Many of these destructive pests are found inside storage buildings and in homes, where they often destroy or contaminate stored food.

Although entomologists (those who study insects) do not agree on how many insects may inhabit the earth at any time, it is known that they make up the largest number of known species of animals found in North America (Figure 7-1). Most of the destructive insects are vegetarians, but some are meat eaters. Other species will eat almost anything that happens to be near. No part of the plant is completely safe from insect attack.

Insects reproduce and multiply faster than any other animal on the face of the earth. Researchers estimate that if one pair of houseflies was allowed to reproduce for about 4 months, they could produce 191,010,000,000,000,000,000 offspring.

Nearly every plant or food product is eaten by some type of insect. Losses run into millions of dollars each year. On the other hand, insects are critical in the food production cycle, particularly in the pollination process. Thus it is incorrect to characterize insects as either good or bad.

Nearly everyone is familiar with the external damage to growing crops, vegetables, and ornamental plants caused by the chewing insects. This kind of damage lowers crop yields and crop quality.

Some insects only reduce the "cosmetic appearance" of fruits and vegetables without damaging the nutritional value. These minor blemishes do not affect the nutritional value, storagability, or taste of the product; but consumers prefer fruits and vegetables that are without blemishes. Treating fruits and vegetables to maintain the fine appearance of the food costs the American producer millions of dollars each year. The American consumer demands such control, however, believing that even an appearance of slight russetting reduces the quality.

CHAPTER GOALS

After studying this chapter you will be able to:

- Identify how insects cause losses in crop production for the American producer.
- Describe the job of an entomologist.
- Describe the general characteristics of an insect.
- Identify the characteristics of insects that enable them to survive so well.
- Explain how insects develop and multiply.
- Describe the feeding habits of insects.
- Describe the benefits of some insects.
- Explain how to control insects.

Figure 7-1. Insect traps are used to determine insect populations. Here a soil insect trap is being prepared for field use. (*USDA*)

CONTROL OF INSECTS

Modern technology has greatly aided producers in their fight against destructive insects. A wide variety of insecticides have been developed to control specific insects. Improved cultural and biological control methods are now available. However, indiscriminate use of control measures may destroy beneficial as well as destructive insects. Therefore, anyone attempting to control insect pests must have a general knowledge of the species to be controlled.

Protecting our plant population should be of concern to everyone. Whatever the purpose of growing plants, whether for beauty, lumber, food, or feed, it is a constant struggle to ward off insect pests. From the moment a seed is placed in the ground until the food has been processed and packaged, the threat of insect damage is present.

Individually, efforts to control insects are minimal, but with everyone working together, an acceptable level of insect control can be reached. The key to combating insects successfully is providing everyone with a general knowledge of potential insects pests and the best methods of combating them.

THE INSECT BODY

An insect is a small animal which belongs to the largest class in the Arthropoda phylum. Of all the animal species known today, over 80 percent are arthropods.

Insects are an important part of the ecosystem. Although a very small percentage of those people who study insects will become professional entomologists, every citizen is directly or indirectly affected by insects. These small creatures may be found almost anyplace at any time. Insects may be found in the large fields of the producers of

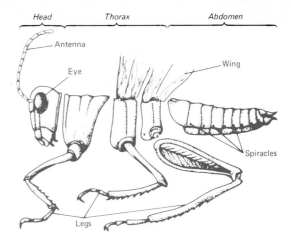

Figure 7-2. The three main body regions of an insect as shown on a grasshopper.

our country and in the small home gardens that are becoming so popular.

Many insects are microscopic in size, while the goliath beetle of equatorial Africa weighs an average of $2\frac{1}{2}$ to $3\frac{1}{2}$ oz [70.8 to 99.2 g]. The body of the mature insect is divided into three distinct sections, head, thorax, and abdomen (Figure 7-2). These three body parts may not be differentiated in certain stages of growth. Most adult insects have one or two pairs of functional wings.

The Head

Each insect has a pair of antennae located on the head, which act as sensory devices. The length, shape, and number of segments in the antennae vary according to the species of insect.

Most adult insects have two compound eyes made up of numerous units called *facets*. The eyes bulge out from the insect's head, enabling it to see in all directions at once without turning its head. This eye characteristic makes it difficult to approach an insect without its being disturbed.

The mouthparts located on the head of the insect vary in structure and function. The mouthparts of different insects are arranged to enable them to feed in different ways. Some insects have mandibles or jaws with which they chew holes in plant leaves or in some cases chew the plant stem. Beetles and grasshoppers have chewing mouthparts. Other insects have piercing-sucking mouthparts. These insects feed by puncturing the surface of a plant with the beak and sucking up the life-giving fluids. It is difficult to determine where insects with piercing-sucking mouthparts have fed on a plant, but one soon notices the plant wilting as its life-giving fluids are sucked out. Aphids, leafhoppers, chinch bugs, and spittlebugs are examples of insects that feed by sucking sap from plants.

The Thorax

The thorax is located immediately behind the head and is typically divided into three parts: the prothorax, mesothorax, and metathorax. The thorax is the heaviest portion of the insect's body. Each division of the thorax has one pair of legs. In most cases the legs are alike, but in some insects, such as the grasshopper, the last pair of legs is very large and designed for jumping.

If an adult insect has functional wings, these wings will be attached to the second and third divisions of the thorax. Only adult insects have functional wings that carry them from place to place, and not all insects have functional wings. Some, such as the dragonfly, have two pairs of wings, while others such as the housefly, have only a single pair. The wings vary in size among the insect species. In some orders, such as the beetle (Coleopera), the upper or front wings have lost their flying function and serve as hard, protective covers for the insect's body; the rear wings are used for flying.

The Abdomen

The insect's reproductive system is found in the abdomen. The number of segments in the abdomen varies with the different insect species.

The spiracles, or breathing organs, are located along the sides of the abdomen; they may be considered the "nose" of the insect. Many insects can open and close these spiracles, thus causing great variation in the usefulness of insecticides that kill by attacking the respiratory system.

The major viscera, heart, and male or female reproductive organs are located in the insect's abdominal region. Some insects, such as wasps, bees, and hornets, have a stinger at the end of the abdomen which is used for defense or in some cases for paralyzing prey.

The Exoskeleton

Insects have no internal backbone like that found in mammals. Their skeleton is on the outside of the body and is called an *exoskeleton*. It is therefore not uncommon to find dead insects that have retained their original shape. The muscles and internal organs inside the exoskeleton dry and decay, but this does not alter the appearance of the dead insect.

The exoskeleton has numerous advantages for the insect. It offers protection from physical damage, slows the movement of moisture in and out of the body, and has a high resistance to insecticides. Finally, it provides a structure for attaching the various muscular systems. The big disadvantage is that in order for the insect to grow, the exoskeleton must be shed or molted. During the molting period, the insect is quite vulnerable to many physical and chemical agents.

THE SURVIVAL AND SPREAD OF INSECTS

The success of insects in the struggle for survival can be attributed to six major assets: (1) their ability to fly, (2) their external skeleton, (3) their small size, (4) metamorphosis, (5) their special reproductive systems, and (6) their ability to adapt to their surroundings.

Many adult insects, when they find conditions unfavorable in one area, can spread their wings and move to a new location quickly. The ability of adult insects to fly enables them to search over a wide area for food and to search actively for mates. Ultimately, this helps spread insect infestations (Figure 7-3).

Many insects migrate over wide areas. Some seem able to fly tirelessly for great distances. Some can breed in one country and fly to another, where they devastate crops. Some species of the locust or grasshopper family may travel 100 miles [161 km] or more during their migrations in search of food. Other insects have regular seasonal flights which often follow rather definite patterns. Some insects overwinter as adults in one part of the country and then move to other parts during the spring and summer.

The tough, hard, flexible exoskeleton that many insects wear provides natural protection. The exoskeleton serves as protective armor against other predatory insects and chemical insecticides. Those sections of the exoskeleton that do not need to be flexible are strengthened by substance called *sclerotin*. This protein material is tough and lightweight and adds further resistance to crushing and to the effects of chemical substances. Waxlike substances may also coat the exoskeleton of some insects. This waxy material provides waterproofing from the outside and helps prevent the escape of

Figure 7-3. The adult Japanese beetle eats many kinds of trees, vegetables, flowers, and fruits, enabling it to flourish in many geographical locations. (*USDA*)

moisture from the insect's body and consequent dehydration. For further protection, the exoskeleton of some insects may have scales, hair, or spines.

Most insects are so small that, individually, they require very little food and water for survival. The small amount of food that is often overlooked by larger animals can be a feast for the tiny insect. A single droplet of water may provide more than enough moisture to sustain the life of the insect. These tiny creatures can find shelter in the smallest niche. Certain weevils spend their life inside seeds, and the leaf miners make their home between a leaf's upper and lower surface. When the smaller insects are ready to move, their size often enables them to be carried from one place to another on wind currents.

Many common insects undergo several changes in size, shape, or form as they develop. Most insects develop through a process called *metamorphosis*. The multiple changes that many insects go through in developing to adults enable the immature and adult insects to feed on different foods, hence there is little or no competition for food. The process of metamorphosis is discussed in more detail later in this chapter.

Insects have flourished for millions of years, partly because of their ability to reproduce. In the insect world, the female is in a dominant position. Many females live for many years and lay millions of eggs. Entomologists say the Australian queen termite can lay as many as 360 eggs in an hour. In a day she could lay 8640 eggs, and in one year, 3,153,000. The queen of this insect species may live for 25 to 50 years.

In some species male insects do not exist, and in other species they play only a minor role in the reproductive process. It is not uncommon for female insects to reproduce without fertilization, a phenomenon called *parthenogenesis*. Aphids reproduce in this manner. Other insects have both a parthenogenetic and a sexual reproductive cycle; many weevils, sawflies, and certain species of bees reproduce in this manner. Other winged adult insects have the ability to delay fertilization of the eggs until the conditions for survival are ideal. A small sac attached to the female's reproductive system stores the male sperm cells. When conditions are favorable, the sperm is released and the eggs fertilized and then deposited. The queen honeybee makes only one nuptial flight, in which she receives sufficient sperm cells to last the rest of her life.

Finally, some species of insects have a marvelous ability to adapt to their surroundings and changing environmental conditions. There is some insect that feeds on almost any plant and many manufactured materials. Some feed on animal tissue. Some chew their food, while others pierce the surface of the plant or skin of the animal and suck out the life-giving fluids.

Besides the usual ways of protecting themselves from their natural enemies by hiding, jumping, or flying away, some insects have the ability to regenerate lost body appendages. Others can protect themselves by using mimicry and protective coloration. Some insects secrete distasteful substances, some sting, and others build special protective structures in which they live.

HOW INSECTS DEVELOP AND MULTIPLY

Different insects follow different patterns of development, but for all insects, life begins with the egg.

Reproduction

Reproduction in insects is of three types. The largest group of insects lay their eggs and are called *oviparous* insects. The second group, *ovoviviparous* insects, retain the eggs in the body until hatched, at which time the female insect gives birth to active young. Some aphids and the citrus mealybug fall into this category. Finally, *viviparous* insects reproduce in the same manner as humans and other mammals. When the embryo is mature, the insect is brought forth alive and active. The sheep tick falls into this category.

Insect eggs differ in size, shape, color, and general appearance. Some insects conceal their eggs, while others lay them so that they blend with the surroundings. Some eggs are laid in masses and others singly. Some insect eggs are so small they cannot be seen with the naked eye. A knowledge of how certain insect species deposit their eggs can often aid in control of the insect (Figure 7-4).

The amount of time required for an insect egg to hatch varies among species. Tempera-

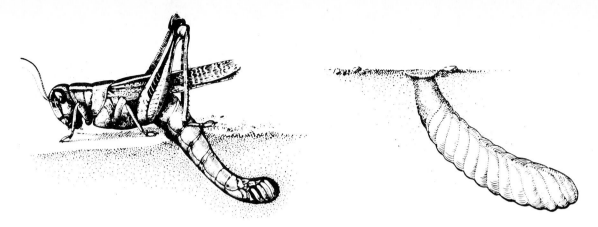

Figure 7-4. An adult grasshopper laying her eggs in the soil. (*USDA*)

Metamorphosis

The change in form, structure, and appearance of insects from the time of their birth until maturity is known as *metamorphosis.* Insects may be divided into one of three groups according to the type of metamorphosis they go through. Insects that emerge from the egg in the same form as the adult except for size are considered to be without metamorphosis. As the insects develop into adults, they go through a series of molts. Each time the insect molts, it increases in size, always resembling the adult form it will ultimately assume. Only a few species, such as the silverfish and the stinkbug, develop without metamorphosis.

A second group of very destructive insect pests, including the aphids, grasshoppers, and leafhoppers, have a gradual or incomplete metamorphosis. When insects in this group are hatched they resemble the adult, but they have no wings. The young insect, called a *nymph,* goes through several

ture has a great effect on the time required. Usually higher temperatures tend to hasten development.

molts. The period between molts is called an *instar.* After several molts, the insect emerges in the adult form. Usually, control is easier during the earlier nymphal stages.

A third group, which includes over 80 percent of insects, develop with complete metamorphosis. When the insect hatches, it has no resemblance to the adult insect. The first wormlike stage is called the *larva.* The larva goes through several molts and instars until it reaches maturity. The larvae feed extensively on plant material and are commonly more damaging than adult insects. Once the larvae have reached their full growth, they go into the resting or quiescent stage, called the *pupal stage.* The pupa is usually concealed because of its helplessness. The pupa, which is usually encased in a cocoon, is quite resistant to insecticides. After a period of time, usually dependent on temperature, the adult emerges from the cocoon as a mature insect, complete with wings and the ability to reproduce (Figure 7-5).

An understanding of the growth pattern of the various species of insects is necessary before an effective control plan can be developed. Different control measures must be used at different stages of the growth cycle.

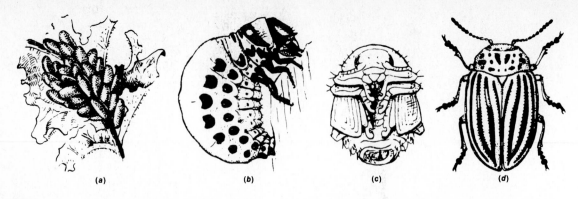

Figure 7-5. The Colorado potato beetle: (*a*) eggs, (*b*) larva, (*c*) pupa, and (*d*) adult. (*USDA*)

In general, for those insects with piercing-sucking or chewing mouthparts, the feeding or larval state is the best time to apply control measures.

THE FEEDING HABITS OF INSECTS

All insects feed on some form of organic material, which may be plant or animal and may be dead or alive. A few insects live on a mixed diet of plant and animal material. Insects that consume a variety of organic food, both plant and animal, are called *omnivorous*.

Some parasitic insects feed on other insects. Some of the parasitic insects select a specific host and lay their eggs inside its body. Others deposit eggs which, when eaten by the other insect, hatch and develop in the body cavity. The larvae feed on the host until they are grown. The parasitic insect may pupate inside the host or emerge and pupate nearby. The wasps are among the better-known parasitic insects that are beneficial in controlling harmful insects.

Predatory insects capture living prey. This group of insects feed on other insects smaller and less active than themselves. The praying mantis is one of the better-known predatory insects; it relishes a wide variety of smaller insect species. The ladybird beetle is now raised commercially and sold to producers, who distribute them in their fields to aid in the control of aphids and other soft-bodied insects.

Within the group of insects that live on plant tissue, some are selective and others are nonselective in their feeding habits. *Polyphagous* insects are those which eat a wide variety of plants. Most grasshoppers fit into this category (Figure 7-6).

A second group of insects, which feed on plant tissue and seem to restrict their feeding to a small number of similar plants, are called *oligophagous* insects. For example, the cabbage butterfly seeks out plants such as cabbage, cauliflower, brussels sprouts, and turnips. These plants have a pungent odor and taste because of the presence of mustard oils. When the plants secrete the oil, the insects smell the odor and are attracted to the plants, where they lay their eggs. The cabbage butterfly caterpillar (the larva) when it hatches will not eat unless it is on the plants to which the cabbage butterfly is accustomed.

The Colorado potato beetle is another example of an oligophagous insect. The potato beetle feeds on common weeds of the potato family as it moves northward from Mexico.

Figure 7-6. The adult grasshopper is an insect with a ravenous appetite that may consume almost any vegetative material in its path. (*USDA*)

The insect is now spread throughout the United States and confines its feeding almost entirely to the foliage of the potato plant, although occasionally it may appear on the tomato or eggplant.

A third group of insects, highly specialized in their feeding habits, are referred to as *monophagous* insects and are restricted to a single species of plant. The boll weevil is one of the better-known insects in this category. The larva of the boll weevil burrows within the flower buds and immature bolls of cotton, destroying the plant's ability to produce mature cotton bolls.

Some adult insects have a specialized siphoning mechanism through which they draw plant nectar. For example, the butterfly lands on a flower, uncoils and straightens its *proboscis,* or siphoning tube, extracts the nectar, and moves to the next flower.

The adult housefly has sponging mouthparts. When the housefly feeds, it secretes saliva onto its food. When the food is dissolved and in suspension, it moves by capillary action into a reservoir and then is drawn into the fly's digestive system.

The chewing-lapping mouthparts of the honeybee are modified in such a way that they can utilize liquid food. As the "tongue" of the honeybee moves up and down, nectar is drawn into the body. The mandibles of the honeybee are not used in the feeding process except to cut the corolla of some flowers to get at the nectar. However, the mandibles are used for defense and to mold wax into combs for storing honey.

An understanding of the feeding habits of insects is necessary if chemical control is to be effective. A more detailed discussion of the feeding habits of insects in relation to chemical control will be presented in a later chapter.

BENEFITS OF INSECTS

Humankind is becoming more aware of the benefits to society of each of the components of nature. It is difficult to characterize any part of nature as all bad or all good, and insects are no exception. Agriculturists are becoming more aware of the beneficial impact of insects on the various ecosystems.

Pollination

The beneficial relationship between insects and flowers has been known for centuries. Insects are attracted to flowers by color, smell, and shape.

Alfalfa seed production in the western states depends upon insects for pollination. The honeybee has been the traditional pollinator, but it is not the most efficient pollinator (Figure 7-7). Seed production is often about 100 to 150 lb/acre [112 to 168

Figure 7-7. A honey bee collecting nectar on an alfalfa blossom. As the bee collects nectar it also picks up alfalfa pollen shown by the white mass packed on the hind leg. (*USDA*)

kg/ha] when honeybees are used to pollinate the crop. Recently, many western seed producers have been using the alfalfa leafcutter bee to increase pollination and raise yield to 500 to 600 lb/acre [about 560 to 670 kg/ha]. The difference in seed production is the result of the leafcutter bee's unique ability to pollinate the alfalfa flower. Producers hope that through proper insect management they will be able to raise their seed production to 1000 lb/acre [1119 kg/ha].

Many of the honeybees used to produce honey in the northwestern states during the summer are shipped to California to pollinate the fruit trees during the winter and early spring. Bees are essential to the citrus industry. Without pollination, the industry would fail. Fifty or more fruit and seed crops depend on the honeybee to produce food or increase yield. Many varieties of apples, sweet cherries, and plums would be barren if it were not for insect pollinators.

The importance of the pollinating insects in agriculture is now recognized. In many locations, however, modern farming methods and the use of insecticides have reduced the numbers of pollinators to a dangerously low level.

Soil Building

Insects are important in the decomposition of dead plant and animal tissue. The most common decomposers are the bacteria and fungi, but insects also feed on residues and burrow into the soil. Insects such as ants and some beetles burrow through the soil, leaving more open space for penetration of air and water. Sometimes as the insects burrow, soil particles from the lower depths are brought nearer to the surface, improving the physical condition of the soil. As insects feed on dead plant and animal tissue, much of the organic material is returned to the soil to provide nutrients for plants.

Human Food

It is estimated that there are over 300,000 beekeepers in the United States. The honeybee is able to survive in almost all areas of the country. If properly cared for, a strong hive of honeybees should produce up to 100 lb [45 kg] of honey per year.

Biological Insect Control

More and more research is being done on the use of certain predatory and parasitic insects in biological insect control. This is discussed in detail in Chapter 8.

Food for Other Animals

Most birds and some mammals include insects in their diet. Warblers, woodpeckers, swallows, and flickers consume large numbers of insects. Normally, birds do not feed solely on insects, but they ingest large numbers of insects each day. Among the mammals, bears, bats, and shrews are specific insect feeders. Insects are important components in the diet of game fish.

Scientific Uses

Insects are being used more extensively for scientific experiments. The fruit fly has been used for genetic studies. Its very rapid reproductive cycle and the low cost of rearing the fruit fly make it useful for scientific studies.

CONTROLLING DESTRUCTIVE INSECTS

Producers have been fighting insects for centuries. But with all their efforts, insects have survived and continue to be a major problem facing today's farmer, livestock producer, food processor, retail food handler, and even homemaker.

The introduction of chemical insect control has made everyone's job easier. However, the indiscriminate use of chemicals is both unwise and illegal. The proper use of chemical insecticides in conjunction with proven cultural practices and nature will help control insects as producers try to meet the demand for food.

Pest Management

The battle against crop pests is a never-ending one. At one time there were those who spoke of eliminating pests. Today, most scientists and producers are talking about controlling pests at a tolerable level which will allow needed crops to be produced economically.

Pest management is rather complicated, because the advantages and disadvantages of all pest control measures must be considered before attempting to control a given pest. Indiscriminate use of broad-spectrum chemical agents is considered unwise, as it destroys both beneficial and harmful insects. A combination of chemical, biological, and cultural techniques is being recommended to minimize adverse effects on the environment while conserving natural control agents.

Pest management will become more and more complex. No single control program will be effective in all situations. The complex relationship between beneficial and harmful insects, weeds, and pathogens must be maintained. Management practices that tend to damage this delicate natural balance must be modified.

Good Farming Practices

Good farming practices can be used to help control insects with little or no extra production cost. For such practices to be effective, producers must understand the life cycles and habits of the insects and the crops that are common in their area. For example, producers who grow hard red winter wheat in the western United States might delay their seeding to help protect the wheat crop against damage caused by the Hessian fly. A few days delay in planting may allow the fall flight of the insect to pass, thus reducing damage. Later planting of corn in Iowa may result in less damage from the larvae of the corn rootworm. Oklahoma and Texas farmers may try to reduce chinch bug damage to their sorghum crop by planting early.

Modifications of the ideal planting date to control insects should be planned to avoid excessive reduction of crop yield. Safe dates for planting vary from year to year in different areas. Reliable information about acceptable dates may be obtained from the local county extension agent or the agricultural college.

Several species of insects reproduce and/or overwinter in crop residues left in the fields following harvest. The European corn borer lives through the winter in the larval form in the stalks of corn or weeds (Figure 7-8). Complete destruction of the stalk by shredding, feeding the entire stalk to livestock, or plowing the stalks under after harvest are ways to control the insect. The sugarcane borer follows a similar pattern in the Gulf Coast area. Again, destruction of the residue in the fall helps control many species of insects.

A number of insects that infest the wheat crop can be prevented from emerging in the spring by plowing under the wheat stubble in the fall. Growers of cotton and other crops find it often helps to destroy infested residue as a means of controlling subsequent insect infestations.

Timely tillage practices throughout the growing season may help destroy insect

Figure 7-8. Popcorn infested with the European corn borer. (*USDA*)

cocoons and eggs or bury them so deeply that the insect cannot emerge. For example, most grasshoppers lay their eggs in the top layer of the soil. Fall or early spring plowing and later tillage may help prevent some of the young grasshopper nymphs from emerging.

Rotating crops may prove to be an effective control measure for those insects with specialized feeding habits. The larvae of the June beetle will feed on the roots of crops in the grass family but does not feed readily on plants of the legume family. The corn root-worm feeds on the roots of the corn plant but will probably starve to death if soybeans are planted in the field.

Some insects can be kept under control by harvesting the crop early. The alfalfa weevil and the lygus bug population can often be reduced by cutting the alfalfa a little early and removing the crop from the field immediately. The exposed larvae, together with eggs and pupae, will then die of starvation or heat exposure before the second growth is large enough to provide food and protection. Wheat infested by the sawfly may be salvaged by cutting it before the insect has a chance to weaken the straw, allowing the heads to fall to the ground where they cannot be picked up by the harvesting equipment.

Any cultural practice that provides an environment in which the crop can grow well will help prevent damage from insects. Preparing a good seedbed, planting at the right time and at the right depth, using insect-resistant cultivars, controlling weeds, and providing proper fertilization will aid in controlling insects.

Insect-Resistant Crop Varieties

Entomologists, agronomists, and other research scientists are constantly working to develop new cultivars that are resistant to insect attack. When one considers the potential environmental effects of excessive use of chemicals, the growing of insect-resistant cultivars is an ideal way of helping to control insects.

When scientists speak of resistant crops, they mean that some cultivars have the ability to avoid an insect attack, recover from an insect attack, or tolerate the presence of insects to a greater degree than other cultivars. One or more of a number of characteristics may make one cultivar more insect-

resistant than another. Some cultivars may have harder, tougher tissue, and the stems or leaves may be covered with excessive hair. The sap of some plants may be bitter and distasteful to the insects.

Developing resistant cultivars is a long and tedious process. Entomologists, agronomists, and geneticists must work many years to produce an insect-resistant variety. Farmers must rely on the plant breeders to do the job for them.

Climatic Conditions

General climatic conditions have much to do with the general distribution of insects. Specific weather conditions may have considerable effect on the insect population within a given area. Because insects are cold-blooded, they respond directly to extremes of heat and cold. Many factors associated with climate and weather, either singly or in combination, have measurable effects on insect populations.

A cold, wet spring will result in fewer grasshoppers in a given area. The high-moisture conditions encourage the spread of fungus diseases that attack many species of hoppers. The abundance of pools and ponds of water, however, creates an ideal breeding place for millions of mosquitoes. The amount and distribution of rainfall in an area may reduce the numbers of some insects while increasing the numbers of others.

Low summer temperature may greatly reduce the numbers of some insects. It has been shown that prolonged cool temperatures may disrupt metamorphosis and prevent many of the European corn borers from becoming pupae and emerging as egg-laying adults. A warm, long summer may allow many of the insects common in the southern areas of our country to move farther north. Generally speaking, an insect species can live through winter conditions in only one stage of metamorphosis. For example, the adult of the chinch bug can live through the winter, while in the gypsy moth only the eggs can survive the cold temperature.

Chemical Control

Some scientists have said that farmers could not produce enough food to feed the American people if they were not allowed to use insecticides to control harmful insects. They warn that without insecticides, food production could be reduced up to 50 percent from current levels. Others warn that the continual use of pesticides will have an adverse effect on the human population. Those who support the latter position point to pesticide poisoning, the reduction of fish and wildlife, losses of susceptible crops, and the destruction of the insect's natural enemies as harmful effects of excessive pesticide use. The purpose here is not to argue the pros and cons of using insecticides but to discuss how registered insecticides can be used to control insects.

A registered insecticide is one that has been approved, following extensive scientific research to prove its reliability, ability to selectively kill insects, and safety to humans when used at recommended rates on the appropriate crop and under proper conditions. An insecticide may kill the insect in various ways, but in most cases a particular insecticide has a chief mode of destroying the insect. A stomach poison is one that attacks the internal organs of the insect after the material has been swallowed. Insects such as the grasshopper and the larval stage of many insects with chewing mouthparts can be controlled most effectively with this type of insecticide.

Contact insecticides kill when the material comes into contact with the insect and is absorbed through the external parts of the insect's body. Insects such as aphids feed by sucking fluids from the plant, so a contact poison is ineffective. The insecticide is sprayed on the plant where the insects can come into contact with the poison.

Fumigant insecticides enter the insect's body through the breathing tubes of its skin as a gas. Fumigants are commonly used by grain producers to control insects that attack stored grains. A number of soil fumigants are on the market and are being used to kill many kinds of insects and nematodes that attack the root system of growing plants.

A relatively new group of insect poisons are classified as systemic insecticides. A systemic insecticide is absorbed by the plant or animal, and the insect is killed when it feeds on the plant or animal. Systemic insecticides do not harm the host plant or animal.

Producers are going to be dependent on agricultural chemicals for the foreseeable future. It is essential that all insecticides be used correctly and that they not be used to try to totally eradicate insect pests. Several reasons can be cited for not trying to kill all insects:

1. The presence of some insects will help maintain an acceptable level of natural enemies.
2. Excessive use of chemicals seems to lead to insect strains that are more resistant to insecticides.
3. Excessive use of chemicals can cause environmental damage.
4. Excessive use of insecticides may damage the human environment.
5. Excessive use of chemicals may increase production costs to a point where total pest control is not economically feasible.

If insecticides are applied correctly they can help protect against insects without harming plants, animals, or humans.

Many insects challenge the crop producer. Some are specialized and seem to be more severe in one crop than in another. The following general discussion relates to those common insects which cause the greatest destruction in a crop. Space limitations will not allow a description of every insect that feeds on a single crop. Even though the insects discussed may concentrate their feeding on a specific type of crop, they may also feed on other crops in the area.

Corn

The European corn borer is one of the most destructive insects of corn and can be found throughout the corn belt. The full-grown larva or caterpillar overwinters in corn stalks, stubble, or other refuse in the area close to the cornfields in which it has been feeding (Figure 7-9).

The European corn borer larva progresses to the pupal stage and finally emerges as an adult moth. The female lays eggs in groups of 5 to 50 on the underside of the leaves and on the stalks. As the eggs hatch, the young growing larvae eat their way down the inside of the stalk, causing tassels to fall off and stalks to break.

Cultural practices used to control the European corn borer include plowing the stalks to expose the larvae to the elements. Planting cultivars that are resistant to the corn borer reduces losses from the insect. Chemical insecticides are also used to control the European corn borer, and several chemical treatments may be necessary in the course of the growing season.

The corn earworm is a familiar insect to anyone who has raised corn. Because the earworm pupae cannot live through the cold winter, they are a more serious pest in the

Figure 7-9. A magnified look at the European corn borer infesting an ear of corn. (*Runk/Schoenberger-Grant Heilman*)

southern states. The moths lay their eggs on the host plant. Each moth may lay from 500 to 3000 eggs. The eggs hatch in 2 to 10 days, and the worms feed first on the leaves and then attack the silk of the corn ear. As the ear develops, the earworms eat their way into the kernels, greatly reducing the corn yield. When mature, they move down the stalk to the ground, where they pupate.

Chemical control is of limited value in field corn. Fall plowing brings many of the pupae to the surface, where they may be killed by low temperature. The planting of resistant corn cultivars will greatly reduce earworm damage.

Corn rootworms are found throughout the corn-growing areas of the United States.

One of two kinds, the northern or the southern rootworm, may attack the corn, depending on the location. The greatest damage to the corn plant is caused by the larvae of the rootworm as they feed on the roots of the plant. The insects slow down growth, and plants are often undersized. Lodging may occur, as the plants may more easily be blown over by strong winds because of the damaged root system.

The northern rootworms overwinter only in the egg stage, the eggs being laid in the ground around the roots. Rotations work well in controlling the insect because the larvae feed only on corn. Fall plowing or disking helps control the insect by exposing the eggs to the weather. When chemicals are used, they either are incorporated into the soil when the soil is disked or plowed or are placed in bands over the row when the corn is planted.

The southern rootworms pass through the winter in the adult stage. The adult can fly, so the use of rotations has limited effect on the spread of the insect. Late planting on land that has been plowed early in the spring helps control the insect. Soil tillage before planting may expose some of the larvae and keep vegetation down. The same chemicals used to control the northern rootworm can be used in protecting the corn against an attack of the southern rootworm.

Wheat

The Hessian fly, wheat stem sawfly, chinch bug, green bug, and grasshopper are among the insects causing economic losses to wheat producers in the major small-grain-growing areas. The adult Hessian fly lays her eggs on the upper surfaces of the wheat leaves. When the eggs hatch, the small larvae move down the plant, where they draw sap from the tender tissue at the base of the young wheat plant. The insects, feeding on the stems, weaken the plant and cause stunting. In the summer the weakened stalks break over and fall to the ground before harvest. The full-grown larvae remain in the stubble and volunteer wheat during the winter.

Good management helps greatly in control of the Hessian fly. Plowing immediately after harvest to cover the stubble and bury the larvae and pupae and destroying volunteer seedlings before seeding in the fall is helpful. Insecticides are not effective, but rotation, maintaining the fertility level, and planting the recommended cultivars at the right time greatly reduce damage from the Hessian fly. Late seeding so that the wheat comes up after the adults have emerged, laid their eggs, and died helps the wheat plants escape injury.

The wheat stem sawfly is found in the northern section of the Great Plains, particularly in Montana and North Dakota. The larva spends the winter just below the surface of the ground. In June and July the adult sawflies emerge from the stubble fields and native grasses. They fly to the growing wheat plants and lay their eggs in the hollow centers of the wheat stems. In a week the eggs hatch and the larvae begin to move inside the stem. The full-grown larva cuts a groove around the inside of the stem at ground level. Losses result from shrunken kernels and the loss of heads when the stems break over in the wind. Early harvesting and good cultural practices are helpful in controlling the insect. However, the best defense is to plant solid-stemmed cultivars of wheat so that the adult sawflies cannot penetrate the stems to lay their eggs.

The chinch bug winters in the adult stage in stubble and trash along the fields. When the spring temperature reaches 70°F [21°C], the adults come out of hibernation and fly to nearby small-grain fields. They begin to feed and lay their eggs on the lower parts of the plant. It takes about 30 to 40 days for the small bugs to complete their development. Throughout the summer they must move to green plants to feed. They mature in mid-summer and start a second cycle. Good farming practices help in controlling the chinch bug. The bugs avoid shade and dampness, therefore uniform, solid stands of vigorously growing wheat escape much of the damage.

The greenbug causes heavy wheat losses throughout the world where small grains are grown. Darkened areas appear in the wheat fields during late winter and early spring when greenbugs are present. The discoloration is caused by a toxic saliva that is injected into the plant when the greenbug feeds. The greenbug gives birth to live young in the southern states. In the north it goes through the winter in the form of eggs.

The eggs hatch in 7 to 18 days, and the females then begin to give birth to live young. Destroying volunteer small grains which green up early in the spring helps hold the greenbugs in check. Rapidly growing plants tend to withstand greenbug attack (Figure 7-10).

Oats, Barley, and Rye

The chinch bug, greenbug, and grasshopper are insects that frequently attack oats, rye, and barley. The extent of the damage caused varies from area to area. The chinch bug is frequently a serious insect pest in barley. Insecticides are effective for the grasshopper and greenbug. Crop rotations and growing resistant crops aid in chinch bug control.

Figure 7-10. Damage to 6- to 10-in wheat in field caused by presence of green-bugs. (*Grant Heilman*)

Cotton

Numerous insects attack the cotton plant. At times destructive insects have threatened the cotton industry. Fortunately, the development of numerous insecticides has reduced greatly the risk of raising cotton in the southern states although control costs still greatly reduce southern acreages. The most famous of the cotton insects is the boll weevil. A single pair of boll weevils can produce millions of boll weevils in a season.

The adult boll weevil overwinters in trash near the cotton field. In the spring, the weevils move into the cotton fields, where they spend the entire summer. The adult insect punctures the boll (fruit) or square (flower bud) and deposits its eggs. In 3 to 5 days the eggs hatch, and the larvae feed for 7 to 14 days before they change into pupae. In another 3 to 5 days the adults emerge and the cycle begins again.

Cultural practices to control the insect include (1) preparing the soil properly, (2) selecting recommended cultivars for the area, (3) planting early and properly, (4) picking the entire crop, and (5) using proper chemical treatment.

The bollworm is a problem particularly in Texas, Louisiana, and Oklahoma. Bollworms feed on tomatoes and corn as well as cotton. The adult insects lay their eggs in rapidly growing, succulent cotton plants. The eggs are laid singly in the squares or in the new plant growth. The larvae feed on the tender buds or leaves or on the inside of the bolls. The larvae pupate in the soil. There can be several broods in a single season.

The bollworm is controlled by using insecticides before the larvae enter the bolls. Treatment must be applied several times during the season. Ladybird beetles are natural predators of the bollworm.

Other insects that attack cotton include the aphid, fleahopper, and leafworm. The aphid and fleahopper feed on the plant juices, causing the plant to appear wilted. The leafworms feed on the underside of the leaf and, when abundant, may strip the cotton plant. Chemical control is the most effective control measure.

Legumes

A large number of the insects that attack legume plants have piercing-sucking mouthparts. As with other crops, some insects prefer one crop over another, but many are common to all types of legume crops.

The potato leafhopper is a serious pest in potatoes but also attacks legumes. The leafhopper sucks the sap from the plant, stunting it and causing reduced yields. Typically, the leaves become yellow and drop to the ground. Plant diseases are often spread from one field to another as the potato leafhopper feeds (Figure 7-11).

Insecticides are quite effective in controlling the potato leafhopper. The population of the insect can often be reduced by delaying cutting of the crop. The delay allows the adults more time to lay their eggs. Thus many of the eggs that cause the population to increase are carried off the field when the hay crop is removed.

The lygus bug is small but can cause serious damage. The plants are injured as the bugs puncture the foliage and suck the sap out of the plant. Insecticides are effective against the lygus bug. As with all insecticides, care must be taken not to spray the plants during the wrong stage or when beneficial insects are pollinating the crop.

The alfalfa aphid, or pea aphid, as it is often called, may kill the small plants and seriously weaken the more mature plants. The adult light-green, soft-bodied insects produce living young each spring. They migrate from field to field sucking sap from the young plants. Chemical control seems to

Figure 7-11. Typical damage to legume plants caused by the presence of the potato leafhopper. (*Grant Heilman*)

be the most effective control measure. The ladybird beetle is a natural enemy of the alfalfa aphid.

The alfalfa weevil is a common insect in the alfalfa fields of the western states. The adult weevils lay their eggs in dead stems or in growing stems. As the larvae develop, they eat the leaves and stems of the plant. When they complete their growth they drop to the ground, where they spin a cocoon. In 7 to 10 days the adults emerge and the cycle repeats.

Chemical treatment of the alfalfa weevil is effective if applied at the proper time. Clean harvesting and quick removal of the hay help keep the weevil population in check.

Soybeans

The velvet bean caterpillar (larval stage) feeds on the leaves, buds, and tender young stems of the soybean. The caterpillar spends the winter in the warm southern regions and moves north as the weather warms. Actually, most of the adult moths fly into the continental United States. The larva is about $1\frac{1}{2}$ in [4 cm] long and eats from the top to the bottom of the plant. Unless controlled, the larvae can strip a field of soybeans in a few days. The caterpillar pupates in the upper layers of the soil, and in 10 days an adult moth emerges. The application of insecticide provides the best protection for the soybeans.

The larvae and the adults of the Mexican bean beetle both feed on the soybean plant. Grubs, grasshoppers, cutworms, and armyworms also feed on soybeans.

Vegetables

The commercial vegetable producer faces a small army of insects that can destroy the vegetables before they reach the market. Only a few of the more common insects that invade fields and gardens will be mentioned. Although not an insect, the spider mite is included.

Spider mites are so small they are hard to find on the plant. The greenish or yellowish mites multiply rapidly, often producing up to 17 generations per year. They feed on a large number of plants, including garden and feed crops and ornamentals. They lay their eggs on the underside of the leaves. As the mites develop, they puncture the leaf and suck out the sap. Heavy infestations cause the leaves to turn red or a rusty brown. The spider mite multiplies most rapidly in hot, dry weather. A heavy rain may slow the infestation. Because they produce several generations in a year, several applications of an appropriate miticide must be applied.

The Mexican bean beetle feeds on the foliage of garden beans, cowpeas, and soybeans. The adult beetles spend the winter in the adult stage. They emerge in the spring, feed on the plants they infest, and lay their eggs on the underside of the leaves. In 20 to 35 days they reach full growth and pupate. In another 10 days the adults emerge and the cycle begins again. The Mexican bean beetle can be controlled with spray and dusts, but one must be sure to treat the underside of the leaves, where the larvae feed. The treatment will need to be repeated every 10 days if there is a bad infestation.

A familiar sight to most gardeners and vegetable growers is the Colorado potato beetle. This striped insect lays its eggs in bunches on the underside of the leaves. In 4 to 9 days the larvae hatch and begin to feed on the leaves. The larvae go through four metamorphic instars, becoming full-grown in 10 to 21 days. The grown larva burrows into the soil beneath the plant and changes to a pupa. The adult emerges in 5 to 10 days, crawls to the surface, feeds on the plant, and begins to lay eggs. Appropriate insecticides applied several times during the season will keep the Colorado potato beetle under control.

The cabbageworm is a velvety green caterpillar that feeds on cabbage, broccoli, and cauliflower. Those who raise these vegetable crops will notice a yellowish white moth making frequent visits to the growing plants. The adult moth lays its eggs usually on the underside of the leaves. In about a week the larvae hatch and feed for about 15 days while they mature. The mature caterpillar changes to an adult butterfly through the chrysalis. Safe insecticides applied to the plants several times during the season will usually control the cabbageworm.

There are many different kinds of cutworm. The cutworm attacks the roots and lower stems of many vegetable crops. The adult moth lays her eggs in the soil. When the eggs hatch, the young cutworms feed greedily, cutting off the stems of the plants. Cutworms pupate in the soil and prefer to feed at night near the surface of the soil. Therefore an insecticide, often in the form of bait, must be applied on the soil surface.

Fruits

Dozens of insects, including aphids, fruitworms, and stinkbugs, attack fruit trees. However, only the Mediterranean fruit fly, the plum curculio, and the codling moth are discussed here.

The Mediterranean fruit fly, or "medfly" (*Ceratitis capitata*), may be found in fruit-growing regions from Florida to California. It was first discovered in Florida in 1929. Over the next 2-year period approximately 120,000 acres [48,600 ha] were treated in Florida in an effort to eradicate it. An outbreak occurred in the Santa Clara Valley of California in 1981.

The adult medfly is about the size of the common housefly. It has yellowish-orange marks on its wings, black spots on its back, and light-gray bands on its yellowish abdomen. The medfly usually remains in trees and bushes near where it has emerged from its pupal case (Figure 7-12).

The female medfly needs food to survive and lay eggs and feeds on honeydew excreted by other insects such as aphids and mealybugs. The females are ready to lay eggs in about a week if the temperature is over 75°F [23.9°C]. The male of the species also feeds on the honeydew until it becomes

Figure 7-12. Closeup photo of Mediterranean fruit flies and the damage they cause to apples. (*USDA*)

sexually active. The males and females mate on the leaves or fruit in the morning when the temperature exceeds 64°F [17.8°C].

The females seek out fruit that is beginning to ripen. They puncture the fruit with their ovipositor and deposit the eggs in a small cavity just below the skin. Each female lays as many as 40 eggs per day, and under optimum conditions may lay up to 1000 eggs over a 60-day period. Temperature has a great effect on the medfly's length of life.

The small eggs hatch in 2 to 3 days. The temperature must be above 53°F [11.7°C], with the ideal temperature for hatching being approximately 79°F [26.1°C]. The cream-colored larvae begin to bore into the fruit pulp. They grow to be about $1/3$ in long [0.83 cm]. The rate of growth is influenced by temperature and the type of host fruit.

When the larvae reach maturity, they leave the fruit and burrow into the soil to a depth of about 3 in [8 cm] or seek shelter under some object on the surface of the soil and pupate. More than one generation may occur per year, depending on the location of the infestation.

The cycle of infestation can be broken by stripping host fruits from trees and removing all fruit lying on the ground. Correct application of recommended insecticides is effective and aids greatly in controlling the medfly.

Biological control relies on the distribution of sterilized males. The males are exposed to gamma radiation in the pupal stage. When the sterilized males mate with female flies, the sperm enter the egg and disrupt its development, and the egg does not hatch.

Figure 7-13. The plum curculio, which is another type of insect which attacks fruit, is shown here burrowing inside a plum. (*Grant Heilman*)

The larvae of the codling moth are off-white or pinkish caterpillars that eat into the fruit. The codling moth favors the apple and is often called the *apple worm*. The moth eats holes leading to the core of immature apples. The hole leading to the core is often filled with dark frass resulting from the internal feeding of the larvae. The moth may also cause damage to pears, English walnuts, and some other fruits.

Plums, peaches, cherries, and apples may be damaged by the plum curculio. The adult beetles feed on the outside of the fruit, causing scars and distortions that destroy its appearance. The larvae cause damage by burrowing inside the fruit (Figure 7-13).

The adults lay their eggs in crescent-shaped cuts. The yellowish-white grubs hatch in about a week. In 2 weeks the grubs are mature and drop onto the ground to pupate and turn into adult beetles. Most of the damage is caused early in the season so that the fruit is deformed or falls to the ground before it matures.

SUMMARY

Eliminating harmful insects is an impossible task. The best that can be hoped for is that these "enemies of the food basket" can be held in check. Scientists continue to study insects and discover new ways to reduce the damage they do.

Good pest management is being followed by most progressive producers. When pest management is followed, producers seek to obtain the most efficient crop production and produce the most economic yield with the least production cost. Such a system requires that growers carefully monitor their fields for pest activity. Control measures must be timely. Nothing can be left to chance.

QUESTIONS FOR REVIEW

1. What are the three specific parts of an insect's body?
2. What are the six insect characteristics that enable them to flourish?
3. What is the process of metamorphosis?
4. What are the different feeding habits of insects?
5. What benefits do insects provide for humankind?
6. How can the producer control insects?
7. What are some common insects that cause damage in small grains?
8. What are some common insects that cause damage to legumes?
9. What are some common insects that cause damage to vegetables?
10. What are some common insects that cause damage to fruits?

Chapter
8

Biological Control of Crop Pests

Controlling pests is necessary if producers are to maintain and increase food production. As chemical agents are used in an attempt to control common crop pests, the control of pests often upsets the balance of nature by destroying natural enemies. Thus the struggle against crop and food pests has become a complicated process. Everyone is affected either directly or indirectly.

Millions of pounds of chemicals are being used annually to control weed and insect pests. In some cases, the uncontrolled use of chemicals has resulted in contamination of our environment. Scientists have estimated that chemicals, to be effective, would need to control 80 to 90 percent of the insects important in agriculture and public health. For such a "kill level," large amounts of chemical agents must be used. New methods of pest control are being studied at our agricultural colleges and research stations and by industrial researchers in an effort to reduce our dependence on herbicides, insecticides, and other agricultural chemicals to control crop pests.

Biological pest control is not new. However, with the increase in concern over the use of chemical pest control methods, more attention is being given to biological control as an acceptable alternative in a well designed pest management program.

CHAPTER GOALS

After studying this chapter you will be able to:

- Describe biological pest control.
- Discuss the importance of biological pest control.
- List the advantages and disadvantages of biological pest control.
- Identify some common biological pest control agents.

BIOLOGICAL CONTROL AGENTS

Several biological controls seem to hold promise for future pest management. These methods use the weed's or insect's natural enemies to reduce the pest population to an economically acceptable level. At the present time, insects are the most often used biological control agents (Figure 8-1).

Biological pest control is not a new method. In the late 1800s and early 1900s, a defoliating moth was used to control prickly pear cactus in India and Australia. In 1920, some 60 million acres [24 million ha] of Australian land was infested with various species of the cactus. Certain defoliating beetles were found to be effective in bringing the cactus under control. As a result, the cactus exists only in isolated locations of that country today.

Some of the most widely used biological agents, namely, insects, are predators. This group of insects complete their growth cycle by feeding on individuals of a particular species, which tend to be general feeders.

The parasitic insects are parasites in that they complete their growth cycle on or in a single individual host of a pest species. The parasitic insects that are used as biological agents tend to be host-specific. Both predators and parasitic biological agents offer hope as scientists and researchers seek new methods of maintaining a more natural predator-prey relationship.

Need for New Methods

Each day, billions of insects and a host of weed pests reduce yields and lower the quality of our food crops. The damage done may vary from year to year and from area to area. Besides lower yields, increased costs result when the harvested crop must be graded, sorted, and cleaned to make the food products fit for human consumption.

Figure 8-1. Hawkmoth larva is a biological control agent for leafy spurge. (*MSU*)

The devastating effect of many of the country's weed and insect pests has been greatly reduced by the use of chemical pesticides. However, the millions of pounds of chemicals used have helped control but have not eliminated a single pest except in very isolated cases. In many cases, certain weed and insect populations have actually increased. Although nearly all insects can be controlled by chemicals, this measure is temporary in nature.

When synthetic chemical compounds were first developed, some scientists felt it would be possible to achieve permanent control of crop pests. However, many did not consider the side effects of the large-scale use of pesticides. It is true that production of food and fiber has increased, but undesirable conditions now exist for certain species of birds and animals. Some pesticides persist in the environment and tend to accumulate to unhealthy concentrations. Some pests have become more resistant to chemical agents, thus the use of more toxic material is required to achieve control. Often, for the producer, the use of additional pesticides has reached the point of diminishing returns.

No single method of pest management will be totally effective under all circumstances. The answer lies in the proper use of a variety of pest control measures. Such an approach will offer the best control with the least adverse effect on our environment and the ecosystem. The use of the pest's natural enemies seems to be a reasonable supplement to selected chemical control measures (Figure 8-2).

How Biological Control Works

Many species of insects feed on a variety of organic material, which often includes other insects. Some insects are quite specific and tend to feed only on selected organic materials. Some insects are parasitic to one or two other insects. Those insects which are host-specific, either to insects or plants, are carefully studied to determine whether they would be effective biological control agents.

Those insects which are host-specific can be reared in large numbers and released in or adjacent to the area where the weed or insect pest is causing damage. As the population of control agents increases, they feed on their host, inhibiting its growth and development. As the population of the unwanted pest begins to decrease, the population of the biological control agent also begins to be reduced because of a lack of food. Usually enough of the unwanted pests remain in the area to maintain the population of the biological control agent. Should the unwanted pest begin to increase, the population of the biological control agent begins to reproduce in greater numbers. Thus a natural balance

Figure 8-2. A quarantine officer packaging approved predator insects for release in strategic locations. (*USDA*)

is maintained and the pest is kept under control.

Once established, the biological control agent begins to feed, consuming the vital parts of the plant or insect. In the case of weeds, the agent may bore into the plant so that the plant structure is weakened. In the case of insects, the agent may eat the pest directly or lay eggs inside the body of the host. The developing larvae then feed on the inside of the insect. If the plant or insect is not entirely consumed, the agent may reduce its

Biological Control of Crop Pests

vigor and thereby affect its ability to reproduce and grow. Finally, the action of the biological control agent may weaken the pest to the point that diseases attack it.

A large number of crop pests have been introduced into this country from foreign countries. Thus it is often necessary to search out natural enemies in the country or area in which the pest originated. The process is often a long and laborious one. But even if a natural enemy is found, successful transplantation is not assured. Often a slight change in environmental conditions or natural habitat will affect the survival of the potential biological control agent. If the biological control agent can live in its new environment, it must then be mass-reared for later release.

The object of using a biological control agent is control rather than eradication. The biological agent reduces and subsequently regulates the pest below the level of economic injury.

Selecting an Agent

A biological control agent must meet a rigid set of criteria before it can be released. It must be proved that the agent will limit itself to one pest, or in other words, be host-specific. To make this determination, researchers must understand the biological agent's natural habits and tendencies. Scientists must know what attracts the agent to a selected host and whether these conditions are unique to the pest to be controlled. Simply stated, the biological agent must not attack other species that may be beneficial to the producer. Finally, the biological control agent should come from a climatically similar area to ensure its survival (Figure 8-3).

Often it takes 3 to 5 years or more of research to determine the acceptability of a biological control agent. Starvation tests are generally used to determine an agent's suitability. If the agent being tested should at-

Figure 8-3. A biological research technician growing plants that will be used for insect feeding tests. (*USDA*)

tack or feed on any beneficial plants during the test, it is no longer considered a potential control agent.

Once a biological agent such as an insect is considered safe, an effort is made to collect or rear enough of the agents to release in an area. It is hoped that the initial group of agents released will survive and become the nucleus of the population. Once there is a sufficient buildup, the agents can be collected and established in another location.

Advantages

Once the control agents are established, they become self-perpetuating. This results in a comparatively economical control method that is highly selective. The biological agents themselves do not introduce toxic substances into the environment, as is often the case with pesticides.

Biological control agents save energy. A large amount of fossil fuel energy is used in the manufacture of chemical pesticides. This saving in cost, plus the long-term nature of the biological agents, makes them relatively inexpensive pest control mechanisms.

In many places it is difficult to apply chemical pesticides. Rough mountains, swampy areas, and other inaccessible places make traditional chemical application equipment almost useless. Biological agents can move into such areas and significantly reduce pest populations without the necessity of ground application or aerial spraying equipment.

Disadvantages

There are limits to the effectiveness of biological pest control. Because it requires time to build up populations, immediate control is not possible. Even when populations are at a high level, biological control does not achieve eradication but only helps to keep

the pest infestation at a tolerable level. Once a population is built up, the control agents are likely to move into a neighboring field. This makes biological control more suited to a large area or region than to an individual farm.

The highly selective nature of the biological control agents can be a disadvantage, because the agents attack only one pest in a group of many that may be equally destructive. In the case of weeds, biological agents may not be able to distinguish between closely related plant species.

EFFECTIVENESS OF BIOLOGICAL CONTROL MEASURES

Under natural conditions, all plants and animals are subject to some natural enemies or environmental factors that control their population levels (Figure 8-4). Without natural control, satisfactory pest management would be virtually impossible. Biological control by itself will not provide adequate protection against the many pests that threaten our food supply. However, it does provide the producer with one more way of controlling crop pests. But there have been and continue to be some important breakthroughs using biological control agents.

Klamath Weed

The Klamath weed was introduced into the United States about 1900, near the Klamath River in northern California. The weed is also called St. John's wort or goatweed. When animals graze on the weed it reduces their appetite and causes general unthriftiness. The weed spreads rapidly in the rangelands of the western United States and western Canada.

Chrysolina quadrigemina, a defoliating beetle, was introduced into California in the

Figure 8-4. Collecting control insects for later redistribution. (*Photo by Gregory Northcutt*)

1940s. The insect greatly reduced the established stands of the Klamath weed in California. In other states the success of the beetle was less dramatic. The success in California can be attributed to the fact that the insect's host specificity and the synchronized relationship between the life cycle of the insect and the growth stages of the Klamath weed are almost perfect.

The adult begins to feed on the leaves of the plant in May and June. The beetle then goes into a resting stage while the weed enters a relatively dormant phase. In the fall, the weed becomes active and so do the beetles. The female beetles lay their eggs in the fall. The adult, larvae, and egg of the beetle survive the winter in California. The larvae feed on the plant during the warm periods of the winter. Thus the weed is under almost constant attack from the beetles (Figure 8-5).

It has been estimated that the biological control of the Klamath weed in California cost between 200,000 and 300,000 dollars. Herbicide savings and the increase of land values, plus the greater land productivity over a period of years, has probably meant benefits of millions of dollars.

Spotted Knapweed

The spotted knapweed has infested some 1.5 million acres [608,000 ha] of grazing land in Montana. In 1973, *Urophora affines*, a gall

Figure 8-5. A Klamath beetle feeding on the Klamath weed. (*Photo by Gerald Johnson, USDA*)

fly native to Europe and western Asia, was released to attack the spotted knapweed.

The gall fly reduces the weed's ability to produce seed. A gall forms as the insect feeds within the receptacle of the flower buds. The fly's effectiveness as a control agent is reduced somewhat because it is a weak flier and spends a large part of its time on the host plant. However, success has been such that further efforts are being made to transplant the gall fly to other areas where knapweed is a problem.

The Gypsy Moth

The gypsy moth was introduced into the United States from Europe. In Europe and Asia the gypsy moth does relatively little damage because of its natural enemies. Un-

fortunately, when it was transported to the United States it left its natural enemies behind.

The gypsy moth caterpillar is capable of defoliating the trees in the forests of the northeastern states. The moth, true to its name, is spreading and is now found in the Midwest and some parts of the south.

The coccyomimus wasp deposits eggs in the pupa, and the parasitic fly *Palexorista* deposits eggs in the caterpillar. Tests are being conducted to find additional species of parasites that will help control this voracious pest of the fruit orchard and the forest.

Aphids

Aphids are a common species of insect pest. They feed on the stems, leaves, buds, flowers, and roots of some plants. Many generations are produced each year. Some members of the species feed on one kind of plant only, but others feed on many different kinds.

Both parasites and predators feed on the aphid. Several wasps are parasitic to the aphid. The female wasp deposits a single egg inside the tiny aphid. The egg hatches in 2 to 3 days, and the wasp larva begins to feed on the aphid. The larva molts several times during this period and finally cuts a hole in the aphid's abdomen. The wasp larva produces a silk substance that cements the aphid to the leaf. In another 5 to 6 days, the tiny wasp emerges to start the cycle over again.

The most common predator of the aphid is the ladybird beetle. The adult lays her eggs on the plant leaves. The black and yellow or orange larvae seek out the aphids and attack them. The adult ladybird beetle also feeds on the aphids. Currently, ladybird beetles are being raised and released in various parts of the United States. In some states the ladybird beetle has become es-

Biological Control of Crop Pests

tablished quite successfully. If sufficient numbers of ladybird beetles can be obtained, this predator should prove an effective biological control for many species of aphids (Figure 8-6).

Mexican Bean Beetle

A pinpoint-sized mite is being tested as a biological agent against the bean beetle. The mites live beneath the wings of the bean beetle and live by sucking the fluids from their host's body, thus lowering the beetle's vitality and reducing its ability to reproduce.

It has been estimated that farmers in the Mid-Atlantic states spend as much as $5.25 per acre [0.405 ha] on insecticides to control the Mexican bean beetle. U.S. Department of Agriculture scientists are continuing to test the effectiveness of the mite. If its effectiveness is confirmed, the bean beetle should be less of a nuisance in the future.

MICROBIAL CONTROL AGENTS

Certain pathogenic microorganisms have the ability to interfere with the growth and development of certain species of plant pests. These organisms, which include bacteria, virus, protozoa, and fungi, are being considered and, on a small scale, being used to bring crop pests under control. The concept is not new, but recent concern about the widespread use of chemical pesticides has increased researchers' experimentation with these disease-causing agents.

Plant pests (insects), like large animals and humans, are susceptible to diseases. As with animals and humans, these disease-causing organisms are quite specific to certain pests. Like biological control agents, microbial control agents are tested carefully to be certain that they can be used safely and, like pesticides, each agent must be registered with the U.S. Environmental Protection Agency.

Figure 8-6. A ladybird beetle feeding on a pea aphid. (*USDA*)

Figure 8-7. Closeup photo of larva on the underside of a tobacco leaf. (*Grant Heilman*).

The use of microbial control agents is not a totally new concept. In the early 1930s, a virus disease was introduced to control the European spruce sawfly. Within 10 years, this harmful insect was virtually eliminated.

How Microbial Insecticides Work

Microbial insecticides must be applied so that the insect feeds on the material containing the organism. The diseases introduced to the pest have an incubation period. The insect will not die immediately upon eating the material. The time lapse between initial acquisition and death will vary according to the insect–microbial insecticide relationship (Figure 8-7).

Some diseases destroy the insect quickly. In this case, the organism develops rapidly and produces a poisonous substance which kills the insect. Other insect diseases take several days; and still others develop so slowly that protection may not be achieved until the following growing season.

Bacillus thuringiensis is a bacterium which produces immediate control in some common agricultural pests. The material must be placed so that the larva consumes it. Insects such as the imported cabbageworm, alfalfa caterpillar, and tobacco hornworm, which feed on the outside surfaces of the plant, are particularly susceptible to the organisms. A larva that feeds on the inside of the plant would be less susceptible to the organism. Thus the European corn borer and the corn earworm, although susceptible to the bacterium, would not be seriously affected by the application of the microbial insecticide.

Many of the diseases caused by viruses, bacteria, and fungi will kill certain insects, but it takes several days before they are affected. Virus diseases have been used to control the alfalfa caterpillar and the cabbage looper. But the producer must remember that the insect will cause some damage before it is killed. On the other hand, when the insect dies and the body decomposes, the fluids of the insect's body will spread over the plant's growing surface, thus infecting other insects when they feed on the plant tissue.

Advantages

The most promising advantage is that the known organisms that cause diseases in insects do not affect higher forms of life. Therefore they are safe to apply and leave no toxic residue on foodstuffs.

EFFECTIVENESS OF MICROBIAL CONTROL MEASURES

The potential of microbial agents as pest control mechanisms has been recognized for many years. It is estimated that more than 1 million lb [453,000 kg] of the pathogen *Bacillus popilliae* (milky spore disease) are sold annually in the United States to help control caterpillars on vegetables and cotton. Plant pathologists and other researchers are continuing to experiment with new microbial agents.

Grasshoppers

Montana researchers are experimenting with two natural microbial agents to control grasshopper populations. One method uses a *Nosema* protozoan. The protozoan occurs naturally in about one-tenth of all grasshoppers and destroys the insect by eating away the fat in its body.

The protozoan is mass-produced by infecting grasshoppers in the laboratory. When they die, the grasshoppers are ground and the organisms extracted from the ground material. The organism is then mixed with bran and fed to other grasshoppers. About $1^1/_2$ lb/acre [1.68 kg/ha] of the mix is spread on the land.

When the control method is perfected, it can be applied to young grasshoppers before they do extensive damage. Moreover, if the grasshoppers do lay eggs, the protozoan is passed on through the eggs and the new offspring will be affected. The major disadvantage is that, by itself, it is a long-term control method.

The second method being developed is designed to reduce the grasshopper's ability to reproduce. Research scientists have found that certain hormones added to the grasshopper's diet will reduce the number of eggs the female grasshopper will lay. Also, many of the eggs that are laid will be sterile and will not produce viable young.

When the hormone treatment control method is perfected, the material can be sprayed on the grass or other plants eaten by grasshoppers. The big advantage will be that the hormones, which occur naturally in some plants, should not have adverse environmental effects.

Canada Thistle

Canada thistle is considered a noxious weed in the western states. It is difficult to control because it spreads by seed and rhizomes. Researchers have observed that some Canada thistles succumb to a rust fungus disease.

When the plant is affected, the rust either kills or severely weakens the plants. It is estimated that rust occurs naturally in about 20 percent of Canada thistles.

The rust fungus must go through several stages to be effective. The fungus spores must germinate and penetrate the leaves of the thistle. In about 10 days another group of spores is produced which can be carried by the wind to other thistles. If the spores have 3 to 4 hours of moist conditions they will germinate. During the winter the spores live in the soil, where they can attack the roots of the plant. The rest of the year, the spores are found on the thistle leaves.

Research is under way to determine why some plants are more susceptible to the rust than others. The rust fungus is promising, and researchers around the country are beginning to study this microbial agent as a control for one of the worst weed pests affecting crop- and rangeland.

SUMMARY

Biological and microbial pest control is a sophisticated and complex venture. Even when natural enemies are found, there is no guarantee that they will be effective in their new environment.

As with chemical control, there are potential hazards that must be considered. The screening procedure requires a considerable amount of time and money. Success of the developmental program will depend on a cooperative effort among government agencies, universities, private industry, farmers, and finally every citizen.

QUESTIONS FOR REVIEW

1. Why is biological control of pests important to agricultural producers?
2. What is biological pest control?
3. What procedure is followed in selecting biological pest control agents?
4. What are the advantages of biological pest control agents?
5. What are the disadvantages of biological pest control agents?
6. Why is biological pest control not effective as the only control measure?
7. What are some examples of biological pest control agents?
8. What is a microbial pest control agent?
9. What are some examples of microbial pest control agents?

Chapter

9

The Agricultural Chemical Industry

Research and technology have revolutionized agriculture, enabling fewer American farmers to produce larger quantities of food, feed, and fiber crops. New and improved cultural practices, new crop varieties, and increased mechanization have helped bring about increased productivity. The agricultural chemical industry, although sometimes criticized, is a vital part of American agriculture. The use of a wide variety of chemicals has helped to make the United States population one of the best-fed, best-clothed peoples in the world, while they spend less than 20 percent of their take-home pay for food—one of the lowest percentages of all the countries in the world.

It is estimated that even with the use of advanced technology and mechanized farming practices, there is still an annual loss of about 30 percent of the world's potential productivity. Most of this loss is caused by pests of all kinds attacking plants at all stages of growth: crops in storage, food products in storage and enroute to the market and even food on the shelves of the supermarket and the home pantry.

Plant diseases, insects, and weeds cause a great deal of the loss in potential crop yield and quality. Some 10,000 destructive species of insects are found in the United States. About 8000 of these species are injurious to crops. Disease organisms kill many plants and often cause rotting of the harvested crop (Figure 9-1).

Weeds cause an estimated 8 billion dollar loss in crop yield and quality. Some 2000 species of weeds and brush are competing with crops for light, water, and nutrients.

Rodents such as rats, mice, and gophers consume large amounts of food. In addition to consuming food, they destroy property and spread disease. The dollar loss caused by rats alone amounts to billions of dollars each year.

Chemicals produced by the chemical industry in this country are necessary to protect plants, animals, and humans from insects, diseases, weeds, and other pests.

CHAPTER GOALS

After studying this chapter you will be able to:

- Describe the ways in which agricultural chemicals help to increase food production.
- Identify the ways in which agricultural chemicals affect our daily lives.
- Describe how agricultural chemicals increase the quality of our food.
- Identify ways in which agricultural chemicals reduce the cost of food.
- Explain why agricultural chemicals are necessary to maintain a high level of agricultural productivity.

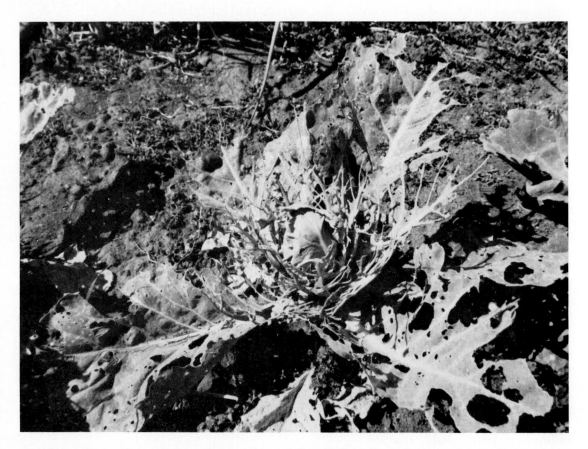

Figure 9-1. This cabbage plant was left unprotected. When left unchecked, insects may completely destroy vegetable crops.

HOW CHEMICALS AFFECT OUR DAILY LIVES

The farmer is not the only person affected by the development and use of chemicals. Chemicals affect each of us every day. The water we drink, the salt we sprinkle on our food, and the carbon dioxide we breathe are simple chemical compounds. Most of the food products we eat are complex chemical compounds.

Many of the chemicals that affect our lives occur naturally in our environment. Others are synthesized by chemists working in the laboratory. Each and every one of us benefits greatly from the work of the chemist and the products of the chemical industry. In this section, a few of the benefits we all receive, directly or indirectly, from chemicals in agriculture are discussed.

Increasing Productivity

Fertilizers make up the largest group of chemicals used in agriculture today. The soil in which crops are grown is like a vast chemical storehouse. As the crops use the nutrients that occur naturally in the soil, chemical fertilizers must be added if the farmer is to maintain a high level of crop productivity (Figure 9-2). Twenty chemical elements are essential for plant growth. Nitrogen, phosphorus, and potassium are the primary elements needed by plants.

It is estimated that 30 to 40 percent of the increase in crop productivity in recent years can be attributed to the application of chemical fertilizers. Each year the fertilizer plants in this country produce millions of tons of plant nutrients necessary to increase crop yields. Because these fertilizers are readily available, the United States farmer is in a better position to maintain our food supply.

Figure 9-2. A farmer spreading commercial fertilizer.

Few soils can sustain high crop yields without the addition of some type of chemical fertilizer. Applying fertilizer without using other improved cultural practices is not likely to increase overall crop productivity. However, proper crop fertilization is an important key to crop productivity.

The chemical plants that produce fertilizers to increase food production are facing many problems. Pollution laws and environmental controls to protect our surroundings must be met. The cost of building new plants is astronomical. Some of the raw materials from which fertilizers are made are in short supply. Natural gas, which is the most economical source of hydrogen for ammonia, the basic ingredient of most nitrogen fertilizers, is being used up at a rapid rate. The high energy requirements needed to produce fertilizer will certainly affect the industry in

The Agricultural Chemical Industry

the future. However, fertilizer is essential to maintaining or increasing food production.

Protecting Food

All efforts to prepare an environment for the plants to produce the greatest yield of food material are wasted unless the crops are properly protected. Each year billions of dollars in crop losses result in increased food costs to consumers.

Pest control is absolutely necessary if feed, food, and fiber production is to be maintained and ultimately increased. Generally, chemical pesticides are used to control insects, plant diseases, weeds, nematodes, and rodents. The common types of pesticides include insecticides, herbicides, fungicides, and nematocides. Other chemicals, including growth regulators, defoliants, and desiccants (drying agents), are in common use in the food production process.

Controversy surrounds the use of chemicals to protect our food, feed, and fiber supply. However, without the use of chemicals (pesticides), it would be difficult if not impossible to maintain our present level of food production and retain a supply of quality food in storage. On the other hand, the misuse and unwise use of chemicals has probably damaged the environment in some areas.

Chemicals used to control pests have received much attention over the past couple of decades. Many of these stories have related only one side of the chemical story and have created a certain amount of distrust on the part of the public for the entire chemical industry. The chemical industry must accept its responsibility to protect the grower, food processor, consumer, and the general public. On the other hand, the public must accept the fact that chemicals are as indispensable to modern agriculture

as the tractor, hybrid seed, and improved crop varieties and animal breeding stock.

Reducing Labor Costs

High labor costs mean higher food costs to the consumer. In conjunction with modern machinery and equipment, chemicals, properly used, can help to reduce labor costs.

In the early 1900s, it took the farmer in the corn belt about 100 hours to produce 100 bu [3.5 m^3] of corn; today, it takes less than 20 hours of the farmer's time. It is true that the reduction in the number of worker-hours needed to produce 100 bu of corn cannot be attributed entirely to the use of chemicals. However, in one hour's time with a power sprayer and a herbicide, the farmer can save hours of cultivation time (Figure 9-3). In addition, reducing the number of cultivations often improves the soil conditions, conserves moisture, and reduces energy requirements.

Other chemicals, called *defoliating agents,* can be applied to the plant shortly before it is ready to harvest. These chemicals cause the leaves of the plant to dry quickly. Getting rid of excess leaves and foliage means that the crop can be harvested more efficiently (and rapidly). Less time is spent in cleaning while preparing the crop for processing. The harvesting of cotton exemplifies the use of a defoliating agent to improve the mechanized harvesting process. The use of a chemical agent can ultimately help keep the cost of a new cotton shirt or blouse at a reasonable price.

Polyethylene film is another labor-saving product used in agriculture. This material is laid on the surface of the soil. Plants such as strawberries are planted in holes punched in the film. In addition to providing protection and warmth, and conserving moisture, the dark film material prevents weeds from growing and thus reduces the labor require-

Figure 9-3. A modern spray rig applying an approved pesticide. (*Tennessee Valley Authority*)

ment for weeding. Harvesting time is reduced because the crop is free from weeds and the ripened fruit is highly visible.

Reducing Energy Consumption

As the world's supply of oil dwindles, agricultural producers must look for ways to reduce energy requirements in farming. Chemicals are being used for this purpose.

No-till or limited-till farming has been made possible through the development of selected chemicals. Selective herbicides are used on a regular basis to reduce the number of times the producer must cultivate the crop to destroy weeds (Figure 9-4). The power requirements for field cultivation are far greater than the power requirements for operating a sprayer with the appropriate pesticide. Chemicals are helping the United States farmer use energy more efficiently.

Preserving and Improving Food Products

Thus far, it has been pointed out that chemicals are used to aid in plant growth, protect growing crops, and save time and labor for the producer. Another important function of chemicals is the preservation of food and the improvement of food flavor. Our early ancestors began to use chemicals in food when they discovered that salt would preserve and add flavor. Smoke was also used to aid in food preservation. Besides preserving food, certain chemicals will improve the nutrient value of food and enhance the quality of a food product or make it more acceptable to the consumer.

Certain chemicals are used to preserve the life of a food product. A chemical coating is often placed on the outside of fresh fruits, vegetables, and eggs to help prevent loss of moisture and slow the spoiling process. The nutrient value is retained, while the quality and appearance of the food product are maintained for a longer period of time.

Any food that deteriorates on the way to the market, while on the market shelf, or in the refrigerator after it is taken home is useless. For example, foods that contain fat or fat products may become rancid if not properly treated. Antioxidants are chemicals that slow fat deterioration. Few people

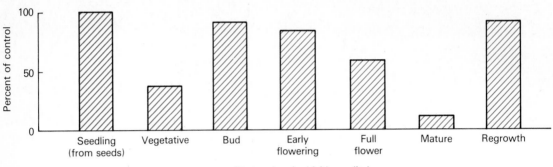

Figure 9-4. Graphs showing percent of weed control for (a) annuals and (b) perennials.

would be willing to purchase foods with repulsive odors or tasting of rancid fat (Figure 9-5). Chemicals used in the preservation of foods are carefully screened and approved as safe by the Food and Drug Administration.

The flavor of cooked foods can often be improved by adding chemicals. Monosodium glutamate is used by the fast-food industry to bring out the flavor of cooked foods. Food and chemicals cannot be separated. We eat food so the human body can assemble the chemicals needed to sustain life.

Synthesizing New Food Products

Almost everything around us is chemical in nature. Chemists employed in the chemical industry are discovering new ways to synthesize natural chemical compounds and produce them in a commercial form. Here

the chemical industry is, in a way, protecting our food supply by actually manufacturing chemicals.

Food-flavoring oils—peppermint, spearmint, and lemon and lime flavors—are readily synthesized by chemists. Oils used to manufacture vitamins A and E are easily synthesized. Thiamine, a vitamin obtained naturally from cereals, can be manufactured readily. Acetic acid found in vinegar can be synthesized easily by chemists. Many chemicals are being used to help preserve and bring out the flavor in frozen foods.

PROBLEMS IN MANUFACTURING CHEMICALS

In recent years, the chemical industry has come under severe attack. Often only one

Figure 9-5. Washing tomatoes with chlorine water before packing helps control soft rot. (*USDA*)

Figure 9-6. This lemon, infested by red scale, shows the result of an insect allowed to go unchecked. (*USDA*)

side of the chemical story is told through newspapers and books. There is always some inherent risk associated with the use of any chemical. For example, penicillin, which is considered to be a wonder drug, can cause serious reactions in about one in every twenty people.

New pesticides are costly to produce. It takes millions of dollars and perhaps 3 to 7 years to produce a new chemical. Federal and state agencies help, but the chemical industry has to take the lead and accept the major portion of the initial developmental cost.

Many new safeguards are being implemented by the chemical industry to protect the American consumer. Fewer and fewer nonselective chemicals are being developed. The pesticide industry is actively engaged in helping to develop pest management systems. Chemists must have a thorough knowledge of insect biochemistry and behavior.

Each chemical pesticide or food additive is carefully labeled. For example, a pesticide label must include a description of all ingredients of the pesticide, along with complete directions on how and when it should be applied. Failure to give such information can result in the levying of severe fines against the manufacturer of the product.

Chemicals are efficient and versatile in controlling pests and, properly used, can help increase crop production (Figure 9-6). The chemical industry has the ability to react rather quickly when a new crop pest appears on the scene. Industry research, along with the efforts of agricultural research stations and research farms, can develop safe chemicals which can help protect growing crops from insect pests, help maintain the quality of stored crops, and lengthen the life of food products being transported to market.

Many unwarranted attacks have been made against the chemical industry by per-

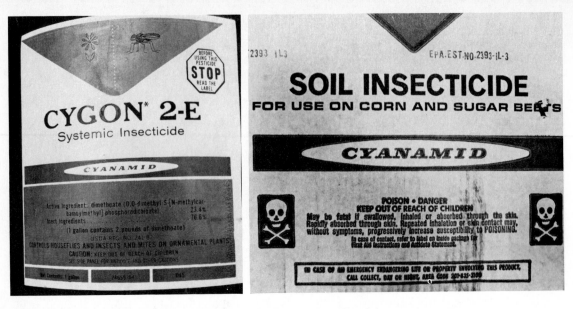

Figure 9-7. Cautionary labels on pesticides. (*Jane Hamilton-Merritt*)

sons who are uninformed on the proper application of chemicals in food production. On the other hand, the chemical industry has an obligation to use every analytical procedure possible to ensure that a chemical is safe for use on or in the food products we consume (Figure 9-7). It must be held accountable for observing the stringent regulations designed to protect the consumer and our environment.

OBLIGATION OF PRODUCERS

In an era of potential food shortages, more, not fewer, agricultural chemicals will be needed to produce the feed, food, and fiber we need. Ultimately, the consumer will benefit from the proper use of agricultural chemicals through quality foods at relatively low cost.

However, if used improperly, chemical pesticides can be toxic. With hundreds of trade-name products on the market, farmers must follow all state and federal regulations to avoid contaminating food, feed, and fiber products. Failure to follow the directions for using each chemical may result in their product being seized and condemned. In other cases, the food processor will not buy the produce if it contains excessive amounts of chemical residue.

Because some chemicals used in agriculture can leave potentially harmful residues, each chemical has its own withdrawal time. This means that the farmer is required by law to stop using a chemical for a required length of time before harvesting the crop for human or livestock consumption. In this way, farmers are helping to protect consumers against excessive drug residues.

To help protect the consumer from the improper application of pesticides, federal law requires that producers and commercial applicators who work with restricted-use pesticides must be trained and certified. Unless the producers or applicators have attended a special training school on the use of pesticides and have received a certificate of completion, they cannot purchase or use restricted pesticides. The pesticide industry has direct involvement in and responsibility for such training programs.

To be sure that they are not violating federal regulations, farmers must choose the right pesticide for the pest involved. This will maximize their production while at the same time protect the consumer. Each chemical has its limitations and should not be used indiscriminately. Furthermore, excessive use of chemicals is very expensive and reduces the producer's profit.

SUMMARY

Never before in history have Americans had more food and a greater variety of food to eat than they do today. Producers and consumers alike must not take this fact for granted. As our society uses more of its productive land for highways, homes, and shopping centers, fewer acres are available for food production. By concentrating agricultural production, this trend intensifies the losses resulting from insects, plant pathogens, nematodes, weeds, rodents, and other pests.

The manufacture and proper use of agricultural chemicals must continue if American farmers are to meet the burgeoning demand for food in the future. However, the government, industry, and individuals must guard against allowing chemicals to be applied without regard for human safety and the long-term effect on our environment.

QUESTIONS FOR REVIEW

1. How do agricultural chemicals increase crop productivity?
2. What problems are faced by the chemical industry in manufacturing chemicals?
3. What safeguards are being followed in the manufacture of agricultural chemicals?
4. What obligations do producers have when using agricultural chemicals?
5. How does federal law help protect the producer and consumer from the improper use of agricultural chemicals?
6. How can agricultural chemicals be used to preserve food products?
7. How can agricultural chemicals be used without harming the environment?
8. In what ways can the use of agricultural chemicals reduce energy costs in crop production?

Chapter
10

Harvesting and Storage

"The crop isn't made until the money is in the bank." This is the attitude that the producer must take. Harvesting, storage, and marketing of the crop are just as important as planting, cultivating, watering, and fertilizing. Harvesting and storage are the steps between producing the crop and starting it on the road to becoming a loaf of bread, a can of corn, a bottle of catsup, or french-fried onion rings. After the crop is harvested, many people become involved in handling it and eventually putting it in the hands of the consumer.

Successful and efficient harvesting of a crop is often the key to a profitable operation. It is desirable to (1) know when to harvest, (2) understand how harvesting equipment operates, and (3) minimize losses during harvest.

After harvesting, it is necessary to store the product and protect it properly until it is processed into a usable product for consumers. Producers will often market their produce shortly after harvesting and move the product on into the channels of commerce. But in many cases the producer will process and store a product for a period of time before it is used. The product may be hay which is processed and preserved for later consumption on the farm or ranch, or it may be corn that it is stored on the farm for marketing at a higher price later in the season.

To protect crops properly after harvesting, producers need to understand how to store and treat crops to preserve quality and to know the storage conditions under which the crop can lose quality.

CHAPTER GOALS

After studying this chapter you will be able to:

- Identify the keys to successful and efficient harvesting.
- Describe the principal methods of harvesting.
- Identify the cause of harvest losses and be able to estimate the size of the losses.
- Describe the conditions needed for suitable storage of crops.
- Explain how to avoid losses or deterioration of crops and quality.
- Explain why crops may need to be artificially dried.

Figure 10-1. Many different types of equipment are used in harvesting. Peanut combine (top left); baler (top right); silage chopper (lower left); combine with corn head (lower right). (*Grant Heilman* [*top left*], *John Deere* [*top right*], *Grant Heilman* [*lower left*], *J. I. Case and Company* [*lower right*])

HARVESTING

Practically all field and vegetable crops are harvested mechanically today. Only a few specialized crops such as cucumbers, watermelons, and cantaloupes are now harvested by hand. Harvesting machines are being constantly improved and new ones developed.

Machines are efficient compared to hand-harvesting. They save time and labor. If machines are properly adjusted, there is less loss of the crop. Uncertainty with regard to weather and markets can also be lessened with machine-harvesting of crops.

The many different types of harvesting equipment include harvester-threshers, diggers, mowers, windrowers, strippers, pickers, choppers, balers, rakes, cubers, stackers, and conditioners (Figure 10-1). The list is endless. The word *combine* is used to describe machines that combine several harvesting operations into one. The most common one is the harvester-thresher combination used to harvest small grains, sorghum, corn, soybeans, and a host of minor crops such as flax, sesame, and guar.

The grain combine is usually adapted to each crop by making small changes in the cutter bar, the reel, or the thresher (cylinder or similar mechanism). For corn, the cutter bar–reel combination used for wheat is replaced by a header which strips the ear of corn off the stalk (Figure 10-2).

The following sections describe the methods used to harvest various types of crops. They include (1) grain crops, (2) root and tuber crops, (3) forage crops, and (4) cotton.

Grain Crops

The harvester-thresher, commonly called a *combine,* is used to harvest grain crops (Figure 10-3). The basic operation of the combine is the same for all crops and includes five separate operations:

1. The grain is cut and conveyed to a thresher.
2. The thresher removes the grain from the head, or stalk, or cob.
3. The grain is separated from the hulls, straw, cob, pods, leaves, or similar plant materials.
4. The grain is cleaned by removing the rest of the chaff, dirt, and remaining trash.
5. The grain is transferred to a grain bin.

Figure 10-2. A picker-sheller combine with a header attachment especially adapted for corn. (*Courtesy Lubbock Implement Co.*)

In operating a combine, one must be sure that the various parts are properly adjusted. The reel and cutter bar should be adjusted to the correct height. Using the correct cylinder speed in the thresher for each type of grain is important. Manufacturers provide instructions on these fine points in the operation manual of the harvester. Conditions of a specific crop, such as moisture content of the grain and weeds in the crop, need to be considered in making adjustments.

When small grain crops have failed to ripen, or mature, evenly or a field is badly infested with weeds, it may be desirable to cut the grain, windrow it, permit it to dry, and then combine. When this is done, a separate pickup attachment is necessary.

Corn today is usually harvested with a combination picker-sheller or a combine with a special header. Both machines remove the ear of corn from the stalk. The ears are then shelled by the machine, the grain is sent to the bin, and the rest of the plant material is discarded back onto the field. Mois-

Figure 10-3. Grain combines can be used to harvest wheat and other grains as well as grain sorghum, soybeans, and similar crops. (*Courtesy Castro Co., News, Dimmitt, TX* [*top*]*; Courtesy Billy Tucker, Oklahoma State* [*bottom*])

Figure 10-4. Grain sorghum was harvested in years past by first cutting, then gathering it into small bundles which were stacked together in a shock.

Figure 10-5. On-the-farm dryers are increasing in use to dry cereal grains. Grain can be harvested earlier and harvest losses minimized if grain can be artificially dried.

ture of around 14 to 15 percent is desirable for harvest.

Today, *grain sorghum* is harvested with a combine (Figure 10-3). This is much simpler than previous methods (Figure 10-4). Sorghum grain is relatively soft, and the speed of the cylinder that threshes, or removes, the grain has to be reduced to prevent grain damage. Grain sorghum should be harvested at 18 to 20 percent moisture if it is to be dried but at around 13 percent moisture if it is to go directly to storage. The grower usually collects a representative portion of the grain and has it tested for moisture at a local elevator or at a laboratory.

Combines are especially well suited to harvest the *small grains*. Moisture percentage of the grain at harvest is an important consideration. In wheat, moisture should be around 12 to 14 percent to be stored safely.

Barley is usually harvested at about 15 percent moisture, but in recent years producers have started harvesting when the grain is at 30 to 40 percent moisture, resulting in a higher yield because of reduced losses at harvest (Figure 10-5). The barley grain is either artificially dried or stored in silage-type bins and then is generally used for livestock feed. Oats should be at around 13 to 14 percent moisture for harvest; however, since oats shatter easily, it is common practice to cut and windrow them when the grain contains about 20 percent moisture. When it dries to a moisture content of 13 to 14 percent, the grain is combined.

Soybeans are harvested with a combine equipped with a grain head (Figure 10-6). Harvesting can begin when the beans are at about 18 percent moisture, but for safe storage moisture should be no higher than

Harvesting and Storage

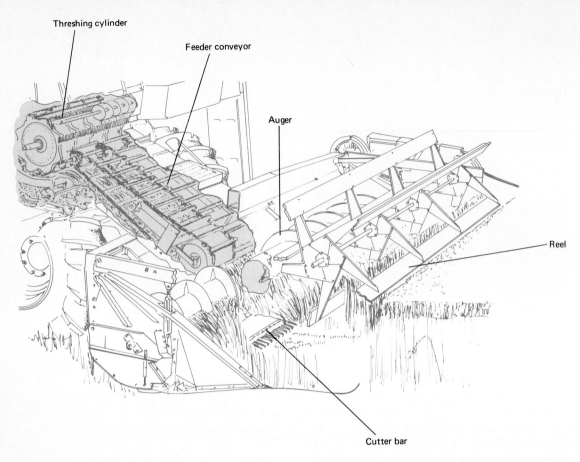

Threshing cylinder

Feeder conveyor

Auger

Reel

Cutter bar

Figure 10-6. Combine header.

13 to 14 percent. The safe-storage level of moisture is usually reached 2 to 3 weeks after all the leaves fall. A relatively slow ground speed is maintained while combining, since mature soybeans shatter easily. The combine is readjusted during the day to compensate for changing moisture conditions in the crop. Even after following the manufacturer's instructions, the operator should check the field for beans on the ground. If four beans per square foot are found on the ground, this represents a loss of about 1 bu/acre [0.087 m³/ha].

Root and Tuber Crops

Potatoes, sugar beets, peanuts, carrots, and other root and tuber crops are harvested with machines that remove the root, tuber, nut or that portion of the plant that is used. The process consists of digging, separating and cleaning, and loading.

These harvesters pull a blade, disk, or shovel underneath the roots, lift them up, and deposit them on a conveyor (Figure 10-7). As the conveyor moves the crop toward the back of the harvester, the soil

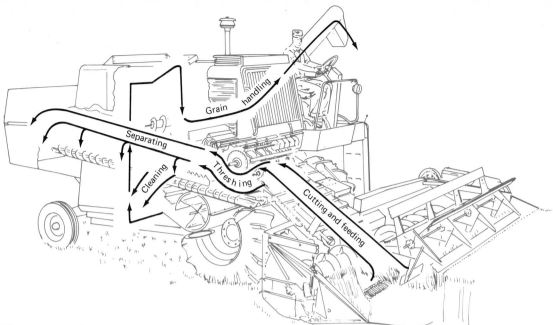

Figure 10-7. Flow of material through a combine.

and/or vines are separated from the produce. From there, the crop is loaded onto a truck or wagon to move it to a processing site. Each different subsurface crop usually requires a special machine for harvesting.

Forage Crops

Forage harvest involves more types of machinery than most other crops: mowers, windrowers (and combination mower-windrowers), hay conditioners, cubers, pelletizers, choppers, balers, stack-forming wagons, self-unloading wagons, and bale handlers, plus the special equipment needed for hauling (Figure 10-8).

The main point to consider in harvesting hay and silage is that quality needs to be maintained. Without the proper harvesting equipment, hay quality can be reduced.

Investment in hay and silage harvesting equipment is relatively high compared to many other crops. A producer needs to plan carefully and be sure the harvesting equipment fits the forage program.

Cotton

Cotton harvesting is done with two principal types of mechanical harvesters: the cotton picker and the cotton stripper. The picker removes only the lint and seed from open bolls of cotton, while the stripper removes the entire boll (as well as leaves if still on the plant)(Figure 10-9). Pickers are normally used in areas with long-season varieties grown in the Mississippi delta, Arizona, and California, while the stripper is used in areas with short-season varieties grown in parts of western Texas and Oklahoma.

A good cotton producer knows that harvest losses should be kept to a minimum. The average cotton grower in the United States leaves over 50 lb [23 kg] of the lint

Figure 10-8. Alfalfa is being harvested by picking up the forage from a windrow, chopping it and blowing into a wagon. This alfalfa will be hauled to a dehydrating plant for making alfalfa leaf meal. (*USDA*)

Figure 10-9. This cotton harvester is being used to harvest research plots in the Southwest. It is a picker rather than a stripper. (*USDA*)

Figure 10-10. Harvest losses in cotton are vividly illustrated above. Cotton was harvested from rows on the right prior to a rain which delayed harvest for a week after which the cotton on the left was harvested. Cotton left due to the delayed harvest was significant. (*Bean, Lubbock, TX*)

cotton in the field. Cotton losses can be easily reduced and yield often increased by 5 percent or more by doing some simple checking or observing of the picker and operator as cotton is harvested.

HARVEST LOSSES

Losses in yield can be due to improper harvesting techniques or harvesting too early or too late. If harvest begins too early, the plant may not have reached full maturity, resulting in a shrinkage of the grain. If harvest is too late, plants can lodge or stalks can break, grain can shatter or ears can drop, and quality will deteriorate.

Producers who are willing to take the time needed to adjust and operate their harvest-

ers properly will reap extra profit from extra yield. Harvest losses from a properly adjusted and operated combine should not exceed 3 percent. Anything greater represents needless loss (Figure 10-10). A number of factors affect harvester efficiency—combine adjustments, field speed, kernel or fiber moisture, and lodging. To determine harvester efficiency, a grower can measure field losses and separate preharvest losses from machine losses. "Rule-of-thumb" methods of measuring field losses have been developed. Directions for measuring are given here for corn and cotton.

Corn

Loss can be estimated by marking off two rows (of 38- to 40-in [about 1 m] row spacings) 65 ft [20 m] in length. Count the number of *ears on the ground* in front of the combine, and then count again behind the combine. The difference between the two is machine loss. Each additional ear on the ground represents a 1 bu/acre machine loss. If machine loss exceeds 1 or 2 bu/acre, make adjustments to the corn header.

Corn kernel loss may be unnoticed because it is hidden by plant material. Clear an area 1 ft [30 cm] wide across the width of the machine. Count the kernels from this 1-ft band. Two kernels per square foot represents 1 bu/acre. In this count, include all kernels remaining on partially shelled or broken cobs. If loose kernel loss exceeds 1 or 2 bu/acre, find where the loss is coming from and make necessary adjustments. Loose kernels may come from improperly adjusted snapping plates and rolls, incomplete threshing (kernels attached to the cob), or inadequate separation. Before making adjustments to the machine, be sure that the losses are not coming from a hole in the conveying components.

Cotton

To check losses of cotton, use the "seed count" method. Pace off and mark 10 ft [3 m] of row. Prior to picking, count the number of bolls that should be harvested. Then count the number of locks per boll (check several bolls from one plant), and count and determine the average number of seeds per lock. Then multiply the number of bolls times the number of locks per boll times the average number of seeds per lock to obtain the total number of seeds. After the picker has passed, pick the remaining locks that have been missed and count the seeds left. Divide the number of seeds behind the picker by the number of seeds before the picker and convert to a percentage of cotton

being lost. If loss is greater than 5 percent, it would normally pay to make adjustments to the machine.

STORAGE

After harvest, many products will go directly to markets and into commercial storage, while some products will be stored on the farm. Regardless of where they are stored, agricultural products must be properly protected until processed into a usable product for the consumer or until fed to livestock. If not properly protected, crops in storage may deteriorate between harvest and consumption. They may lose valuable nutrients through natural aging processes. Even when a crop has matured and has been harvested, it is still "alive"—not in the usual sense of the word, but changes are taking place in storage. Dried products such as corn and small grains undergo minimal changes, but high-moisture produce such as potatoes, sugar beets, and fruits can undergo considerable changes.

A crop in storage can also be contaminated by insects and other pests. Molds, rodents, insects, and similar organisms often cause more damage to stored crops than the aging process.

The conditions for proper storage will depend on the type of product stored and its moisture content. Proper storage conditions are desirable in order to keep a product such as grain from losing quality or to maintain the quality of a product such as hay or silage (Figure 10-11).

Stored crops are in constant danger of deteriorating from insects or heat and moisture and may even be destroyed. Each crop has an optimum level of temperature and moisture for best storage. Most grains store best in a cool, dry environment, while root

Figure 10-11. Storing grain on the farm or delivering it to a "country" elevator is the first step in the marketing process.

and tuber crops require cool temperatures but higher levels of humidity.

Grains and Edible Legumes

These crops are normally stored in bins which are equipped for air circulation and humidity control. It is fairly common today for grains and edible legume seeds (hereafter in this chapter the word *grains* is used to include edible legume seeds) to be harvested at a moisture content well above the safe storage level. This is done to decrease losses before or during harvest when the grain is at around 15 percent moisture. As a result, the grain must be dried before it is put into storage.

Temperatures above 50°F (10°C) and moisture content in excess of around 15 percent are conducive to the growth of fungi, or molds, in grain. Such conditions are also more favorable for insect growth and damage to the grain.

Rodents, particularly rats and mice, can be a serious problem around stored grains. Not only do they eat large quantities of grain, but they contaminate the grain with urine, hair, feces, and other debris.

Insects not only eat significant quantities of grain but decrease the quality by the presence of insect fragments, feces, and unnatural odors.

One key to the control of insects and fungi is to provide areas with cool temperature and low moisture in the storage unit. Proper air circulation is needed. Fumigants can also be used to kill insects and rodents. Clean conditions around the storage area will often help keep rodent numbers down.

Keeping foreign matter out of stored grain is also helpful. Straw and trash brought in with harvested grain tend to accumulate along the edges of bins. Pockets of trash and foreign material will serve as breeding

Figure 10-12. Grain inspectors check stored grains for quality and grade. (*USDA*)

grounds for insects and disease organisms (Figure 10-12).

Root and Tuber Crops

These crops, because of their high moisture content, are more perishable than the grains. The storage facility must have high humidity to prevent drying and loss of weight. A low temperature is necessary for best storage of root and tuber crops. Each crop has a specific desirable range of temperature and humidity while in storage. Proper storage conditions are much more critical for these crops than for grains.

Root and tuber crops are usually stored in special sheds and cellars designed to provide the correct temperature and humidity. Some crops, such as sugar beets, are stored outside in huge piles without covers. Even though deterioration occurs, it is not sufficient to warrant the cost of providing better storage conditions.

Root and tuber crops are much more susceptible to molds and other organisms because of their high moisture content. It is important to maintain good storage conditions to prevent the growth of these undesirable organisms.

Other Crops

All agricultural crops undergo some type of storage, and special types of storage and storage conditions are usually needed for each crop. It is beyond the scope of this book to describe the various conditions needed for proper storage of all crops, but the principles of storage described in this chapter are essentially the same for all crops.

Crop storage is a critical part of a farming operation. Successful production of a crop is obviously essential to profits in farming, but proper harvesting and storing of the crop form the capstone of a profitable farm operation.

SUMMARY

Harvesting and storing are integral parts of a production program. Harvest losses must be kept to a minimum in order to take full advantage of other good management practices and to help ensure a profit.

A good producer will check constantly for harvest losses. Proper adjustment and operation of the equipment will help minimize harvest losses. Each crop grown requires special adjustments for the equipment designed for that crop.

Storage of crops after harvest, whether commercially or on the farm, is critical. The correct storage conditions must be considered for each crop in order to minimize losses and deterioration in quality. Temperature and humidity are two conditions that need to be considered, while pests such as insects, diseases, and rodents need to be controlled.

QUESTIONS FOR REVIEW

1. In what ways do harvest losses affect farmers' profits and consumer prices?
2. What is the impact of time of harvest (early or late) on market prices and potential field losses?
3. What are the three keys to successful and efficient harvesting of a crop?
4. What are the four keys to controlling insects and diseases in storing grains and edible legumes?

Chapter
11

Marketing Plant Products

Marketing is a complex, integrated process which can be simply expressed as "getting the product to the consumer." In this age of specialization, just producing plant products is not enough. The cotton produced in Texas does not make a shirt available for the Seattle businessmen. Neither does durum wheat delivered to the elevator in North Dakota provide the spaghetti and other pastas at an Italian restaurant in New York City. The products that farmers grow and sell must be transported, processed, stored, handled, financed, and delivered to the consumer in the proper form and quantity and at the appropriate time and place.

The movement of huge quantities of grain, fiber, oil, fruits, vegetables, ornamentals, and other plant products from the producer to the ultimate consumer requires the services of a vast array of middle-level firms and businesspeople. In addition, storage, transportation, and processing facilities must be available at the right place and at the right time. For orderly marketing, financial arrangements and procedures require the services of banks, production credit associations, the Farm Home Administration, and other agricultural credit institutions and their highly trained professional staffs. Also included in the marketing chain are the agencies and personnel involved with measuring and recording quality and setting quality standards.

CHAPTER GOALS

After studying this chapter you will be able to:

- Identify the activities that interact to permit the marketing process to move plant products from the producer to the consumer.
- Identify marketing institutions through which plant products may flow.
- Understand how marketing and sales agencies function.
- Understand how futures markets are used to provide a mechanism for price discovery.

FUNCTIONS OF MARKETING

A number of activities interact to permit the marketing process to move the product to the consumer. The following are functions of marketing:

Grading and Standardization

When a buyer in Texas is communicating with a seller in Michigan, it is essential that the terms used to describe the quality and quantity of the product be understood by both parties. A uniform grading system is needed to facilitate interstate and international trade in agricultural products. In countries without product standardization and grading, marketing is stifled and inefficient.

Grading standards have been a part of the marketing process in the United States for so long that most people take them for granted. *The Official United States Standards For Grain,* published by the U.S. Department of Agriculture (USDA), gives the characteristics for each of the official grades of the commonly traded grain crops in this country. The United States Grain Standards Act requires that anyone selling grain by grade and shipping it in interstate commerce (across state lines), must have it inspected for grade (Figure 11-1).

Each crop is divided into market classes, and grades are assigned within each class. The factors used to determine grade vary from crop to crop, as do grade designations, but the procedure for evaluating is similar for most plant products that are graded. Table 11-1 shows the standards used in determining United States grades of corn.

Corn grown in the United States is divided into three market classes: yellow corn, white corn, and mixed corn. Six grades describe the various quality factors: U.S. No. 1 to U.S. No. 5 and Sample Grade. The

Figure 11-1. This inspection crew is gathering grain samples from a rail grain car for grade determination. (*USDA*)

highest quality is U.S. No. 1, and Sample Grade is the lowest.

Determination of both total and heat-damaged kernels is made on the basis of the grain remaining after removal of the broken corn and foreign material. All other determinations are made on the basis of the grain as a whole.

Percentages are determined on the basis of weight. Moisture is determined by an approved moisture meter. Test weight is the weight of a specific volume of grain equal to 1 bu. High test weight indicates plump, full kernels and high quality.

Plant products other than grain are also graded to facilitate marketing and to protect the consumer. When hay is shipped in interstate commerce, grading is required only if requested by one or more of the interested

TABLE 11-1 STANDARDS USED TO DETERMINE UNITED STATES GRADES OF CORN

Grade	Minimum test weight per bushel (lb)	Moisture (%)	Broken corn and foreign material (%)	Damaged Kernels Total (%)	Heat-damaged (%)
				MAXIMUM LIMITS	
Class A					
U.S. No. 1	56.0	14.0	2.0	3.0	0.1
U.S. No. 2	54.0	15.5	3.0	5.0	0.2
U.S. No. 3	52.0	17.5	4.0	7.0	0.5
Class B					
U.S. No. 4	49.0	20.0	5.0	10.0	1.0
U.S. No. 5	46.0	23.0	7.0	15.0	3.0
U.S. Sample Grade*					

*U.S. Sample Grade shall be corn which does not meet the requirements for any of the grades from U.S. No. 1 to U.S. No. 5, inclusive; or which contains stones; or which is musty, or sour, or heating; or which has any commercially objectionable foreign odor; or which is otherwise of distinctly low quality.
Source: Adapted from "The Official United States Standards for Grain," January, 1978, USDA Federal Grain Inspection Service.

parties. Selling and buying hay by grade is relatively rare.

Grades have been established for fruits and vegetables, using appropriate measures of quality, freshness, and wholesomeness. Other inspections of plant products destined for human consumption are performed to assure that the food is wholesome and free from excessive levels of harmful chemical residues. In addition, livestock feed and feed additives are under scrutiny. Cotton is classified by staple length and grade.

In all the grading and standardizing of plant products, the USDA acts as a referee in the marketplace to assist in the marketing process and to provide a service and protection for the buyer, the seller, and the ultimate consumer (Figure 11-2).

Assembly

Another function of marketing is the assembly or the gathering together of small lots of products into larger quantities for more efficient transportation and processing. For example, the wheat farmers in Kansas take their grain to local elevators, where it is assembled into homogeneous units by grade. It is then shipped to terminals where it is assembled and blended to meet the specifications for export or sent to millers and bakers who process it and distribute it to consumers.

Storage

In most regions of the United States, plants grow and mature seasonally, which means

Marketing Plant Products

Figure 11-2. This inspector is looking over apples to rate their color and to check for bruises and other defects. (*USDA*)

charges are accruing. For example, a farmer sells his soybeans at a local elevator and is paid for them. The elevator must pay interest on this investment until the beans can be moved to the terminal or to the processor. The elevator manager either has to use funds which could be invested and earning interest or must borrow money from a credit source to pay the farmer. Either way, there is an interest charge. The time between transactions, then, becomes a significant factor in marketing costs. The more time it takes to move the product to the consumer, the greater are the marketing costs and the higher the price of the commodity.

For a marketing system to operate efficiently, it is essential that reliable information on the quantity and price of commodities being traded be available to buyers and sellers. This information is made available by trade associations, newspapers, commodity publications, and the marketing services of the USDA.

Merchandising

It is necessary to distribute products to consumers, make the consumers aware that the products are available, and provide a way the goods can be purchased. This process is called *merchandising*.

Consumers today expect more in the way of processing of agricultural products than in the past. There is also a considerable difference in the amount of processing required with different products. For example, moving soybeans from the producer to the consumer involves much more processing than moving pinto beans. The oil must be extracted from the soybeans, and the oil and meal may have to be further processed before reaching the consumer. The pinto beans are harvested, cleaned, distributed, and sold to the consumer. The amount of processing determines to a large extent the

that plant products must be stored and distributed from storage through the rest of the year. This function of marketing permits a steady flow of products to the consumer throughout the year. Costs are involved in storage, and the longer a product is stored, the greater is the risk of spoilage. The costs of assuming this risk and the accumulation of interest on money tied up in the produce being stored add to the cost of storage. Storage is a service that ultimately benefits both the producer and the consumer.

Financing

The marketing process takes time, and capital is tied up in the plant products along the way. Money is invested, therefore interest difference between the price the farmer receives and what the consumer pays. This price differential is referred to as the *marketing spread*.

Packaging of food products also adds to their cost. The public demands frozen fruits and vegetables, packaged and prepared meals, and other convenience foods that require special processing and packaging. The extra service costs money, and these costs are passed on to the consumer (see Chapter 12).

Transportation of products takes them from the farm through the various processing steps to the distributor and finally to the consumer. This may be accomplished by trucks, railroads, or air transportation or, with bulky, nonperishable products, by water. Transportation rates vary greatly depending on the mode, the distance involved, perishability of the product, and amount of load to be hauled. Rail freight rates can vary greatly from location to location, which gives rise to seemingly inequitable differences in rates. For example, some longer hauls from the Midwest to the West Coast cost less than shorter hauls from the intermountain region.

Another merchandising function that also adds to product cost is advertising and promotion. This may seem wasteful, but it is necessary to inform buyers what products are available. Producers may actually benefit from promotion efforts to motivate buyers to purchase their products.

INSTITUTIONS FOR MARKETING PLANT PRODUCTS

The market system includes a number of marketing institutions, although not all marketed products pass through each of these institutions. These marketing institutions include:

Figure 11-3. Fruit being loaded onto a ship for transport from the Port of Seattle to an overseas destination. (*Port of Seattle*)

1. Local assembly and processing markets such as the country grain elevator, which serve as collection points for product until sufficient quantities are accumulated for shipment to district markets.
2. District markets, which further assemble and process the products for shipment to central markets.
3. Central, or terminal, markets, which store, process, and assemble products from several local district markets.
4. Seaboard markets such as the Port of Seattle and other ports, which serve as the embarkation point for overseas shipments (Figure 11-3).
5. Secondary processing markets, through which pass products that require additional processing to meet specifications of millers or other prospective buyers.

6. Wholesale distribution markets, through which products that have been assembled and processed are dispersed to the consumer. These markets include brokers, commission merchants, jobbers, speculators, buyers for chain stores, and service wholesalers.

7. Retail markets, the final institutions, represented by the stores that sell the products to consumers in local communities throughout the country (Figure 11-4).

Marketing and Sales Agencies

Hundreds of business firms are involved in the movement of plant products from farms to consumers throughout the country. Examples of marketing and sales agencies include:

1. Commercial plants such as canneries, bakeries, cotton gins, creameries, elevators, flour mills, packing plants, and chain and independent grocery stores.

2. People such as buyers, assemblers, wholesalers, salespersons, exporters, jobbers, and factory representatives.

3. Transportation systems such as railways, barges, and truck and air cargo firms.

Cooperatives

The member-user-owned cooperative is a business corporation that has most of the characteristics of any other corporation. It differs from other corporations in that it is organized to serve its members, who are also its owners. A corporation is controlled by stockholders, who are owners in proportion to the number of shares of stock they own.

Figure 11-4. Retail outlets, such as this supermarket, bring the farm products to the consumer. (*USDA*)

The cooperative is controlled by its members, who are also the owners, either on a one-member–one-vote basis or in proportion to the amount of service provided the member. A board of directors and officers elected at an annual meeting of the members manages both business organizations in accordance with their charter and bylaws.

Cooperatives are formed to increase the advantages of buying inputs—for example fertilizer or petroleum products—and selling outputs—wheat or corn or other products. Their success depends on the members working together and supporting the cooperative. For example, marketing cooperatives sell what the members produce (Figure 11-5). All or some of the following services may be included: harvesting, receiving, storing, processing, sorting, packing, transporting, and financing.

In addition to marketing cooperatives, there are supply cooperatives that supply needed production inputs such as seed, fertilizer, machinery and parts, hardware, and other production needs and credit cooperatives that serve the credit needs of members through the farm credit system created by the United States Congress. Local production credit associations are cooperatives that make production loans to agricultural producers and farm-related businesses. Other examples of cooperatives include consumer, energy, wholesale, housing, health, and other special-purpose cooperatives.

Cooperatives have been an important part of the agricultural business section of our economy since the mid-1800s, when farmers started to join together to carry out some of the marketing functions and to increase their share of the profits. The success of these cooperatives, as with other business organizations, is dependent on efficient management. Some of the most successful cooperatives have been those among citrus fruit

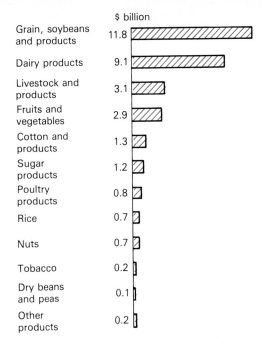

Total net marketing business = $32.1 billion. Wool and mohair included in livestock and products. Fiscal year 1977.

Figure 11-5. Major farm products marketed by farmer cooperatives. (*Agriculture Handbook No. 574 USDA, 1980*)

growers, walnut producers, Idaho potato farmers, and the dairy industry.

Futures Markets

A primary function of the futures market is to provide a mechanism for price discovery. Producers, processors, and merchants use these prices to make important decisions as to what and how much to buy as well as what, how much, and when to sell. Through the buying and selling that takes place the futures market demonstrates the present price of grain or other commodities for which delivery is assured as much as 15 months into the future.

The futures market provides for the buying and selling of contracts for goods such as

Marketing Plant Products

wheat, barley, corn, and other products to be delivered in the future. These transactions take place at trading centers such as the Minneapolis Grain Exchange or other futures markets (Figure 11-6). *Speculation* is the term that describes the buying and selling of futures contracts with the objective of making a profit. The speculator seldom sees the goods being traded, because the contract is liquidated before it is necessary to accept or make delivery.

Speculation in the futures market is a high-risk venture, and those who are casual dabblers usually end up losing money. Successful speculators depend on careful and continuous study of the market, systematic trading over a long period of time, cool judgment, and emotional detachment. They do not count on luck as a factor in marketing.

A second function of futures markets is the transfer of risk through a procedure called *hedging*. Speculators are a necessary part of the futures marketing process. They provide financing and easy entrance into the market by hedgers. This permits the hedger to shift the risk of price fluctuation to the speculators. A farmer may wish to hedge, to establish a position in the cash market before the crop is harvested. This reduces the risk of a drop in the price of the product. For example, Jim, an Iowa farmer, estimates that he will harvest about 50,000 bu of corn. He determines that he must have $3.40/bu to cover his production costs and give him a reasonable profit. If he does not want to take the risk of a drop in price in October when he is ready to sell the crop, he may sell ten futures contracts for 5000 bu each to deliver corn in November at $3.70 ($3.40 plus transportation costs of $0.30/bu). If the price of corn drops $0.12/bu, he can buy his futures contracts back for $3.58/bu and recover the $0.12/bu loss he would have taken because of the drop in price. It is imperative that at the same time traders close out the hedge position by buying back the futures contracts, they sell the commodity on the local market. Careful

Figure 11-6. Traders buy and sell futures contracts in trading centers such as the Minneapolis Grain Exchange shown here. (*Minneapolis Grain Exchange*)

timing of these transactions will determine the success of the hedge.

However, if the price of corn goes up, Jim will not have gained from the price increase. He reduces his risk of a price decline, but he also reduces his opportunity to gain if the price goes up. This is an example of a perfect hedge, where gains exactly offset losses. The changes in prices of futures contracts do not always exactly parallel the actual cash price of commodities being traded, but the tendency is for the prices to move similarly.

Hedging can also be used by a buyer to protect against a sharp increase in price at a later date when the buyer plans to purchase the commodity. The same principle applies as in the above example but in the opposite direction. It is a little like buying price insurance. If the hedge will allow the trader to "lock in" a reasonable profit, it is often an advantage to do so. This is especially true if the trader needs to be assured of a cash flow to service a debt and supply normal living expenses.

SUMMARY

It is basic to the marketing and sale of plant products that the output of farms and ranches be harvested, stored, transported, processed, and merchandised in the form wanted by consumers, at the time and place they want it, and at acceptable prices. The difference between the price paid to the producer and the price paid by the consumer—the marketing spread—has increased dramatically in recent years as consumers demand and are willing to pay for more intermediate services in the form of processing, packaging, and other marketing services. Other factors that affect the price to consumers are perishability of the product, bulkiness relative to value, and the seasonableness of the produce.

It is important that both producers and consumers understand the complexities of the agricultural market and the way the marketing system operates. It is this system that determines what should be produced, how it is produced, the amount needed, where it can be efficiently produced, and when the products are needed. The well-being of the people of America is dependent on an efficient and effective marketing system.

QUESTIONS FOR REVIEW

1. Why is a uniform grading system so important to orderly marketing of plant products?
2. Briefly describe the importance of the other functions of marketing: assembly, storage, financing, and merchandising.
3. Using wheat or corn as an example, trace the marketing institutions that would probably be used between the producer and the consumer.
4. What is a cooperative?
5. What is the purpose of hedging on the futures market?

Chapter

12

Food Processing and Packaging

Many agriculturists and historians attribute the Industrial Revolution, which started in the mid-nineteenth century, in part to increasing mechanization of crop production: as fewer people were required to produce food, more were freed to pursue other occupations. Looking back farther into the history of the human race, the processing and packaging of food were necessary steps preceding much organized exploration of any continent. Prehistoric people were limited in their travels by sources of readily available food—wild game, fruits, berries, nuts, and such. Only when methods of preserving essential foods were discovered was extensive travel, including oceanic travel, made possible.

Today, advanced nations have a wide variety of foods available throughout the year. This is due, in part, to advances in food processing and packaging. However, in spite of these advances, up to 30 percent of the food produced worldwide is lost due to inadequate or inappropriate processing or packaging. This waste must be reduced and, at the same time, energy efficient methods of processing and packaging must be identified. The broad area of food packaging offers many exciting, important career opportunities.

CHAPTER GOALS

After studying this chapter you will be able to:

- List the major reasons why food processing and packaging are important to human welfare.
- Recognize the magnitude of food loss from inadequate processing or packaging and the causes of such losses.
- Explain the basic principles and methods of food processing and packaging.
- Identify some of the career opportunities for well-trained persons in this general scientific field.

FOOD PRESERVATION

The earliest methods of food preservation go back to the roots of history, perhaps to around 15,000 B.C. when wheat was recognized as a valuable food crop. The preservation was simple—allow the grain to dry and keep it dry so that it did not spoil. The processing was a matter of grinding the grain into a coarse flour, a precursor to modern milling.

Early civilizations developed in areas where food production was successful because of a favorable climate, fertile soil, adequate moisture, and the presence of suitable plant species for domestication, such as wheat, *Triticum aestivum,* in ancient Mesopotamia. Ancient Egyptians grew a variety of crops, from cereals (wheat) to vegetables and fruits. The cereals were preserved as dried grain, flour, or baked bread; fruits were dried in the sun—grapes became raisins or, when fermented, wine. Some plant products were pickled, and meat, poultry, and fish were preserved by salting and smoking.

TODAY'S PROBLEMS

The miseries of human hunger, malnutrition, and starvation are well known. We must, in the name of humanity, make every possible effort to reduce these tragic losses. Food processing as part of the total agricultural food system represents a major step; but even if we produced enough food worldwide, starvation would remain a serious problem. Not only are there serious problems related to food distribution (some nations report so-called surpluses, while people elsewhere starve), but also we have an unacceptable amount of food wastage, much of which can be reduced, if not eliminated, by appropriate processing and packaging techniques.

A long-term, permanent solution to human hunger, malnutrition, and starvation is extremely complex. It involves agricultural issues of producing more per acre or hectare, moral issues concerning the control of population growth, and political and economic considerations. Currently, up to 66 percent of the starvation worldwide could be significantly alleviated by reducing food wastage through proper processing and packaging of agricultural products. In a day when the world needs all the food that can be produced and when we cannot afford to waste energy, losing food after it is harvested is the most tragic type of waste. No benefits are derived from the time and energy devoted to producing a crop that is wasted, and people who need the food suffer more from its loss. The actual amount of food lost due to wastage is difficult to determine. All too frequently the true loss is masked because the food substance remains apparently unchanged although the nutritional quality has been greatly reduced because of lack of processing or inappropriate processing. This is readily illustrated by the common practice of overcooking vegetables and discarding essential vitamins with the cooking water. The amount of food wastage varies with the type of food and local customs. In some instances up to 50 percent may be wasted, in others as little as 10 percent. An acceptable general estimate is that about 20 percent of the food produced is wasted.

Wastage

Although we may justifiably brag about our highly productive crop and livestock production technology, the loss, due to wastage, of 1 out of every 5 lb of food produced is a disgrace. Even the minimum level of 1 out of 10 lb is shameful. These losses can and must be reduced.

Food wastage is generally the result of one or more of the following factors; contamination, consumption by pests, inefficient utilization, and spoilage.

Contamination can come from several sources. There is increasing concern about contamination from pesticides. This type of contamination can be minimized by careful use of such materials in growing crops and by following recommended procedures in washing crops such as fruits and in maintaining clean storage areas (Figure 12-1). When the food is free of contaminants, packaging it so as to avoid contamination or recontamination is very important. Various types of pests, frequently insects or rodents (e.g., mice) contaminate foods and render them unfit or unacceptable for human consumption. Here again, proper storage and appropriate packaging can essentially eliminate losses.

Consumption of food materials by pests can be a problem in the field as well as in storage. In both instances, losses can be significant. As a general rule, those steps which ensure cleanliness in storage and proper packaging to minimize losses due to contamination in storage will also minimize losses due to direct consumption. Losses in the field due to direct consumption may be more difficult to minimize. Such losses may range from deer feeding in fruit orchards, where extremely expensive fencing appears to offer the only lasting solution, to rabbits in a vegetable field, birds in grain sorghum, and of course multitudes of insect pests that consume plants directly. In practice, to minimize these losses, the farmer must first determine the cause of the damage or loss and next decide whether the level of loss justifies the expense of any control measure. In spite

Figure 12-1. Proper storage of fruit helps ensure high quality and extends the market period for fresh produce. (*Byron Schumaker/USDA*)

of the fact that we need all the food we can produce, it must be remembered that many management decisions in crop (and livestock) production are based on economics as much as or more than on biological considerations. Also, if a pesticide is to be used, the danger of contamination must be determined and weighed against expected benefits. *If* a pesticide is used, all safety and use directions and regulations should be followed.

Wastage due to *inefficient utilization* may arise from many causes. With many fruits and vegetables, vitamins are lost as a result of overcooking by the individual homemaker. Losses are also associated with untimely harvest schedules, premature harvesting before fruits are fully developed and have attained their maximum nutritional level, or late harvesting, when shattering of grain is excessive. In livestock production, common examples are poor storage of hay as a result of exposure to rain and losses due to excessive sun bleaching. To minimize these losses, the producer and the processor must understand not only the production practices for the crop but also the ultimate crop product and how the crop is processed to yield this product. The entire sequence from seedbed preparation through producing the crop to processing the product should be viewed as a continuous process, with the goal of producing the maximum quantity of a food product. Not all losses can be avoided. So much of the actual production phase of the continuum depends on the weather that on occasion a crop must be harvested (or planted) at other than the ideal time. Once again, understanding the *entire* sequence in producing the food product will help all concerned reach a compromise that should minimize wastage. For example, it may be most desirable to harvest tomatoes when they are red and plump, but to avoid harsh weather, earlier harvest when some fruits are green may be a compromise. This may be fully acceptable when the tomatoes are used in pastes or juices, not as fresh fruit.

Spoilage is a constant threat to all types of food. Spoilage is the result of the activities of microorganisms, generally bacteria and fungi. The raw plant product may be exposed to the organism in the field or at nearly any stage up to packaging the food product. Losses are minimized by a variety of methods of reducing the number of organisms or slowing their growth or reproductive rates. Once the microorganisms are removed, packaging to minimize reinfestation or contamination is important. Unfortunately, much food is lost by spoilage at home. This waste can be greatly reduced by refrigeration and proper restorage of food. Fruit requires special handling immediately after harvesting, as is illustrated in Figure 12-2.

FOOD PROCESSING AND PACKAGING

The primary goal of food processing is to minimize wastage. Packaging also contributes to meeting this goal. Modern processing and packaging also are the basis of many so-called convenience foods, from frozen french fries and concentrated juices to dehydrated meals for hikers and campers.

To expand on the primary goal of food processing: we seek to minimize wastage by preventing undesirable changes in the food while retaining its wholesomeness, nutritive value, and sensory qualities (color, smell, texture) by economical means. A wide variety of methods are available for processing foods. Generally these methods are designed to control the growth of microorganisms and to minimize chemical, physical, and physiological changes of the food substance.

The particular method used to preserve a given food depends on various factors. Food materials differ greatly in their relative perishability. Generally, foods with a high water content are more perishable than those with lower water content. Meats, seafood, and soft foods are very perishable, potatoes and apples are moderately perishable, and grain of various types is relatively nonperishable. The method of food processing to some extent depends on the water content of the food material. Other factors that may affect the processing method include the length of time the food is expected to be preserved, the equipment available, and the way the food finally will be packaged.

Figure 12-2. After harvesting, most fruits are sent to packing houses where the fruit is washed, dried, and graded. Most packing houses use a system of conveyor belts, as shown here, to move fruits from one work station to the next. (*Florida Department of Citrus*)

Methods of Food Processing

Food processing is a major industry in the United States and in other highly developed nations. In spite of the development of specialized, sophisticated equipment, the general methods and principles of food processing used commercially are in principle the same as those used by many homemakers. Three general methods of food processing are recognized: chemical methods, biological methods, and physical methods.

Chemical methods involve the addition of some substance to the food. Usually the substance is salt, sugar, or an acid (as in pickling or smoking). The addition of such substances retards or stops microbial growth; frequently it serves to reduce the water content of the food material. Although chemical methods are quite effective, frequently fairly simple, and represent some of the ancient methods of preservation, they are far from perfect. Most chemical methods alter flavor and may also alter the texture and color of the initial food product. In most instances these changes are desirable (e.g., making pickles from cucumbers). Sometimes chemical methods yield a modern lux-

Figure 12-3. Some foods, such as the smoked pork shown here, are delectable products of chemical food processing. (*Jane Hamilton-Merritt*)

ury, such as smoked meat, fowl, or fish. Some modern smoked products are shown in Figure 12-3.

Biological methods of preservation usually involve a microbiological fermentation process that leads to production of alcohol or acid. The preservation of grape juice (and other fruit juices) as wine is a good example. The fermented product must be packaged and/or stored properly to avoid spoilage. A fine wine may turn to vinegar rapidly if it is not properly stored in an airtight container. The production of cheeses and similar dairy products also frequently involves microbial activities, and prolonged storage demands appropriate (airtight) packaging and storage at a cool temperature. A good example of cheese storage is illustrated in Figure 12-4.

Physical methods are frequently employed in the more highly developed nations, such as the United States. Such methods on a simple scale are used by many homemakers and are employed extensively by leading food-processing companies. Physical methods are extremely effective when linked with appropriate packaging methods. Unfortunately, such methods require the input of large amounts of energy. These methods include the use of heating or cooling (including freezing) and dehydrating.

As energy conservation continues to be a serious concern, food-processing methods must be refined and made more efficient. From farm to dinner table, more than 50 percent of the total energy required to produce some foods is used in processing. In the United States, the entire food-supply system uses about 12 percent of the energy we produce. Overall, food processing uses about one-third of this (4.5 percent). A significant amount of energy is devoted to convenience foods, which may become more of a luxury if energy becomes too expensive or scarce (Figure 12-5).

The first method results in killing microorganisms and is thus a highly desirable method of preservation. Energy inputs are high, however, and caution is essential to minimize recontamination. In addition, the energy input (heating or irradiation) frequently has undesirable effects on the product, varying from subtle changes in texture (firm flesh may become spongy) to actual destruction of some vitamins. For this reason, these methods are not best for many fresh fruits and vegetables.

Figure 12-4. Preserving the quality of cheeses demands airtight packaging and storage at cool temperatures. (*Jane Hamilton-Merritt*)

With highly perishable foods, the greatest concern is with microbial growth. Physical methods tend to either kill microorganisms or greatly retard their growth. Packaging to avoid recontamination is essential or the benefits of processing will be lost and the energy used wasted.

There are numerous specialized physical methods of preserving foods. Sometimes several are used in combinations for specific types of food or for a given type of packaging. In general, the physical methods involve one (or more) of the following procedures: (1) heating or irradiating the product; (2) controlled reduction of temperature; (3) controlled reduction of water content, which may also include heating; and (4) freeze-drying.

Figure 12-5. Convenience foods require lots of processing and energy which adds to their cost. (*Jane Hamilton-Merritt*)

Food Processing and Packaging

Chilling or freezing does not necessarily kill microorganisms, but low temperatures retard the growth of these organisms, hence reduce wastage. Freezing requires energy, and once a product is frozen it must be kept frozen or used fairly rapidly if thawed. In many instances, refreezing results in serious losses in texture plus losses from renewed microbial activity. Also, the freezing itself harms some products. A temperature of about 17°F [−8°C] is suitable to hold many frozen foods; unfortunately, vitamin C may be lost at this temperature.

Dehydration, or the controlled reduction of water content, is perhaps the oldest method of food preservation. Early wanderers settled when they discovered that grain (wheat, corn, and rice) could be dried and stored for future use, either to mill into flour or for planting. Sun-drying of fruits has been recognized since the dawn of recorded history. Raisins are dried grapes, and in many areas sun-dried apricots and peaches are considered luxuries. The technology of producing a wide variety of dehydrated foods was expanded during wars to facilitate shipping supplies. It has grown to support space-exploration missions, and many campers and backpackers enjoy luxury meals prepared from dehydrated products.

In addition to extensive dehydration, removal of water produces concentrates. Good examples are the familiar frozen, concentrated fruit juices and drinks—orange juice, grape juice, and lemonade. Prior to the mid-1940s, if fresh fruit was not available, the only way to enjoy the products was to purchase bulky, relatively expensive cans of the various juices. Concentration methods, linked with appropriate packaging technology and the widespread availability of home freezers, have made these ''fresh-frozen'' items popular and available to many people.

Freeze-drying is a relatively new procedure. It involves both reducing the tempera-ture of and removing water from the food material. Commonly, freeze-drying is done under vacuum conditions. Water removal is facilitated, because cool tissue retains less moisture than warm. The procedure requires fairly substantial energy inputs and special equipment, but many products, once preserved, do not have to remain frozen to stay preserved; this is a real benefit. Also, the combination of reduced temperature and reduced water content leads to fewer undesired changes in many food products. The most successful product of freeze-drying currently found on grocery shelves is freeze-dried instant coffee. Expense has limited the number of products preserved by freeze-drying techniques (Figure 12-6).

Methods of Food Packaging

Few, if any, raw agricultural products are purchased directly for human consumption. The vast majority are processed in some manner, and essentially all are packaged. For fresh fruits and vegetables, processing may be as simple as washing the produce, and the packaging may involve merely bagging products in convenient sizes or weights.

There are two major reasons for packaging food: protection of the food product and convenience of the consumer.

The protective functions of packaging are by far the most important. The type of packaging used depends on several factors: the aspect, such as the type of food, of it that is to be protected, and the method by which the food is processed. A simple example of packaging according to the aspect of the food to be protected is the development of foam-like containers for fresh eggs, the purpose of which is to minimize breakage (Figure 12-7). These containers also help to preserve freshness by partially controlling temperatures or insulating the eggs. Reflecting the effect of the type of food on the packaging

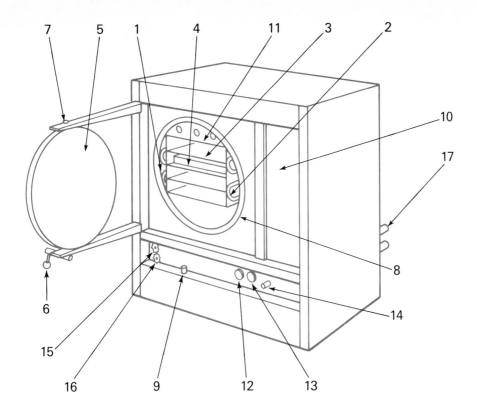

Figure 12-6. Diagram of small freeze-dryer used in research. In commercial units, the vacuum chamber holding shelves and trays is a bulk area.

1. Vacuum chamber	7. Door adjustment	12. Vacuum release valve
2. Condenser coil	8. Door gasket	13. Vacuum gauge valve
3. Shelf	9. Heat liquid level indicator	14. Sterile air inlet
4. Sample tray	and filler tube	15. Chamber drain valve
5. Door	10. Control panel	16. Condensate drain valve
6. Door latch	11. Probe jack panel	17. Water line connection

method, fresh fruits and some vegetables are packaged in simple mesh bags, a type of container wholly inappropriate for a liquid product. Plastic bags have become acceptable for many products, including some liquids, but they are not fully acceptable for frozen materials. The method of processing influences the packaging, and the reverse is also true. In the case of metal cans, heat sterilization is easier because scientists can measure heat within the can very accurately. Thus a raw product can be placed in a can, sealed, then sterilized. The procedure is one of both processing and packaging.

Many packaging materials are available. For convenience they can be classified in one

Figure 12-7. Foamlike containers, which keep eggs fresh and resist breakage, are an example of packaging tailored to a specific product. (*Ken Karp*)

of six broad types: (1) glass, (2) metals, (3) paper, (4) plastics, (5) wood, and (6) combinations. In selecting a packaging material, the product to be preserved and the characteristics of the product to be protected are of major concern. The packaging material must not add toxic substances or otherwise affect the quality of the food it is supposed to protect. Ideally, it would be light, durable, and inexpensive. Some packaging materials are illustrated in Figure 12-8.

Glass containers have been used for centuries, and pottery containers have been used for food storage throughout the history of civilization. They are suitable for all types of food and are commonly used with heat sterilization and radiation methods. They are not suitable for frozen products. Sometimes problems with airtight seals in lids are encountered. When the seal is broken and air contacts the food, spoilage may occur.

Two general types of metal containers are in common use. The "tin can" is familar to people worldwide. It is acceptable for nearly all foods and processes. There are different types of cans for specific types of food. The critical differences are associated with the type of metal from which the can is produced. Very acid foods are stored in L-type cans, which resist corrosion and the possible contamination of the food. MR-type cans are used for moderately corrosive foods, and the MC type of can is used only for noncorrosive foods. Modern technology has provided food scientists with a wide array of metals from which containers for special purposes can be made.

In addition to cans, in recent years metal-foil packages have become extremely popular. They can be used for many foods, are light, and are relatively inexpensive. The most popular use of these containers seems to be for dehydrated, highly processed food products such as instant breakfast drinks and the dehydrated meals carried by many campers. Aluminum foil is by far the most common metal foil used. The major drawback to these packages is their relatively poor durability.

Paper containers are of great importance. Besides the simple, baglike containers used for many fresh products, other paper containers are treated in various ways to give them special properties. Many frozen foods are packaged in paper containers; in fact, the foods are first packed in these containers, then frozen. Various types of paper containers are also used for milk and a multitude of dairy products.

The impact of modern technology is clearly evident in the wide array of plastic containers used in food packaging. Some containers are essentially metal-like, others are in the form of edible plastic films. Cellophane is popular for many fresh products, and polyethylene containers are now of great importance. Plastics differ in their reaction to high and low temperatures,

Figure 12-8. The selection of food packaging materials depends on the product and the characteristics of the product to be protected. (*Jane Hamilton-Merritt*)

strength, and reaction to other chemicals and to light. The type of container must be carefully matched with the food product and with the type and location of ultimate storage. Looking to the future, it seems likely that certain plastics will be replaced because their manufacture depends on the use of increasingly limited supplies of fossil fuels.

Wood is used more as a package for shipping than as a container of immediate interest to the individual consumer. However, an increasing number of Americans are purchasing wooden cases of common vegetables and fruits for home canning.

Combination packages may use any of the five previously discussed materials and may include, for instance, the plastic-capped glass jar common for freeze-dried coffee and the paper "tin" with a foil inner lining and metal ends in which concentrated fruit juices are packaged. Combination packages are developed to take advantage of the best properties of several packaging materials and to avoid deficiencies of those same materials or of alternatives.

Convenience has become a major factor in packaging. The American consumer has a choice among a staggering number of excellent food products that have been developed for the convenience of the consumer. The surge of products started in the late 1940s with the advent of TV dinners and now includes essentially all food products from frozen apple pies to pizzas. The cost in terms of energy and other resources suggests that perhaps the expansion of this industry has reached a peak.

CAREER OPPORTUNITIES

The food-processing and food-packaging industries are major employers, and career opportunities frequently require college de-

grees in a wide spectrum of subject areas. Development of packaging materials is a specialized career area. A solid background in crop and livestock production is helpful, because this helps in understanding how the raw product is produced.

Various aspects of this complex industry hire graduates in plant science (agronomy and horticulture) and in the animal sciences (livestock, dairy, and poultry). These people find careers as buyers for the company. They may deal directly with the farmer in purchasing produce, milk, or meat, for example. Large supermarket chains hire graduates, as do the packing and processing firms.

Graduates in food science or food technology are in demand in all phases of processing and packaging. These graduates must recognize and understand the aspects of the plant or animal that affect its quality once it is processed and packaged. In-depth understanding of on-farm production is of less importance because of the skills needed in processing. Food scientists emphasize chemistry, biochemistry, microbiology, and physics. Frequently college curricula in dairy science and poultry science consider both on-farm production of raw products and processing. These curricula parallel some specialized aspects of food science curricula. A rural background is frequently in demand by employers, but in the food-processing and food-packaging areas this is not essential. Even if you were born and raised in the city, you can find exciting agricultural careers in this general area.

Food processing and food packaging require specialized, sophisticated equipment. Energy conservation will become of increasing concern in this giant industry. Some colleges and universities have special curricula in food engineering; in other areas, this subject is taught as part of the curriculum in agricultural engineering. Thus, if you are a solid student in the physical sciences (math and physics) and are thinking about engineering as a career, as an agricultural engineer your skills may be applied in developing equipment to handle and preserve perishable foods. This is an important field of study, worth your serious consideration.

Biochemists and microbiologists are in demand in the food-processing industry. Graduates are needed to work in quality control to determine the effectiveness of processing and packaging and to develop and evaluate new methods of processing and packaging. Along with food science graduates, these graduates may also find careers in various types of inspection and sanitation positions. As with the food science college graduate, a farm background is not essential, but understanding how crops and livestock are produced is beneficial.

SUMMARY

Food packaging and processing are important steps in the total food chain from the farm to the consumer. A significant reduction in human suffering could be realized if wastage could be reduced by proper processing and/or packaging of foods. Food processing is a highly technical field, as is packaging. An understanding of chemistry, microbiology, and physics is important to the real understanding of many processing and packaging procedures. There are excellent career opportunities in the field of food processing and packaging for well-trained people. Understanding how crops are produced adds to a thorough understanding of processing and packaging, but a farm background is not essential for a career in this growing field.

QUESTIONS FOR REVIEW

1. What is the earliest record of food preservation?
2. What are the benefits of food preservation?
3. What is wastage? Name four factors that contribute to it and describe each one.
4. What is the purpose of food processing?
5. Name four methods of food processing. For each give an appropriate type of food, an advantage, and a disadvantage.
6. What is the main purpose of food packaging?
7. Name the six common methods of food packaging. For each, give an appropriate type of food, an advantage, and a disadvantage.
8. Discuss food processing and packing in the light of the current shortages of fossil fuels.
9. Describe some of the career opportunities in the food-processing and food-packaging industry and the education needed to enter these careers.

Chapter
13

Cereal Grains

Grain was the basic food crop for the first farmers more than 10,000 years ago. Today, grain is still the most basic food crop. Every year over 1 billion tons of rice, wheat, and corn are produced worldwide. This is enough grain to fill a train long enough to go around the world more than six times.

Grain is important for two reasons. First, it is the major source of food for the world's population. Second, it is used to feed livestock, which provide meat, dairy products, wool, and eggs.

Grain is easy to store and will not spoil if properly stored. It is easy to convert into food. Grains are excellent sources of needed nutrients, particularly the carbohydrates which provide energy. They are also easy to grow in many different parts of the world.

The cereal grains consist of wheat, corn, rice, grain sorghum, barley, oats, rye, and millet. The name *cereal* comes from the Latin *cerealis,* which was originally used to describe the smallest grains of the European continent exclusive of corn, grain sorghum, and rice. The latter three crops, introduced into America from places other than Europe, were later included in this group of crops.

Wheat and rice are the two leading crops in providing food for people worldwide. (Rice will not be discussed in this book because most of it is produced outside of the United States.) The other small grains, in order of importance as food sources, are barley, rye, and oats.

Grain sorghum and millet are the two principal food sources in most African countries. Grain sorghum is grown in significant quantities in the United States, but it is used primarily as a feed grain. Corn, too, is grown in the United States as a food grain but used principally for feed for livestock.

In the United States, cereals make up about 25 percent of the average diet. In other parts of the world they may account for 50 to 90 percent of the average diet. Cereals are widely adapted, with wheat being the most widely produced.

The leading producers of the cereal grains are in the temperate climates of the world. The United States is one of the leading producers, along with Russia, China, and Canada. Table 13-1 shows the ranking of the five leaders in the production of the principal crops discussed in this chapter.

TABLE 13-1   LEADING PRODUCERS OF CEREAL GRAINS, IN ORDER OF RANK

Wheat	Barley	Rye	Oats	Corn
USSR	USSR	USSR	USSR	USA
USA	China	Poland	USA	China
China	France	W. Germany	W. Germany	Brazil
India	Canada	E. Germany	Canada	S. Africa
Canada	UK	Turkey	Poland	Yugoslavia

CHAPTER GOALS

After studying this chapter you will be able to:

- **Identify the cereal grains and explain their importance as sources of food and feed.**
- **List the countries of the world that are leading producers of the cereal grains.**
- **Describe the various cereal grains and their general characteristics.**
- **Describe the general production practices required for the cereal grains.**

DESCRIPTION OF CROPS

Cereal crops are members of the grass family of plants called Gramineae. They have the common characteristics of an upright stalk with nodes and internodal spacing and a pithy stalk. The leaves are elongated, being narrow relative to their length, with parallel veins. The grasses have fibrous root systems. These crops have a *determinate* growth pattern, which means they grow primarily in a vegetative stage and then go into a reproductive stage. Stalks and leaves grow first and then start producing the grain. Vegetative growth still continues during the reproductive stage, but at a much slower rate than prior to the start of reproduction. The other type of growth is called *indeterminate*. Plants of this type grow both vegetatively and reproductively at the same time; examples include cotton, some soybeans, and tomatoes.

The cereal crops are used for both grain and forage, but their principal use is for grain. The head of grain is normally produced at the top of the plant. The one exception is corn, which produces one or more ears at the side of the stalk at about the fifth and sixth node. Most corn varieties have the capability of producing two or more ears. The second ear is often responsible for additional yield, but the third or additional ears seldom produce grain.

These crops are used as a forage for livestock feed as grazing, as hay, or as silage. The use of these crops as forage is discussed in Chapter 18.

Grains have been used to produce alcohol for human consumption down through the centuries. A relatively new use of alcohol from cereal grains is for blending with gasoline into "gasohol," thereby providing an alternative energy source. Not only the grain of the plant but also the vegetative portion can be used. Sugars from sorghum stalks and sugarcane can be used. The stubble from the small grains, sorghum, and corn can also be used for this purpose. Because of potential energy shortages and costs, the fermentation of grains and their residues to produce alcohol will most probably increase in the future.

Wheat

Wheat, Triticum aestivum, is divided into classes on the basis principally of kernel characteristics and growth habits. The classes are hard red winter, hard red spring, soft red winter, white, durum, and club wheat (Figure 13-1).

Wheat is used principally as a food because of its relatively high protein content, but it is also used as feed for livestock. Feed usually consists of wheat of mixed classes and is further mixed with other species that are not as suitable for food purposes. Occasionally wheat is grown specifically for livestock feed and not intended for the food market.

Wheat originated in Asia and was first grown in North America in the seventeenth century. It is now the leading small cereal grain in acreage and production in the United States (Figure 13-2). It is grown in every

Figure 13-1. Wheat varieties have different sizes and shapes of the grain head.

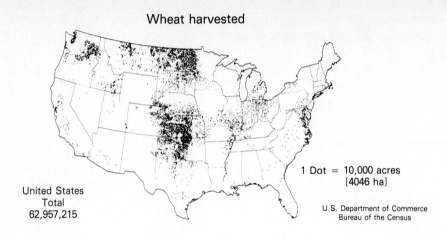

Wheat harvested

United States
Total
62,957,215

1 Dot = 10,000 acres
[4046 ha]

U.S. Department of Commerce
Bureau of the Census

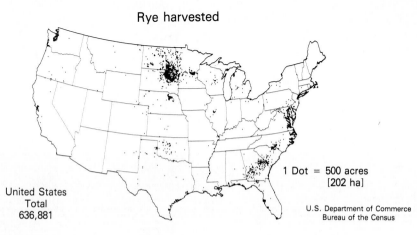

Rye harvested

United States
Total
636,881

1 Dot = 500 acres
[202 ha]

U.S. Department of Commerce
Bureau of the Census

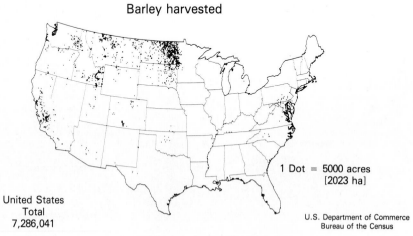

Barley harvested

United States
Total
7,286,041

1 Dot = 5000 acres
[2023 ha]

U.S. Department of Commerce
Bureau of the Census

Figure 13-2. Distribution of areas of production of the cereal grains.

Oats harvested

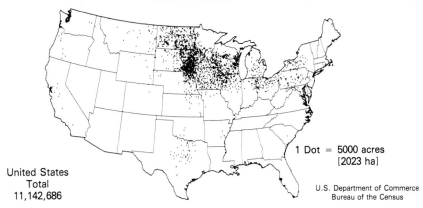

1 Dot = 5000 acres
[2023 ha]

United States
Total
11,142,686

U.S. Department of Commerce
Bureau of the Census

Corn harvested for all purposes

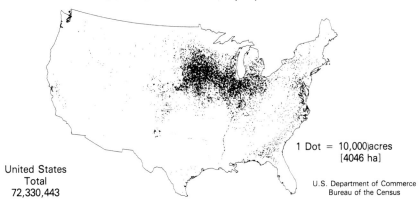

1 Dot = 10,000 acres
[4046 ha]

United States
Total
72,330,443

U.S. Department of Commerce
Bureau of the Census

Sorghums harvested for all purposes, except syrup

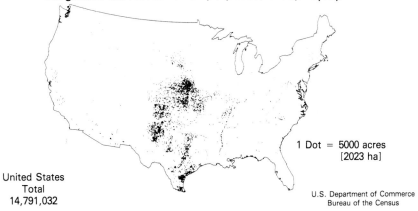

1 Dot = 5000 acres
[2023 ha]

United States
Total
14,791,032

U.S. Department of Commerce
Bureau of the Census

state in the nation except for the New England states. The wheat-producing areas of the United States are shown in Table 13-2.

Winter wheats are seeded in the late summer or fall when soil temperatures drop below the 50 to 55°F [10 to 13°C] range.

Winter wheat must go through a prolonged period of low temperature, 25 to 30°F [−4 to −1°C] during the seedling stage—a process called *vernalization*—before it will produce a head of grain the following spring. This period of cold temperatures is necessary for the formation and development of flowering parts (heads). If it is not vernalized, no head is formed and only vegetative growth is produced. Spring wheat, planted in the spring and maturing in late summer, does not have to go through the vernalization process.

Barley

Barley, Hordeum vulgare, originated in the Near East and was brought to America by the early English settlers. Barley is relatively more important than other grains in certain parts of the world such as Korea, the Middle East, and northern Africa. The principal growing areas in the United States are the states of the upper Great Plains, Pacific Northwest, and West Coast.

Barley is grown as both winter and spring types, although the differentiation between the two is not as distinct as for wheat. Spring barley is planted in areas with severe winters, because winter barley might be killed. Winter wheat is much more hardy than winter barley.

Barley is used in the United States principally as a feed grain for cattle. It is also used in processed foods and, when pearled, as a rice substitute. A small amount is used in the baking industry. About one-quarter of the barley crop is used in making malt. Production for malting purposes requires special management efforts. The grain of malting barley must be plump, uniform, high in quality, low in nitrogen or protein, and bright in color. It should be free from broken and skinned kernels. The grain should contain no more than 13 percent moisture for storage. Certain cultivars of barley are preferred for malting, and if the protein content is low enough, a premium is paid for malting barley.

TABLE 13-2 USES AND AREAS OF PRODUCTION OF THE VARIOUS TYPES OF WHEAT

Class of Wheat	Use	Area of Production
Hard red winter	Bread flour	Southern Great Plains
Hard red spring	Bread flour	Northern Great Plains and Canadian provinces
Soft red winter	Pastry flour	Midwest
White	Pastry flour	Pacific Northwest and Northeast
Durum	Spaghetti-type flour	Northern Great Plains and Canadian provinces
Club	Starchy flour	Pacific Northwest

Rye

Rye, Secale cereale, is a small grain which is more important in other parts of the world than in the United States. The leading producers and consumers of the grain are in Europe, with Russia and Germany leading the way. It is used in the European countries as a leading source of flour, second only to wheat. Other uses include bread, cereals, and baby foods.

In addition to its use as a grain food, rye is grown in the United States as a winter forage crop. Very little rye is used as a feed grain in the United States. Rye often poses a problem when grown in wheat-producing areas. When rye is harvested, some seed will remain in the field and produce grain the following year. In a wheat field, and mixed with wheat at harvest, the rye mixture lowers the quality of the wheat. Rye is especially susceptible to ergot, a fungal disease which reduces the quality and value of the crop (Figure 13-3).

Oats

Oats, Avena sativa, originated in Asia, then moved to Europe and eventually to the United States. The crop is grown principally in the Midwest and the upper Great Plains. Oats are used as a food for humans, in the form of oatmeal, baby foods, and the like, but over three-quarters of the crop produced in the United States is used for feed for livestock. It is an excellent feed for horses. Oats are often utilized on the farm where they are produced. The crop is also used as a forage crop for hay.

Triticale

Triticale is a relatively new cereal grain. It is a hybrid cross between wheat, *Triticum,* and rye, *Secale,* hence the name *triticale.*

Figure 13-3. The fungal disease ergot, shown here on barley, lowers the quality of such cereal grains as rye, wheat, and barley, and is potentially toxic to livestock as well as humans.

The hybrid combines the quality of wheat with the hardiness of rye. Only limited amounts of triticale are produced in the United States, and markets and uses for the grain are limited.

Corn

Corn, Zea mays, is grown in every state in the continental United States. It is the leading feed grain. The United States produces approximately 50 percent of the world's supply of corn on less than 22 percent of the acreage planted to corn! South Africa is another major corn producer.

The principal corn-growing area in the United States is in the Midwest, commonly called the *corn belt.* Figure 13-2 shows the distribution of corn production by states.

Many different types of corn are grown. Most corn is classed as field corn. There are both yellow and white varieties. Dent corn and flint corn are types of field corn. Dent, identified by the depression in the crown of the seed, has a relatively soft endosperm; flint has a hard endosperm. Dent corn is the principal field corn grown in the United States, while flint is grown most in South America and Europe.

Yellow field corn is used primarily as a feed for livestock, but it is also used for cornmeal and snack foods such as corn chips. White field corn is used primarily for food products such as corn chips and similar snack foods and for hominy and grits.

Other special types of corn are popcorn, sweet corn, waxy maize, flour corn, podcorn, and decorative corn. Popcorn is a special corn which has the ability to "explode" to 20 to 30 times its original size to produce the very tasty snack that is enjoyed at the theater. Sweet corn is a type bred primarily for its sweetness and tenderness and is used as a human food. Waxy maize is a special corn bred for a specific type of starch in the grain. It is used in tapiocas and puddings and also has a few industrial uses. Flour corn is used for the purpose its name implies—flour for use in making breads and pastries. The ears of podcorn are enclosed in a husk as with other types of corn, but each kernel is also enclosed in a pod or husk. Decorative corn has multicolored grains ranging from black to red to yellow to white. Acreage for this special-purpose corn is relatively small.

Grain Sorghum

Grain sorghum, Sorghum bicolor, is grown principally in the Great Plains, with the largest concentration of acreage in the southern part. Texas, Kansas, and Oklahoma are the leading producers (Figure 13-2). It is grown in this area principally because of its relatively low water needs as well as its efficient use of moisture and drought-resistant characteristics. These characteristics include an extensive root system, waxy leaf covering, few *stomatal openings,* and an ability to roll its leaves when moisture is limited. The Great Plains area is normally characterized during the growing season by low rainfall and low humidity. The crop is normally grown either under dry-land conditions or with limited irrigation.

Grain sorghum originated on the African continent. In its native habitat and original form, it was a relatively tall plant (up to 12 to 14 ft [366 to 427 cm] tall) with very little grain. The varieties now grown in the United States (and also in Africa) are much shorter (usually not over 4 ft [122 cm] tall), and the grain head is much larger. Present varieties will often have a grain/stover ratio on a dry-weight basis of close to 1:1.

Grain sorghum is referred to by several common names: milo is the most commonly used, and maize, hegari, and kafir are also used.

Because it is grown under limited rainfall conditions, yields may be relatively low, with a normal dryland yield being 1500 to 2000 lb/acre [1678 to 2238 kg/ha]. With adequate irrigation water or rainfall, yields of 6000 to 7000 lb/acre [6713 to 7832 kg/ha] are common, with potential yields of up to 10,000 lb/acre [11,189 kg/ha] under ideal conditions.

Grain sorghum is used almost exclusively as a feed for livestock in the United States, but in other parts of the world, especially Africa, it is made into flour for human food use.

Millet

Millet, Setaria italica, is seldom grown in the United States as a food or feed grain but is grown throughout the South as a summer pasture forage. Small acreages are grown for specialty purposes such as birdseed. Millet is of considerable importance in other parts of the world, particularly Africa, as a flour source. Millet is a short-season crop and an efficient water user. It is therefore grown in many areas of the world which have limited rainfall. It is planted in the spring or early summer and harvested in the fall. Most varieties mature in 90 to 110 days.

PRODUCTION PRACTICES FOR SMALL GRAINS

Land Preparation

Land preparation for the cool-season small grains—wheat, rye, barley, and oats—is quite similar for most producing areas. Variations depend on soil type and texture, management of residue from the previous crop, need to control weeds, fertilizer application methods, and moisture management. Sandy soils may receive only two diskings from one wheat crop to the next, while the finer-textured soils may need several diskings followed by plowing, and, in irrigated areas, preparation of beds for rows.

Land preparation starts with the incorporation of the residue of the previous crop into the soil so that decomposition will start. In low-rainfall areas, a portion of the crop residue may be left on the surface as a stubble mulch, to increase water infiltration into the soil and reduce evaporation and wind erosion. A moldboard plow, a disk, or chisels are usually used to incorporate the residue (Figure 13-4). Subsequent tillage prior to planting is usually only for weed control. The field may be crossed several times with sweeps or chisels for this purpose. There is normally at least one tillage operation to control weeds prior to planting. In fine-textured-soil areas where row irrigation is practiced, the land is formed into beds with lister plows so that irrigation water can be applied in the furrows.

Planting and Stand Levels

Wheat, barley, rye, and oats are usually produced in narrow rows in a drilled pattern or are broadcast at random. Drilled rows are usually 7 to 9 in [18 to 23 cm] apart, but the space between rows may range from 4 to 16 in [10 to 40 cm] (Figure 13-5). Relatively large numbers of plants per acre are produced. The planting rate will be determined by the potential yield (principally determined by average annual rainfall). The planting rate in semiarid areas is around 20 lb/acre [22.4 kg/ha] for wheat and barley and increases with increasing rainfall or under irrigation up to 80 to 100 pounds per acre [89.5 to 111 kg/ha]. Seeding rates for oats range from 30 to 70 lb/acre [33.6 to 78.3 kg/ha]. For rye, the rates are from 20 to 30 lb/acre [22.4 to 33.6 kg/ha]. Precise planting rates for small grains are not critical. Most small grains have the ability to *tiller* or *stool,* that is, produce additional stalks from one plant. Plants tend to tiller where the stand is somewhat thin and conditions are good for plant growth. Planting depth for small grains is normally 1 to 3 in [2.5 to 7.5 cm].

Water Needs

A characteristic of the grasses is the fact that actual yield or percent of yield potential is greatly influenced by available moisture. A large portion of the small-grain acreage is located in semiarid, or low-rainfall, areas of the nation. Yields are usually limited by moisture in most of these areas. Fall-seeded small grains can survive with surprisingly little moisture during the winter until

(a)

(b)

(c)

(d)

Figure 13-4. Several different types of equipment are used in the preparation of land for planting: (a) moldboard plow; (b) chisel plow; and (c) rod weeder. While the (d) cultivator is normally used for weed control.

rapid growth starts in the spring. At that time, moisture levels must be adequate for good production. Fertilizer rates need to be tied closely to moisture levels.

Fertility

Crops in the grass family are relatively high nitrogen users and have a moderate need for phosphorus, potassium, magnesium, and all the other essential elements. If the soil cannot supply a sufficient quantity, the necessary nutrients must be applied as fertilizers. The amounts of the various nutrients that need to be applied range widely.

Most states have soil-testing laboratories to which soil samples can be sent for analysis to determine the amount and type of fer-

tilizer to apply. Plant-tissue tests can also provide some guidance. Plant parts that can be used would include the leaves or the entire above-ground portion of the plant. Guidance on soil and plant sampling for testing to determine fertilizer needs is available from vocational agriculture teachers, county extension agents, local farm advisors, and fertilizer dealers.

Several points should be kept in mind about fertilizer use. A balanced program of nitrogen, phosphorus, potassium, and other elements is desirable. This does not mean that all the nutrients should be applied in equal quantities but that the nutrients the soil cannot supply should be applied. For some soils, this may mean only nitrogen—for example, 120 lb/acre [134.3 kg/ha] of ni-

Figure 13-5. Most small grains are seeded with a grain drill with row spacings of 7 to 9 in [18 to 23 cm] apart. (*John Deere*)

trogen for wheat. Or it may mean significant quantities of phosphorus and potassium plus some zinc and iron in addition to an adequate level of nitrogen.

The fertilizer rates should be adjusted to the yield potential and the ability of the soil to supply nutrients. If yield potential is low because of a lack of moisture, perhaps no fertilizer is needed. But if yield potential is high, fertilizer rates should be relatively high. This same concept was discussed in regard to water and stand level.

Most fertilizer nutrients, with the exception of nitrogen, normally should be applied prior to seeding and incorporated into the soil. Nitrogen can be applied as needed. For fall-seeded grains, it can be applied as a topdressing in the fall and again as a topdressing in the spring, just prior to initiation of rapid growth or jointing. Nitrogen should be top-dressed on spring-planted grains prior to initiation of rapid growth and jointing.

Pest Control

Pest control has been discussed in Chapter 8. Cereal grain yields can be greatly reduced by weeds, insects, and diseases. The loss annually from these pests runs into the millions of dollars. It has been estimated that 25 to 35 percent of the potential crop yields for wheat and other small grains is lost each year in the United States because of inadequate management measures. Good management for production of cereal crops dictates that appropriate methods of pest management be used. Weeds can be effectively controlled in most cases by chemicals or by mechanical means. Many cultivars have "built-in," or genetic, resistance to insects and disease. When resistant varieties are not available, chemicals often can be beneficially used on cereal grains.

Harvesting

The harvesting process is the final step in the production of the crop. Proper handling of the crop after maturity and correct harvesting procedures may often be the difference between profit and loss (Figure 13-6).

Cereal grains reach a specific maturation stage at which maximum dry weight has been achieved and no further growth takes place. Drying continues even though growth processes have ceased. The producer should observe the moisture content of the grain and be ready to start harvesting when the moisture has reached a level at which the grain can be harvested and stored.

Figure 13-6. Harvest of the matured grain is the final step in the production process. (*John Deere*)

Harvesting equipment should be properly adjusted to minimize the loss of grain during harvest. Periodic checks should be made behind the combine to be sure that losses are at a minimum. Management practices used during the season can greatly influence the efficiency of the harvest. If weeds are not properly controlled or if a certain variety tends to lodge, efficient harvesting may be difficult.

Processing

Small grains are used in a large number of food products, but the principal use is for flour for various types of breads, pastries, and pasta. After being transported to the elevator, the grain is graded and subsequently sold.

Wheat provides an example of the processing of small grains. First, the *bran* (seed coat, or covering over the kernel) is removed. The bran is used mostly as livestock feed but also in cereals and other foods. The inside of the kernel, composed of the endosperm and the germ, is ground to make flour. Various types of flour are produced, including whole wheat flour, which contains the bran and germ plus the endosperm, and white flour, containing only the endosperm.

PRODUCTION PRACTICES FOR CORN AND GRAIN SORGHUM

Land Preparation

Tillage operations for corn and grain sorghum start with handling of the residue of previous crops. The residues may be shredded and incorporated into the soil (Fig-

Figure 13-7. Shredding stalks prior to disking is the first step following the harvest of the previous crop of corn.

ure 13-7). In some cases stalks may be only partially incorporated, leaving some on the surface. This practice, stubble-mulching, reduces evaporation and wind and water erosion and increases water infiltration. These two crops are normally grown in rows with a distance between the rows of 38 to 42 in [97 to 107 cm] apart, but in humid areas there is a trend to closer rows, 20 to 30 in [51 to 76 cm] apart. The land is often formed into beds on which the crop can be planted to provide a more favorable environment for the plants.

Planting and Stand Levels

The planting rate for corn and grain sorghum should be adjusted for the yield potential of a field. The planting rate, particularly for corn, should be more precise than for small grains, especially when other production practices, such as use of irrigation, extra fertilizer, and control of pests, are more precise.

The planting rate and the resulting stand level for corn are usually dependent on yield potential. In areas with the highest potential, such as irrigated areas, stand levels range from 24,000 to 30,000 plants per acre [60,000 to 74,000 per hectare]. In areas such as the corn belt, which are mostly rainfall-dependent, stands will normally be lower, in the range of 18,000 to 24,000 plants per acre [45,000 to 60,000 per hectare]. "Precision" planters are now in use which properly space seeds and plant the "precise" number of seeds per acre which will result in a desired stand level.

Stand levels for grain sorghum range from about 15,000 to 50,000 plants per acre [37,000 to 123,500 per hectare]. The lower stand level should be used in low-rainfall areas with low yield potential, and stand level should be increased as normal moisture and yield potential increase.

Stand levels and fertilizer rates for corn and grain sorghum should be adjusted to normal moisture conditions. The supply of moisture, except under irrigated conditions, cannot always be known in advance, but normal rainfall levels generally are known, and the producer should manage accordingly.

Corn varieties vary in the number of days needed from planting to maturity. Long-season varieties take up to 180 days to mature, while short-season varieties mature in around 120 days. Some short-season varieties mature within 90 days after emergence. Long-season varieties normally have a higher yield potential than varieties which mature in a shorter length of time. One key to good management in corn production is to choose for a given area a variety that will use the entire growing season (Figure 13-8).

In addition to maturity, other varietal characteristics need to be considered. These might include resistance to insects and diseases, uniformity of ear height, harvest ability, and standability (resistance to lodging).

Corn and grain sorghum planting start in southern Texas in mid-February. Planting occurs in late March and April in the mid-South and Southwest and in May in the corn belt of the Midwest. The latest planting dates are around June 1 in the northern states. Grain sorghum is normally planted 2 to 3 weeks later than corn.

Depth of planting ranges from 1 to 3 in [2.5 to 7.5 cm] for both crops. Soil moisture content will influence depth of planting; seed needs to be planted deep enough to be in contact with moist soil.

Figure 13-8. Seed corn companies establish demonstration plots to provide the producer an opportunity to compare varieties.

Water Needs

A continuous adequate supply of soil moisture is essential for highest production of corn and grain sorghum. The latter, however, has certain plant characteristics which permit it to withstand drought, such as fewer stomatal openings, less leaf area, a more extensive root system, and a waxy covering on the leaves. A short period of limited moisture or lack of moisture will have a relatively greater detrimental effect on yields of corn than on yields of grain sorghum.

For corn, the water needed for a yield of 10,000 lb/acre [11,189 kg/ha] is 20 to 24 in [51 to 61 cm] of moisture, as either rainfall or irrigation, provided that it is properly distributed throughout the growing season.

Fertility

The fertilizer needs for corn and grain sorghum are generally the same as for the small grains, i.e., principally nitrogen plus other nutrients which the soil cannot supply. Soil tests or plant-tissue tests are used as a basis for determining fertilizer needs.

Rates of fertilizer should be adjusted to the potential yield. Rates of nitrogen can be reasonably predicted for various yield levels of these two grains. Local farm advisors can provide guidelines for nitrogen and other nutrient use. Local conditions will dictate the best time to apply nitrogen.

Pest Control

When conditions are ideal for good corn and grain sorghum production, weeds tend to flourish also and, if not controlled, will cause a reduction in yield of 10 percent or more. Weeds can now be effectively controlled or practically eliminated with chemicals and good cultural practices.

Other pests—insects and diseases—are more difficult to control. Spider mites and the southwestern corn borer are two common insect pests on corn. For sorghum, the principal insect pests are midges, greenbugs, and chinch bugs. Corn earworms and cotton bollworms can also be a problem. Diseases of corn are minimal, with southern corn leaf blight being one of the most prominent. For grain sorghum, maize dwarf mosaic virus (MDMV) is a problem disease. Insects and diseases can be controlled by chemicals, use of resistant varieties, and proper crop rotation.

Harvesting

Proper harvesting is necessary as the final step in the production program to ensure maximum yield. As with small grains, corn and grain sorghum reach a stage when the grain has reached its maximum dry weight. For corn, this is approximately 52 to 54 days after silking. Drying processes then take over. When moisture content reaches a certain percentage, the crop should be harvested.

New storage technology has been developed which permits the storing of grain with a moisture content higher than the normal storage moisture. Some livestock feeders buy grain at moisture levels above normal and feed it before spoilage can occur.

Grain drying has become a common practice, particularly for corn. If grain can be harvested at a relatively high moisture level, for example, 25 to 30 percent, and artificially dried, the producer can harvest much earlier and often avoid considerable losses of grain due to dropped ears, fallen stalks, or perhaps damage from natural disasters such as hail or delayed harvesting due to wet weather. The decision on whether to harvest high-moisture grain and dry it is principally an economic one. If enough addi-

Figure 13-9. Iowa shelled corn is being loaded into a grain drying bin where it will be artificially dried before storage. (*Bill Gillette*)

tional grain can be harvested to offset the cost of drying, it will pay to dry it. Rising energy costs, however, may slow the trend toward increased use of this practice. Solar energy for drying is under investigation and will possibly offer a cheaper source of energy to use for drying grains (Figure 13-9).

Corn is harvested by a mechanical picker-sheller. Grain sorghum is harvested by the combine—the same type of machine used to harvest small grains. To be sure that harvest losses are at a minimum, the producer should check behind the combine and make adjustments to minimize losses.

SUMMARY

Cereal grains are the major sources of food in the world. They account for about 25 percent of the average diet in the United States and 50 to 90 percent of the average diet worldwide. In addition, they provide feed for livestock and will be increasingly used as a source of energy.

Even though the crops in this group of cereal grains have many similarities, each one is unique and is grown to suit the environmental conditions in an area for a particular need.

Management practices are also similar for various crops, but each crop requires specific managerial skills for best production.

The uses of the cereal grains are not restricted to flours and livestock feed; many specialty products are made from them. Corn may be processed for its oil or syrup or tapioca.

QUESTIONS FOR REVIEW

1. What percentage of the average U.S. diet is made up of cereal grains? What percentage of the diet of the world population is cereal grains?
2. What are some possible reasons why cereal grains are grown on almost 75 percent of the agricultural land of the world?
3. Who are the principal producers of the cereal grains worldwide?
4. Why is the principal production of grain sorghum in the Great Plains section of the United States, while the principal production of corn is in the Midwest?
5. Why is grain important to humankind?
6. Describe vernalization.
7. Describe the principal differences between corn and the small grains in row spacing and stand level.
8. Describe the general fertility needs of grass plants.

Chapter
14

Edible Legumes

For many people of the world, edible legumes, cereal grains, and nuts provide the major portion of calorie and protein needs. There is every indication that the importance of legumes in human diets in the United States, as well as other parts of the world, will increase significantly in the future.

Cereal grains and legumes complement each other nutritionally. Almost all cereal grains are deficient in one or more of the essential amino acids, which are found more abundantly in legumes. Other essential amino acids are more abundant in cereal grains than in legumes. It follows, then, that a diet containing both cereals and legumes can provide a reasonably good balance of the dietary protein needs for human beings.

The complementary relationship between cereals and legumes is extremely important, especially for people who may not be able to afford a high-meat diet. Examples that illustrate this complementarity are found abundantly in some of our neighboring countries, where corn (tortillas) and beans (often refried) or rice and beans are a major part of the daily diet. In other areas of the world, chickpeas (garbanzo beans) and wheat combine to make an almost perfectly balanced diet as far as calories and protein are concerned (Table 14-1).

CHAPTER GOALS

After studying this chapter you will be able to:

- Explain why edible legumes are a significant protein source for human diets.
- Identify the major edible legume crops.
- Describe the principal uses of edible legumes.
- Describe the cultural practices applied in producing edible legume crops.

TABLE 14-1 PROTEIN CONTENT OF PRINCIPAL CALORIE AND PROTEIN SOURCES

Food	Protein Content (g/100 g)
Rice, brown or husked	7.5
Maize, grain or whole meal	9.5
Millet, grain	9.7
Oats, meal	13.0
Rye, whole meal	11.0
Wheat:	
Whole grain	12.2
Germ	22.9
Bran	13.6
Bean (*Phaseolus vulgaris*)	22.1
Broad bean (*Vicia faba*)	23.4
Chickpea (*Cicer arietinum*)	20.1
Cowpea (*Vigna spp.*)	23.4
Peanut (*Arachis hypogaea*)	25.6
Lentil (*Lens esculenta*)	24.2
Peas (*Pisum sativum*)	22.5
Soybean (*Glycine max*):	
Seed	38.0
Cake	46.0

Source: From International Development Research Center, [IDRC = TS1] Food Legume Processing and Utilization, Ottawa, Canada, 1976.

THE LEGUME FAMILY

Edible legumes refers to the seeds of plants belonging to the Leguminosae family, which is one of the three largest families of flowering plants, containing 700 genera and 18,000 species. The Leguminosae family is divided into three subfamilies, the largest of which is Papilionoideae, which includes the edible legumes that are widely distributed across the world. The members of this sub-

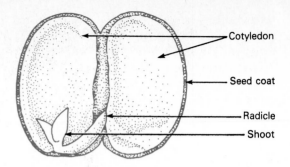

Figure 14-1. Parts of a dicot seed. (*FAO*)

family are shrubs and herbs which have characteristic flowers resembling the flowers of sweet peas. The flowers give rise to pods which contain one to several seeds, depending on the species and environmental conditions. There is wide variation in the size and shape of the pods and seeds. Each seed is attached to the pod at one point, which permits easy removal of the seeds from the pods.

The seeds are made up mainly of the cotyledon and the seed coat. The seeds separate into two halves, each of which is a cotyledon; because there are two, the species are classified as dicotyledons or "dicots." The cotyledons contain the stored nutrients that sustain the embryonic plant at the time of germination. These stored nutrients are the part that is consumed as human food. Legumes have no endosperm, which distinguishes them from the cereal grains, made up primarily of endosperm (Figure 14-1).

Pulses and Oilseeds

The food legumes can be further classified according to use. Edible legumes can be divided into two such groups, the oilseeds and pulses. The oilseeds consist of legumes such as soybeans or peanuts, which are grown for their oil as well as the meal. The oil is extracted either by pressing or by solvent extraction. There is a high-protein meal produced by these processes which can be used for human food or for animal feed. Pulses are non-oil-producing cultivated legumes grown for their edible seeds. Historians record the use of pulses in the human diet for thousands of years. For example, dry beans and lentils were common in the diet of the ancient Romans, Greeks, Jews, and Egyptians.

Pulses have been regarded by many societies as "the meat of the poor." It is unfortunate that this stigma is attached to such a valuable protein source. By developing new processing techniques and with some creativity in the kitchen, this image is being changed and legumes are becoming more common in the diets of all people.

Edible legumes have another significant advantage which is becoming more important as energy costs continue to rise. These plants, grown in association with bacteria of the *Rhizobium* genus, are able to reduce atmospheric nitrogen to ammonium nitrogen, which can be used by the plant to produce protein. Legumes also tend to improve the yield and quality of other crops grown with them as well as the crops that follow in the next season.

There is no doubt that world demand for edible protein will increase dramatically in the future. In view of the comparative inefficiency of providing this protein from animal sources, it will be necessary to give more attention to increasing the availability and acceptability of edible legumes for direct consumption in human diets (Table 14-2).

PROCESSING

Traditionally, edible legumes have been processed in the home as part of preparing meals for families. A significant portion of food legumes are still prepared and used in

TABLE 14-2 WORLD PRODUCTION OF MAJOR LEGUMES, 1972

Common Name	World Production (1000 metric tons)	World Legume Production (%)
Soybean	53,024	49.8
Peanut	16,887	15.9
Field bean	10,899	10.2
Garden pea	10,218	9.6
Chickpea	7,618	7.2
Faba bean	5,326	5.0
Cowpea	1,260	1.2
Lentil	1,182	1.1
World Total	106,414	100

Source: From International Development Research Center, [IDRC-TS1] Food Legume Processing and Utilization, Ottawa, Canada, 1976.

the home. Some small- and large-scale industries for processing them are developing, but not on the scale of the milling and baking industries with wheat, for example.

Home and Small-Scale Industrial Processing

Many of the processes that originated in the home and small-scale industries have been adapted and refined to permit large-scale commercial milling, frying, germination, and fermentation of edible legumes.

Peas and lentils are sometimes split or made into grits or a paste to make soups or served as a paste with crackers or bread.

Many of the pulses are harvested and eaten in the immature state. For example, garden peas are usually harvested while

Figure 14-2. Chickpeas, also known as garbanzo beans, are among the most important edible legumes providing much needed plant protein. (*California Dry Bean Advisory Board*)

still young and tender. They are processed commercially by canning and freezing. String beans or snap beans and baby limas are other examples.

Pulses such as pinto, great northern, lima, navy, and faba beans are usually harvested when dry and mature. They store easily and will not readily spoil if reasonable care is taken to keep them clean, dry, and free from pests.

A common and popular food for people whose cultural background is the Middle East is humus, which is prepared by crushing cooked garbanzo beans (chickpeas) and adding lemon juice, garlic, and salt to make a puree that is nutritious and flavorful.

People who migrated from Central and South America commonly include in their diet beans that have been soaked overnight and boiled. They are eaten either whole or mashed or are included in the preparation of other dishes. Crushed beans are often fried and served with rice, meat, or other foods.

Some leguminous seeds can be roasted or parched and eaten as snacks, including the peanuts that are shelled and eaten at a ball game or the circus. In other parts of the world chickpeas, mung beans, and soybeans are also consumed this way. Roasting improves the flavor and texture and also improves the nutritive value of the seeds as human food (Figure 14-2).

A method has been developed for extracting the oil from peanuts to produce a product which can be ground into a bland, white flour containing a highly soluble protein that can be used to make a very palatable milk. This peanut flour can be blended with wheat flour for bread and biscuits. It can also be used to make ice cream.

Germination is another way of processing legumes which allows the whole seed to be eaten. In this process the seeds are placed in a warm, moist environment and allowed to germinate until the hypocotyl has grown to about 1 to 3 in [about 3 to 8 cm] in length. Sprouted mung beans, peas, and soybeans are a common constituent of the diet of Americans whose culture has its roots in the Far East and the Orient. In recent years, legume sprouts have become a popular constituent of salads in homes and gourmet restaurants. Research has shown that the germination process actually improves the nutritive value of the seeds by making them easier to assimilate in the human digestive process (Figure 14-3).

One of the oldest known ways of processing edible legumes is by fermentation. In the Orient, fermented legume products have been eaten for centuries. Fermented legumes in human diets have been common in Asia and Africa also. Fermentation occurs when microorganisms are allowed to break down the carbohydrates to form acids in a limited oxygen environment. Many products are produced by this process, but the one most Americans are aware of is soy sauce, which is made from an equal mixture of soybeans and wheat. The flavor of the fermented product can be varied by using different strains of yeasts, bacteria, and molds.

Commercial Processing and Utilization

One of the most commonly used methods of preserving edible legumes for human food is canning. The canning industry in the United States processes huge amounts of food legumes, especially green (snap) beans, garden peas, and the picnic favorite, baked beans. Lima or butter beans, black-eyed peas, and kidney beans are also preserved in this way. Canning makes these products available the year around in a safe, easy-to-store form.

A recently developed food-processing method, quick-freezing, has created a large industry in the United States employing thousands of people. Frozen peas and green beans are among the many foods that are preserved in this way to retain a flavor

Figure 14-3. Legume sprouts, such as the alfalfa seed sprouts shown, are a valuable source of protein in the human diet. (*Jane Hamilton-Merritt*)

resembling that of products freshly harvested from the garden.

We often take these advances in food-processing technology for granted. Our early ancestors, however, were not able to open a can of green beans in the middle of winter to enrich and vary the diets of their families.

Travel to some other parts of the world soon brings into focus the advances made in the United States in processing and preserving food. There are many opportunities for employment in this agriculturally related industry, including everything from production and harvesting to transporting, processing, and marketing the products.

PRODUCTION PRACTICES

The cultural practices for the production of edible legumes vary to some extent according to the species, climate, type and fertility of the soil, latitude, equipment available, size of the production enterprise, and other factors. A number of general practices, however, are common among the various food legume species.

Seedbed preparation for the production of most of the food legumes is much the same as for corn or cotton. They are planted in rows, with the width between rows varying according to fertility level, water availability, equipment availability, and species involved. Modern seeding equipment can be adjusted to space the seeds in the row and adjust the width between rows to meet individual environmental and species needs. With management practices in recent years, row spacing has narrowed and plant population per unit of area has increased. This practice, along with better weed control and higher rates of applied fertilizer, has led to increased yields.

In general, edible legume seeds are larger than the cereal grains (except corn), so planting depth is usually greater. Depth of seeding can be controlled by adjusting the planter. Variability of soil characteristics makes it necessary to check the seeding depth by digging down to the seeds in the row behind the planter to see if the desired depth is being achieved. In general, the larger the seeds the deeper they can be planted. Most of the edible legumes are planted from 0.5 to 2.5 in [1 to 6 cm] deep, depending on seed size and moisture level of the soil. A soil test is helpful in determining the fertility level of the soil and the need for applying additional plant nutrients. It should be remembered that most legumes, if properly inoculated, will not require supplemental nitrogen fertilizer. However, a good response can often be expected from the proper application of phosphorus, potassium, and sometimes calcium. The soil test will also reveal the pH of the soil, and lime, if needed, can be applied on acid soils (see

Chapter 3). The requirements for other minerals vary widely with the crop species and the soil in a particular field. However, it has been shown conclusively that application of needed fertilizer can not only increase yields but also improve the quality of the crop.

The edible legumes are subject to the ravages of many diseases and pests which vary from species to species (see Chapters 6 to 9). Pesticides can be used to control harmful insects and diseases, but producers must be extremely careful when applying pesticides to prevent accumulation of harmful residues in the harvested crop which is to be consumed as food. It is important that, if needed, approved pesticides are applied according to the directions on the label to assure the production of food that is wholesome as well as nutritious.

Weeds can seriously reduce the yield and quality of edible legumes, as with other cultivated crops. Widely spaced rows generally provide space for mechanical cultivation to remove the weeds between the rows. Chemical herbicides are sometimes used, but caution must be exercised as with other chemical pesticides to assure that the crop is not damaged or toxic compounds in the plants allowed to accumulate (see Chapter 5).

Edible legumes, in general, require more intensive management and care than the cereal crops. The value of these crops is usually somewhat higher in the marketplace, which justifies a greater investment of time and resources to produce them. A large proportion of these crops is consumed directly as human food. Therefore it becomes extremely important that producers use optimum field management practices to produce the best possible quality of crop.

Soybeans

Although soybeans (*Glycine max*) are used directly as human food to some extent, this

Figure 14-4. Mechanical cultivation of corn in soybean residue. The widely spaced rows allow the equipment to remove harmful weeds. (*Gene Alexander/USDA-Soil Conservation Service*)

crop is produced primarily for its oil. No other crop produced in the United States has increased so greatly in importance and in area under production in the past 25 years. Soybeans are now challenging corn as the most important crop in this country. The fact that they have the same environmental requirements as corn and that much of the same equipment can be used to produce both corn and soybeans has enhanced the popularity of this crop (Figure 14-4).

The soybean is one of the oldest known cultivated crops. It was first recorded in Chinese literature in 2838 B.C. and was undoubtedly an important crop in the Orient long before that. It is only in recent years that this crop has developed into a major economic influence in the United States. In the early 1800s, a few farmers grew soybeans as a hay and green manure crop. Later, the technology and machinery for extracting the oil was developed; and in the

early 1900s soybeans began to increase in importance as a source of vegetable oil and protein. In recent years, the high-protein meal has become a more important component as methods of processing it into textured vegetable protein and other human food products have developed. It is now used to make less-expensive meat-substitute foods that may be a partial solution for the problem of inadequate diet and undernourished people.

The soybean is an upright, bushlike plant that branches from the nodes of the main stem. At maturity, the plants vary in height from 1 to 6 ft [30 to 180 cm] depending on environmental conditions, variety, and planting rate. The leaves are trifoliolate and vary in shape and size among different varieties. The flowers, similar in structure to most other legumes, are small and unnoticeable and are borne in racemes in the leaf axils. When the pods mature, they usu-

Edible Legumes

Figure 14-5. A high-yielding cultivar of soybeans. (*USDA*)

ally contain two or three seeds, as can be seen in Figure 14-5. The seeds are usually round but vary from spherical to somewhat flattened disks in some varieties. They are usually yellow but may vary from pale green to yellow to dark brown.

The soybean is an annual warm-season crop, best adapted to areas favorable for corn production. Seeds germinate more rapidly and seedling growth is more vigorous when soil temperatures are above 60°F [16°C], and the minimum temperature for effective growth is about 50°F [10°C]. Varieties differ in the length of growing season required for maturation. In general, a minimum of 120 frost-free days, with a mean summer temperature above 70°F [21°C], is needed to produce acceptable yields.

The soybean plant is very sensitive to photoperiod (day length), and varieties differ in terms of photoperiod required to initiate the formation of flowers. Soybeans are considered to be short-day (or long-night) plants, because most varieties start to flower as the days begin to shorten. Scientists have found that it is actually the lengthening of the dark period that triggers the flowering response. Soybean cultivars are grouped by their photoperiodic response and the time they take to mature. Varieties have been developed by plant breeders that are adapted to different latitudes with their varying day lengths. A county extension agent, vocational agriculture teacher, or reliable seed dealer can recommend varieties that are adapted in a particular area.

Optimum yields of soybeans require from 30 to 40 in [76 to 102 cm] of moisture in the form of either precipitation or irrigation. Adequate available moisture at the time of germination and during flowering and immediately thereafter is especially critical.

Seedbed preparation is similar to that previously described for corn. It is important to have a seedbed that is free from weeds, to provide a good environment for germination and early rapid growth of the seedling. During the early growing season, weeds are usually controlled by sweep cultivation or a rotary hoe. Herbicides are sometimes applied to reduce the number of cultivations required. Weed control is considered to be one of the major management problems and challenges in producing soybeans.

Soybeans grow on nearly all types of soil, from light, sandy soils to heavy clay and highly organic soils, but they are most productive on fertile loam soils that are relatively high in organic matter. A pH of 6.0 to 7.3 is optimum, and lime should be applied to highly acid soils. Fertilizer and lime rates should be determined from soil-test results, which can be obtained from state soil-testing laboratories. Soil-fertility specialists are employed by most fertilizer companies, and independent consultants also are available to make fertilizer and other management recommendations.

In the fall, when the seedpods turn yellow and dry down to about 14 percent moisture or less, the beans are ready for harvest. The crop is cut and threshed by a combine. It is important to assure that the combine is properly adjusted to reduce threshing losses and minimize damage to the seed.

Peanuts

An important cash crop of the humid southeastern and south central United States is the peanut (*Arachis hypogaea*). This plant originated in Brazil and Paraguay. It was later introduced by traders into Africa, India, and Asia and finally was brought to America by slave traders. Production of peanuts as a cash crop increased significantly in the early 1900s; about 1.5 million acres [679,500 ha] are now harvested annually (Figure 14-6).

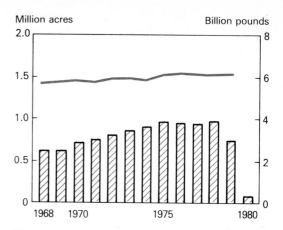

Figure 14-6. Peanut acreage and production. (*Source: Agricultural Handbook, No. 574, USDA, 1980.*)

Peanuts are an annual warm-season crop. Other names for this crop are groundnut, goober, or pinder. The seed is actually classified as a pea, even though it looks and tastes like a nut. The plant has a central stem with numerous branches bearing many pinnately compound leaves. There are both upright or bunch types, such as the Valencia and Spanish types, and runner types that are more prostrate in growth habit, such as the Virginia runners.

The plant has a deep taproot with numerous lateral roots; and if properly inoculated, many nodules will form. The typical legume flowers are borne in the axils of leaves either above or below the surface of the soil. Usually the flowers are self-pollinated. The peg, a stalklike structure that is technically referred to as a *gynophore*, grows downward into the soil, pushing the ovary beneath the soil surface. The pods, which usually contain from three to six seeds, develop underground on the pegs (Figure 14-7).

Peanuts are adapted to environments that are favorable for cotton and tobacco. A frost-free growing season of 200 days is desirable, although some varieties will mature in a

Figure 14-7. The pods of peanuts are borne on pegs beneath the soil surface. (*USDA*)

shorter growing season. This crop grows best in areas receiving 40 in [102 cm] of precipitation annually, but it can be grown under irrigation in drier areas. The moisture must be distributed throughout the growing season, but adequate moisture is especially critical during the fruiting or pegging period.

Well-drained, sandy-loam soils produce the best yields. The crop is not well adapted for high-organic-matter soils, which harbor harmful pests and reduce the yield and quality of the peanut. Fertilizer should be applied as determined by soil tests. If properly inoculated, peanuts do not need nitrogen fertilizer.

Planting dates vary according to location and season from mid-March in the deep South to early May in Texas. The soil temperature should be between 62° and 70°F [16 to 21°C] for optimum germination and seedling growth. A well-prepared, friable seedbed is desirable to permit penetration of the pegs on which the pods form. The rows are usually spaced from 24 to 40 in [60 to 100 cm] apart, and planting can be done with a corn or cotton planter by making some adjustments for seed size and shape. The seeds should be treated with an approved fungicide prior to planting. The rate of seeding varies from 60 to 110 lb/acre [67 to 123 kg/ha], with bunch types generally being seeded at the heavier rates.

As with most cultivated crops, early weed control is essential to allow the seedlings to become successfully established. Caution must be exercised when cultivating peanuts to avoid damaging the pegs with shovels or sweeps. Herbicides can be used to reduce the need for cultivation.

If peanuts are harvested too early, they shrink and lose weight. At full maturity, which can sometimes be determined by ex-

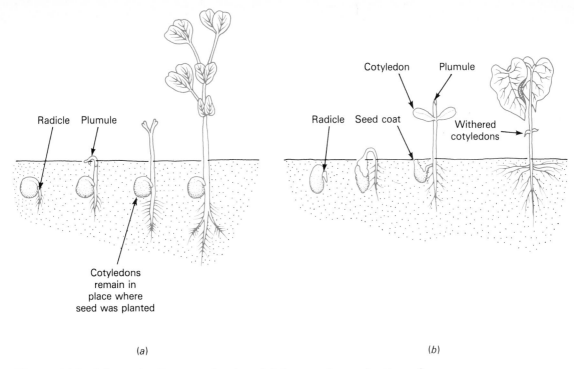

Figure 14-8. Schematic diagram showing (a) hypogeal germination of pea seedling versus (b) epigeal germination of bean seedling.

amining the inside of the shells for characteristic dark veins, the peanut plants are usually lifted from the soil with a digger-shaker which removes most of the soil from roots and pods. They are windrowed or placed in small stacks and allowed to dry before being combined to separate the pods from the plants. Harvesting is usually accomplished in the fall before the first killing frost.

Field Peas

As early as 3000 B.C., the Swiss lake dwellers were cultivating field peas (*Pisum arvense* and *P. sativum*). Field peas, introduced by colonists who migrated to America, are grown for food and livestock feed. Current production is concentrated in Idaho, Washington, Colorado, Montana, Wisconsin, and Michigan. Peas for canning and freezing, *Pisum sativum,* differ somewhat from the field pea, *P. arvense,* that is harvested dry. Field peas would not be classified as a major crop in this country, but in some areas they are an important component of the economy.

The field pea plant is an herbacious annual that ranges in height from 2 to 4 ft [0.6 to 1.2 m]. It germinates in a *hypogeal* manner; that is, the cotyledons do not rise above the surface of the soil (Figure 14-8a). The leaves are pinnately compound, with the exception of the terminal leaf, which forms a double tendril. At the base of each leaf petiole is a pair of large stipules. The flowers are typical of other legumes and are usually borne singly in leaf axils. They are white to reddish purple in color. Four to nine seeds are borne in a typical pod.

As a cool-season crop, field peas require moderate temperatures during the growing season. Field peas can be grown successfully in cool, semiarid regions but generally are grown in areas with over 20 in [50 cm] of annual precipitation. To attain high yields, adequate moisture at the time of flowering is most important. Well-drained clay loam soils are best for high yields. Fertilizer should be applied as determined by soil-test results. Proper inoculation eliminates the need for nitrogen fertilizer.

Field peas are generally planted as early as possible in the spring in northern regions. In southern and coastal areas, fall planting is customary, to allow the crop to mature before hot weather begins. Seeding can be done with a standard grain drill. A stand of three to four plants per square foot, or about 175,000 plants per acre [432,250 per hectare] can be attained by varying the seeding rate from about 110 lb/acre [123 kg/ha] for small-seeded types to 180 lb/acre [202 kg/ha] for large-seeded types.

When pods are dry and mature, the crop is ready for harvest. Usually it is combined directly, but if the pods are not uniformly mature or the field is weedy, it may be necessary to windrow and allow time for drying before combining.

Field peas are sometimes grown as a forage crop. When harvested for hay, they should be cut when the pods are well formed but still green. They may also be grown as a green manure crop to enrich the soil.

When garden peas are produced for canning or freezing, a field person usually advises the producer on time of planting and date of harvest. It is necessary to stagger these dates to give an even flow of properly mature peas at the processing plant over as long a season as possible. The public is demanding more fresh-frozen peas, and this industry has increased significantly in recent years.

Most pests of this crop can be controlled by crop rotation and the planting of resistant varieties. Peas are especially susceptible to powdery mildew, for which there is no known practical control.

Field Beans

The various types of field beans (*Phaseolus* spp.) apparently evolved in South America and Mexico, where there is evidence of domestication and cultivation for over 4000 years. They are still an important protein source in the diets of the people of that region today.

There are several types of field beans, including navy, great northern, pinto, and kidney beans, *P. vulgaris;* and large lima beans and baby limas, *P. lunatus.* They are all annuals with pinnately compound trifoliolate leaves. They have the typical legume flower, and the seeds are borne in pods with several seeds per pod.

In contrast to the hypogeal germination of peas, field beans have *epigeal* germination; that is, the cotyledons rise above the surface of the soil (Figure 14-8*b*).

Field beans are warm-season crops. They grow best at temperatures of 68 to 78°F (20 to 25°C). They will not tolerate frost in either the spring or the fall and also are sensitive to prolonged exposure to near-freezing weather. A 120- to 130-day frost-free growing season is needed for most types, but red kidney beans need 140 days to mature. Very high temperatures, e.g., 100°F [38°C], at flowering will cause floral blasting. Also, large limas will not tolerate heat, but baby limas will.

The moisture required varies with soil type and other climatic factors, but field beans are generally grown in areas receiving 12 to 24 in [30 to 61 cm] of precipitation annually. Supplemental moisture in the form of irrigation water is frequently applied. Adequate moisture throughout the growing season and dry harvest weather usually give a high yield and a high-quality crop. Field beans seem to grow well on a wide range of soils, from somewhat acid in humid regions to slightly alkaline in the dry West. However, a relatively neutral, sandy–clay loam soil that is well drained produces the best yields.

Field beans require about the same seedbed preparation as corn and soybeans, and they can be planted with the same type of planter with proper plates and adjustment. They are generally seeded in rows 22 to 42 in [56 to 107 cm] apart. Seeding rates vary from 40 to 80 lb/acre [45 to 90 kg/ha], depending on row spacing and size of seed. Both narrower space between rows and larger seed would result in a need for higher seeding rates to attain the desired number of plants per acre. Normal planting depth is about 2 in [5 cm]. Seeds should be planted in moist soil if possible. If wind is a problem, it is recommended that strips of corn be planted at intervals across the field to act as a windbreak.

Cultivation during the early growing season not only controls the weeds but also hills up the rows to ease the operation of the puller at harvest. Two cultivations are generally needed, with soil being thrown into the row to cover weeds and hill up the rows. The beans should not be cultivated when the foliage is moist, to avoid spreading bacterial blights through the field.

Harvesting of dry beans usually involves pulling the vines, windrowing them, and, when they are dry, combining from the windrow. The plants are normally lifted from the soil by blade-type pullers, which then place them in windrows for drying. The bean plants should be left in the windrow only long enough for the lower stems and root parts to dry enough for easy threshing in the combine. The beans should be harvested and handled at about 15 percent moisture to reduce splitting and damage to the seed coat; if they are too dry, excessive damage can occur. Harvest begins after a killing frost in the fall, when the pods are dry and yellow. Excessive rain at harvest can cause discoloration of the seed, which reduces the value of the crop. Most grain combines can be equipped with bean attachments which reduce the cylinder speed and provide special sieves and screens.

The beans should always be handled carefully. Dropping them from great heights in unloading can cause unnecessary damage, which can sometimes be avoided by cushioning or deflecting the beans as they fall into bins or wagons.

Diseases of field beans can reduce the yield and quality of the crop. Examples of some common diseases include white mold, common bacterial blight, rusts, bacterial halo blight, bacterial wilt, fusarium root rot, and virus mosaics. Cultural practices and pesticides for control vary from region to region. The local county agent, vocational agriculture teacher, or a reliable agricultural consultant can assist with recommended control measures.

Lentils

Lentils (*Lens culinaris*) have been an important food crop since ancient times. According to Genesis 25:34, Esau sold his birthright for a "pottage of lentils." They are grown in Europe, North Africa, Western Asia, and to a limited extent in the Pacific Northwest in the United States. The fruit is borne in pods containing several seeds. The seeds are red-

Figure 14-9. Faba bean plant. (*Walter H. Hodge*)

dish brown, gray, or black, are shaped like a lens, and are smaller in size than peas. The glass lens as we know it today was so named because of its resemblance to the shape of the lentil seed.

Lentils thrive best in sandy loam soils. Fertilizer should be applied according to soil-test recommendations.

Faba Beans

Faba beans (*Vicia faba*) are also commonly called horse beans, broad beans, pigeon beans, tickbeans, and Windsor beans. In the United States they are called either faba or fava beans. It is believed that they, too, have been cultivated in Europe and North Africa since prehistoric times. The plant is upright, with strong stems 2 to 5 ft [0.4 to 1.2 m] in height. Many flowers are produced on each plant, but only a few produce seeds. Pods containing three or four seeds are borne on the stems. The flowering habit is indeterminate or continuous, and at any given time during the production period pods can be found at various stages of maturity (Figure 14-9).

Faba beans are a cool-season crop and should be planted early in the spring during the cool, moist spring weather. The plants are able to withstand the heavy frosts which can occur in May or June at northern latitudes. The crop is best adapted to fertile silt loam soils that are well drained. The seeds should be planted in a well-prepared seedbed at a depth of 2 to 3 in [5 to 8 cm]. In the germination process, the hypocotyl does not elongate, so the cotyledons remain below the soil surface. The first internode elongates, either pushing or pulling the terminal bud to the surface. If weeds are controlled, the highest yields are obtained in rows spaced 12 in [30 cm] apart, but wider spacing is sometimes used to permit cultivation.

Faba beans are high in protein and can be used as human food, livestock feed, or a green manure crop. When consumed as human food, dry beans can be soaked and baked. Immature beans can be shelled and eaten like fresh garden peas. The beans contain from 26 to 33 percent protein and less than 1 percent oil. They are high in the amino acid, lysine, and low in methionine.

Mung Beans

The mung bean, *Phaseolus aureus,* is an ancient and widely used pulse crop. In some parts of the world it is referred to as *green gram* or *golden gram.* It is grown extensively in south, southeast, and east Asia, east Africa, and India. In the United States, production is mainly in California and the south central states. It is a relatively drought-tolerant crop, and varieties that mature in 60 to 70 days are available.

The mung bean crop is in demand because of increased use as dry beans, bean sprouts, and noodles. The bean offers 21 to 28 percent protein and can be processed in many ways for human food.

Chickpeas

The chickpea, *Cicer arietinum,* is also known as the garbanzo bean, Bengal gram, or gram. It is an important food crop in Asia, the Middle East, and northern Africa and is grown commercially in California and the southwestern United States. It is an herbaceous, annual, semierect plant that grows to a height of 8 to 20 in [20 to 50 cm]. It has a taproot. The plant is pubescent over all its parts. The leaves are pinnately compound and serrated. The seeds are borne in pods and vary in size and shape from round to semiround and from wrinkled to semiwrinkled. Germination is both epigeal and hypogeal.

Chickpeas are another nutritious food legume which could well become a much more important high-protein food crop in the future.

Cowpeas

The cowpea, *Vigna sinensis*, is believed to have evolved in central Africa and was introduced into the United States in colonial times. It is presently grown in the southern states and California for human food and in some cases as a forage or green manure crop. This is a warm-season annual with a vine-type stem and numerous trifoliolate, smooth, shiny leaves (Figure 14-10). The flowering habit is indeterminate, so flowers and seeds continue to be produced until the plants are killed by frost in the fall.

The seedbed is prepared as for corn and other row crops, and seed can be planted with a regular corn planter. Seeding rates vary from 30 to 50 lb/acre [34 to 56 kg/ha] when seeded in rows. The plants are sensitive to frost, so seeding should be delayed until danger of frost is over and the soil is warm.

The mature, dry cowpea crop is harvested with a combine. Because of its indeterminate flowering habit, windrowing may be necessary to allow the plants to dry out before combining. Cowpeas are sometimes harvested as green pods and green beans for canning.

Figure 14-10. Dwarf cowpeas with pods held above the foliage. (*USDA*)

SUMMARY

The edible legumes in general are becoming an important source of protein in human diets as well as a protein supplement in livestock feed. When the availability of meat is limited, these legume crops can provide the needed amino acids for good health. When combined with cereal foods, the edible legumes can become a way of coping with the continuing rise in population around the world.

QUESTIONS FOR REVIEW

1. What characteristics do edible legumes have that make them an increasingly important source of human food?
2. Describe some of the plant characteristics that are common among most edible legumes.
3. List some food products that are commonly made from edible legumes.

4. What environmental conditions are needed for optimum production of soybeans?
5. Describe the peanut plant and how the peanuts are formed.
6. What part of the country produces most of the field peas?
7. How do field peas and beans differ in their manner of germination?
8. Describe the practices a producer would follow to successfully grow soybeans, peanuts, field peas, and field beans.

Chapter
15

Oil Crops

Thousands of products taken for granted day by day are made partly or entirely from vegetable oils. Among these products are the margarine spread on bread, the creamy white shortening used in cooking, the bar of soap for washing hands, the chocolate coating on a candy bar, the paint used on houses and cars, medicines, cosmetics, lubricants, and illuminants. These are merely a few of the many uses for vegetable oil. In considering alternative energy sources, the prospect of using vegetable oil, a renewable source of energy, in place of diesel fuel, a nonrenewable source of energy, looks promising. Even though oilseed crops have not played as significant a role in the cultural development of humankind as the cereal crops and edible legumes, there are a number of important species of oil-bearing plants grown throughout the world. Several of these make a significant contribution to the economy and well being of the United States.

CHAPTER GOALS

After studying this chapter you will be able to:

- Describe the major sources of vegetable oils and fats.
- Understand the various procedures used to extract the oil from the meal.
- Identify the major commercially grown oilseed crops produced in the United States.
- Identify the characteristics of the major oil crop plants.
- Describe how to apply basic cultural practices for growing oil crops.

SOURCES OF OIL CROPS

Oils and fats are substances derived from plants and animals. The terms *oil* and *fat* are interchangeable, except that a fat is solid or semisolid at room temperature and an oil is liquid at room temperature. Fats and oils are important to all forms of life. They are derived from five general sources:

1. *Seeds of Annual Plants* This is the largest source and includes such crops as soybeans, cotton, peanuts, sunflowers, safflowers, flax, rapeseed, mustard, and other species. The fact that they are annuals allows for relatively rapid adjustment of the supply of these oils to meet changing market demands. They are adapted primarily to temperate environments.
2. *Oil-bearing Trees* Oil from these trees falls into two categories; edible and inedible. Examples include coconut and olive oil, which are edible, and tung and palm, which are inedible and are used for industrial purposes. In general, the oil-bearing trees are adapted to warm and tropical environments, and production is limited in the United States.
3. *Animal Body Fat* This important source of fats and oils includes tallow from cattle, lard from swine, and, of less significance, the fat from other animal species.
4. *Dairy Products* Included in this category is the fat in milk, which is consumed directly or made into a myriad of dairy products such as ice cream, butter, yogurt, and cheese.
5. *Fish and Marine Mammal Oil* A number of small fish, such as sardines and herring, as well as whales are the main sources of these oils.

This chapter is confined to category one, the seeds of annual plants, which are the predominant plant-derived source of oils and fats in the United States. The other categories are important also and are mentioned to ensure that the sources are kept in proper perspective. Figure 15-1 shows production of oils and fats in the United States from 1975 to 1980 by category.

CONSUMPTION

The annual consumption per capita of fats and oils for food in the United States increased about 8 percent from 1970 to 1980—from 79 lb [35.5 kg] in 1970 to 85 lb [38.5 kg] in 1980. The increased use of vegetable oils for food accounted for a significant portion of the increase in consumption. When considering the consumption of oils and fats on a worldwide basis, the United States is a major consumer. Table 15-1 shows our high consumption rate in comparison with some other countries.

PRODUCTION

The production of five major oilseed crops in the United States in 1979–1980 was about 71 million t, which was an increase of nearly 20 percent over the previous year. This continues an upward trend for recent years, as shown in Figure 15-2. The increase in production results from higher yields and an increased acreage. It should also be noted that soybean crops make up about 85 percent of the total. The production of cottonseed is increasing significantly also. Sunflower production increased rapidly up to 1979; but in 1980, a reduced price resulted in a 28 percent drop in acreage. However, sunflowers will probably continue to be an important oilseed crop in the United States. The other oilseed sources are relatively small but important to the economy in local areas where they are being grown.

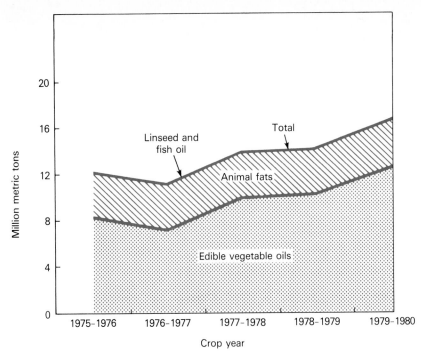

Figure 15-1. Production of oils and fats in the United States (excluding tall oil) from 1975 to 1980. (*Source: ESCS/USDA*)

TABLE 15-1 ESTIMATED PER CAPITA CONSUMPTION OF OILS AND FATS*

Country	Consumption (kg)		Population (millions)	
	1970	1980	1970	1980
China, P.R.	2.8	3.5	774	940
India	6.2	6.8	560	673
U.S.S.R.	18.4	21.7	243	266
United States	35.5	38.5	202	220
World	N.A.	N.A.	3,700	4,400

*Food and nonfood.
N.A. = Not available.
Source: "Fats and Oils Situation," USDA No. 298, February 1980.

FUEL FROM VEGETABLE OIL

The use of vegetable oil as a substitute for diesel fuel in engines is receiving increased attention. One reason this potential energy source is being considered is that it is renewable; it can be produced on the farm by the producer for his own use. It could reduce the risk of unavailability of fossil fuel (diesel) at critical planting and harvesting times in the energy-intensive farming systems that are predominant in the United States.

Scientists have shown that vegetable oils can be substituted for diesel fuel as a backup power source. Researchers have tried mixtures of 25, 50, and 75 percent vegetable oil with diesel fuel, as well as 100 percent vegetable oil, and have shown that there is little

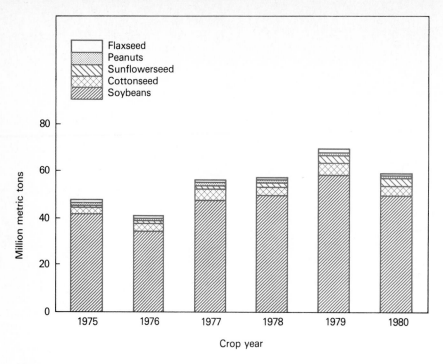

Figure 15-2. Oilseed production in the United States from 1975 to 1980. (*Source: ESCS/USDA*)

variation in horsepower up to 75 percent. Mixtures above 75 percent vegetable oil result in a slight reduction in horsepower. This could be caused by the fact that vegetable oil is thicker than diesel fuel and may slow the flow through the filters and fuel lines to the engine. The economic feasibility of substituting vegetable oil for diesel fuel will depend on the relative price of the two sources. However, if diesel fuel should suddenly become unavailable, there is no question that vegetable oil could be used to run tractors, combines, and other agricultural implements (Table 15-2).

THE VEGETABLE OIL INDUSTRY

It is most likely that the first fats used by humans were animal fats separated from other tissue by heating or boiling with water. In general, it is more difficult to extract the oil from plant sources, but there is evidence that vegetable oils have been used for thousands of years. The Phoenicians and Egyptians may have been among the first to develop the technology for extracting vegetable oils. Historical references show that the Egyptians used olive oil as a lubricant in moving building materials and statues and used fats and oils to grease the axles of their chariots.

The early extraction methods left a high percentage of the oil in the meal. It was not until the hydraulic press was developed in 1795 that the technology for more efficient extraction was available.

The continuous screw press was developed about the turn of the twentieth century, and again the efficiency of extraction was im-

TABLE 15-2 ESTIMATED VALUE OF SUNFLOWER OIL LESS PROCESSING COST*

	Market Value of Sunflower/cwt				
	$7.00	$8.00	$9.00	$10.00	$11.00
Value of oil/cwt seed (32 lb)	3.60	4.60	5.60	6.60	7.60
Cost per gallon sunflower oil	0.88	1.12	1.36	1.61	1.85
Cost per pound sunflower oil	0.11	0.14	0.17	0.21	0.24

*Above chart based on:
  Sunflower = 40% oil by weight
  Extraction rate = 80%
  Sunflower oil = 7.7 lb/gal
  Sunflower oil meal = $100/ton or 5¢/lb
  100 lb sunflower seed ⟶ 32 lb oil ⟶ 4.1 gal oil
  ⟶ 68 lb sunflower oil meal
  Source: Adapted from North Dakota State University, CES-13AENG-5, Fargo, North Dakota, 1979.

proved, leaving a meal containing only about 4 percent oil when processing soybeans, for example.

But processors were not satisfied with leaving a meal with 4 percent oil. The answer to more efficient extraction was found in a different concept, organic solvent extraction. This process was developed in the 1840s in England, but it did not become extensively used until the continuous solvent extraction process became operational in the 1930s. This method has now become the predominant extraction procedure in the United States, and a huge industry employing thousands of people has developed.

The solvent generally used by processors in the United States to separate oil from, for example, soybeans is called *hexane*. The residual meal contains less than 1 percent oil.

The crude vegetable oil must be processed further before it is suitable for food or industrial uses (Figure 15-3). It can be sent through the edible oil refining process, or it can be treated to furnish a substance called lecithin, which is used in food products and pharmaceuticals.

After being refined, the oil is bleached to improve the flavor stability and the color of the oil. Following bleaching, the oil flows to the hydrogenator and then to the deodorizer for conversion to edible products. The hydrogenation and deodorization processes form margarine-base oils, shortening, cooking oils, and an all-purpose salad oil.

After the oil has been deodorized, it is filtered and stored. Antioxidants may be added to increase its shelf life.

The soybean oil meal is processed into soy flours and grits and high-protein-concentrate food and feed products. See Chapter 14 for further discussion of whole soybean and soybean meal products.

Other oilseed crops are processed in a manner similar to soybeans, with dif-

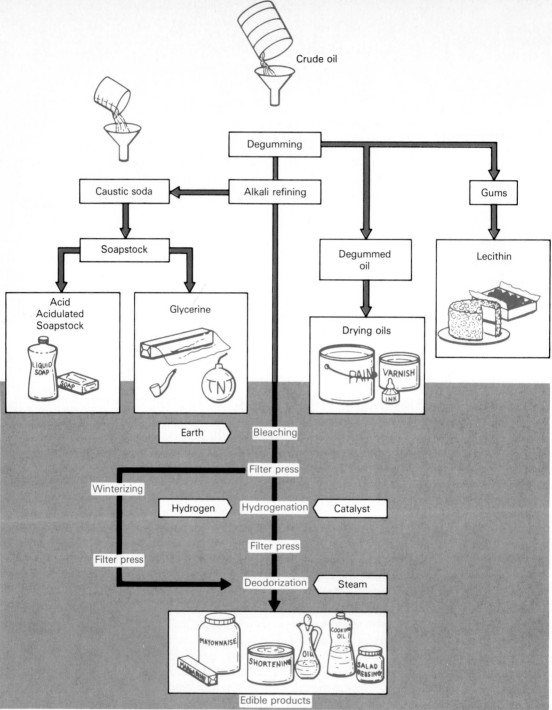

Figure 15-3. The refining processes and end uses of crude vegetable oil. (*Source: Economic Research Service/USDA*)

ferences to accommodate differences in the characteristics of each species.

OIL CROPS OF THE UNITED STATES

The predominant oil crop produced in the United States is soybeans. It is also classified as an edible legume, so a full discussion of the characteristics and production practices of this crop can be found in Chapter 14, "Edible Legumes."

Another major source of vegetable oil is cottonseed. Cotton is the primary fiber crop of the United States, and Chapter 16 is devoted entirely to this crop.

A third important vegetable-oil crop is peanuts. Peanuts are also classified as an edible legume as well as an oil crop, so the description, characteristics, and production practices for this crop are found in Chapter 14.

Sunflowers

The state of Kansas made the sunflower, *Helianthus annuus,* its state flower. It is a plant that is native to North America and grows wild in many parts of the United States today. It was used by American Indians for food prior to the colonization of the new world. Sunflower seed was taken to Spain in the sixteenth century and found its way to several countries of the Mediterranean, Middle East, and North Africa. It spread to China and Russia, and agriculturists in Russia developed it into an important seed crop. Sunflowers began to become an important oilseed crop in the United States in the 1960s.

The sunflower, a member of the thistle (Compositae) family, is an erect annual plant. It grows to a height of 4 to 20 ft [1 to 6 m], depending on the variety and date of planting. The hairy stems are woody and thick. The leaves are large, sometimes measuring 12 in [30 cm] in length, and are borne on petioles. The flowers are borne at the end of each stem in a large-head-type inflorescence. The sunflower is well known for its unique response to light. The head faces east each morning, and as the day progresses, the head turns and follows the path of the sun until it is facing west at sunset. During the night it rotates until it is facing the east again the next morning (Figure 15-4).

Sunflowers are adapted to the northern and western fringes of the corn belt, where corn and soybean production is marginal because of the short growing season or low rainfall. They are also grown in the prairie provinces of Canada. Some sunflower production is found in Texas and California. The sunflower is considered a warm-season crop, but it will tolerate light frost in the early seedling stage. After it has reached the four- to six-leaf stage, it will not tolerate frost without damage to plant tissues.

Sunflowers are widely adapted to most soil types. They require less water than corn and will produce relatively well on low-fertility soils. It requires a shorter growing season than soybeans, about 120 days to reach maturity.

Scientists have recently developed hybrid varieties of sunflower which have resulted in significantly higher yields and higher oil content. The hybrids also display superior resistance to rust. They tend to mature more uniformly than open-pollinated types, which facilitates the harvesting process.

The sunflower is usually planted with a corn planter in rows spaced 24 to 30 in [60 to 76 cm] apart to permit cultivation for weed control. The seedbed is prepared as it would be for corn. Seeds are planted about 2 in [5 cm] deep. Planting time varies with latitude from March to May. Late planting in north-

Figure 15-4. A field of sunflowers being grown in North Dakota. (*USDA*)

ern areas results in lower yields and lower oil content.

Several diseases and insects attack the sunflower. The most serious diseases are downy mildew, rust, sclerotinia rot, verticullium wilt, phoma black stem, and alternaria leaf spot. Treating the seed with fungicides has not normally been shown to be economical. Growing resistant varieties and using good pest management practices will keep pest losses to a minimum.

Two types of sunflower are produced: those for oilseed production and those for nut and birdseed production. Those grown for oil are generally black-seeded and have a thin hull which adheres to the kernel. The non-oil sunflower seeds are larger and striped, with thick hulls that thresh free from the kernel. The non-oil type is used for human food as a snack or in salads and is also sold as birdseed. The sunflower meal and hulls that remain after the oil is extracted are used primarily for livestock feed.

The hulls can also be used as fuel to generate steam at processing plants. Sunflower oil appears to have potential as a substitute for diesel fuel. A large proportion of the sunflower crop is exported to Europe, where it is processed to extract the oil for human consumption. Sunflower-processing plants are being built in the United States, which will reduce the need to export it for processing in the future.

Safflower

Safflower, *Carthamus tinctorius,* is believed to have originated in northern India. It has been grown in India, North Africa, and the Middle East for centuries. It was used as a source of dye by the Egyptians to tint the wrappings of mummies over 3500 years ago. It was also produced for its edible oil.

Safflower was introduced into the United States in 1925 as an experimental crop. It was first grown commercially in about 1944

Figure 15-5. A safflower plant. (*Walter H. Hodge*)

to supply needed oil during World War II. By 1965, there were about 320,000 acres [128,000 ha] planted. It is a minor oilseed crop on a national basis, but in parts of California, North Dakota, South Dakota, Montana, Nebraska, and several other western states it is a significant crop.

The safflower plant is a member of the thistle family (Compositae) (Figure 15-5). It is a taprooted, drought-tolerant, long-season annual with stems ranging from 2 to 5 ft [60 to 150 cm] in height. The stems branch near the top, with a flower forming at the terminal end of each branch, forming a head-type inflorescence with green bracts. The flowers are usually yellow or orange in color. Each plant produces several heads, each containing 15 to 50 seeds. The leaves are oblong and waxy in appearance, with pointed, toothlike margins. In most varieties there are spines on the leaf margins, but plant breeders are attempting to develop spineless types. The fruit are smooth, shiny, white, angular, and

somewhat wedge-shaped, about 0.4 in [1 cm] in length. The seed weighs 37 to 38 lb/bu [4.8 to 4.9 kg/m³]. The hulls account for about one-third to one-half of the weight. The oil content of the seed varies from 32 to 40 percent and the protein from 11 to 17 percent.

Safflower is adapted to the northern Great Plains, the intermountain west, California, and, under irrigation, the Southwest. It is considered to be a cool-season crop. Spring-planted safflower requires about 120 frost-free days to reach maturity. Seedlings can withstand temperatures down to 20°F [−7°C], but plants become less tolerant of frost as the plant develops beyond the four- to five-leaf stage.

This crop requires a minimum of 16 to 18 in [40 to 45 cm] of moisture annually, but for maximum yield 25 in [63 cm] is needed. Excessive moisture has resulted in the development of damaging diseases. Because of its deep, extensive root system, safflower can use moisture deep in the soil. It produces best on deep, fertile, relatively neutral soils with good water-holding capacity. Good drainage is important for healthy plants.

Seedbed preparation for the safflower is about the same as for wheat or barley. Fall plowing is recommended; and a firm, friable seedbed that is free of weeds should be prepared. The soil should be tested and the recommended fertilizer applied.

Safflower is usually planted with a standard grain drill, plugging the openings to obtain the desired row width. Under irrigation, narrow row spacing of 6 to 15 in [15 to 35 cm] is most common. Seeding rates vary from 15 to 40 lb/acre [14 to 45 kg/ha] on dry land to rates of 25 to 60 lb/acre [28 to 67 kg/ha] under irrigation. Seeding depth varies with seed size and the level at which moisture is found, but generally a seeding depth of 1 to 2 in [2.5 to 5 cm] is recommended. Safflower is ready for harvest when the plants are completely dry. The crop is usually combined directly from the standing crop, but if green weeds are a problem, it can be windrowed, allowed to dry, and then combined with a pickup attachment.

Yields of safflower are seldom reduced significantly by insect damage. Wireworms, army cutworms, thrips, and lygus bugs can be destructive, and occasionally other insect pests will invade the crop. The most prevalent diseases are bacterial blight, caused by *Pseudomonas syringae,* alternaria leaf spot, rust, and sclerotinia head rot and root rot. Practices that reduce loss from diseases include planting treated, disease-free seed and following proper crop rotations to break the life cycle of the organisms causing these diseases.

Safflower is grown almost exclusively for its oil, which is popular as a cooking oil and in the manufacture of mayonnaise and margarine. Its popularity is enhanced by the fact that the oil has high levels of unsaturated fatty acids, which some medical evidence has shown to be more beneficial to human health than foods containing saturated fatty acids.

Flax

Flax, *Linum usitatissimum,* has been grown as long as history has been recorded, as a source of both oil from the seed and fiber from the straw. It is believed that flax evolved in the Mediterranean area and was first used as a fiber crop for making the fine cloth called *linen.* The Egyptians used this fabric to preserve mummies. There are records of the use and drying qualities of linseed oil in the Roman Empire as early as A.D. 230. Flax has been grown for both fiber and oil in the United States; but in recent years, it has been grown exclusively for its oil and high-protein meal.

The flax plant is an herbaceous annual that ranges in height from 1 to 4 ft [30 to 122 cm] (Figure 15-6). In thick stands, one main stem develops, but in thin stands, four or more branches may form. It has a short taproot with many fibrous branch roots that penetrate as deep as 4 ft [122 cm] in light soils.

The flowers are usually blue but may be white, pink, or violet, depending on the variety. They open shortly after sunrise, and the petals usually drop off by noon. Flowering usually continues for about 3 weeks. It is usually a self-pollinated plant, but some cross-pollination takes place, with insects as vectors. The seeds form in capsules, each of which contain about ten seeds. The seeds are reddish to deep brown or smoky yellow in color and are covered with a mucilaginous material that gives a glossy appearance and is sticky when wet.

Flax is a cool-season crop that will tolerate frost in the seedling stage. It requires

Figure 15-6. A flax plant. (*International Linen Promotion Commission*)

from 18 to 30 in [46 to 76 cm] of moisture annually for good yields. It is better adapted to loam, silt-loam, and clay-loam soils than to sandy soils. Generally, flax does not respond to fertilizer as well as cereal crops but responds well to application of nitrogen when needed. The results of a soil test should be observed before applying added fertilizer.

The seedbed should be prepared by using a plow or one-way disk that will control weeds and leave a firm seedbed. If the soil crusts, it may be necessary to harrow to facilitate emergence of the seedlings. The seed is planted at a depth of 1 to $1^{1}/_{2}$ in [$2^{1}/_{2}$ to 4 cm], usually with a standard grain drill. Seeding is usually done in late April or early May as soon as the soil is moderately warm. Seeding rates vary from 28 to 40 lb/acre [31 to 45 kg/ha] on dry land to 45 lb/acre [50 kg/ha] under irrigation. The seed should be treated with an approved fungicide before planting. Weeds are a serious problem with flax. Good cultural practices must be used, and herbicides are available that are very beneficial in controlling weeds.

Flax should be harvested when the bolls are dry and the stems yellow. When dry, the seeds give a dry rattle within the boll. Direct combining of the standing crop is most common, but it can be cut and windrowed to allow green material to dry before combining if necessary. Flax stems are tough, so it is important to keep the cutter bars clean and sharp. The combine must be carefully adjusted to avoid cracking and damaging the seed.

A number of diseases have been known to attack flax, including rust, wilt, seedling blight, root rot, pasmo, heat canker, and aster yellows. To reduce losses from diseases, only clean, disease-free seed should be sown in a firm seedbed, and flax after flax should be avoided in rotations.

Various insect pests that can be troublesome in flax include cutworms, wireworms, aphids, leafhoppers, armyworms, flax bollworms, beet webworms, and grasshoppers. The fields should be examined regularly and control measures applied promptly if pests become a problem. Recommended control measures vary from region to region; information on control measures is available from local agricultural advisors.

Flax is used almost exclusively to produce linseed oil and high-protein oil meal for livestock feed. Linseed oil is best known for its use in protective paints that dry to a durable finish. It is a fast-drying oil. The seed usually contains from 35 to 45 percent oil. The meal is about 35 percent protein that is 85 percent digestible and appears to have a regulating affect on the digestive system of livestock.

Rapeseed

Not much is known about the origin and early culture of rapeseed, *Brassica napus* or *B. campestris*. There are records of what is believed to be rapeseed being grown in India and China up to 2000 years ago. The oil from rapeseed was used in lamps during the Middle Ages. The oil, high in erucic acid, was not considered suitable for human consumption. Canada now is the major producer of rapeseed; the crop is sometimes called *canola* there. Small amounts are produced in the United States, but marketing can be a problem and producers should have a market contract before producing this crop. Low-erucic-acid varieties have been developed, and the oil can now be used as a human food oil as well as being a high-quality lubricating oil for marine engines and other uses.

There are two types of rapeseed, the Argentine type, *B. napus,* and the Polish type, *B. campestris* (Figure 15-7). The Argentine type matures about the same time as wheat in the northern states. It grows to a height of 2 to 4 ft [60 to 120 cm]. The stems are coarse and have many branches. The flowers and pods are typical of those in the mustard family. The flowers have four sepals and four petals which open into a crosslike pattern, giving rise to the family name—*Cruciferae*. The pods tend to dehisce (split open) at maturity. Each pod contains 15 to 40 seeds, which may vary from black to yellow in color.

The Polish type matures 1 to 2 weeks earlier than the Argentine type, and the plants generally are somewhat shorter. They also yield less. Low-erucic-acid varieties are available in both types.

Rapeseed is a cool-season crop and grows well on most soil types. It grows best, however, on clay-loam soils. Rape is not as drought-resistant as cereal grains and does best where adequate moisture is available throughout the growing season.

A well-prepared, firm seedbed is essential for successful establishment of rapeseed plants. The seeds are small and must be planted at a uniformly shallow depth of not more than 1 in [2.5 cm]. The seed is planted with a grain drill at a rate of 5 to 8 lb/acre [6 to 9 kg/ha] for Argentine types and 4 to 7 lb/acre [5 to 8 kg/ha] for Polish types. Special seed treatment is not generally recommended.

Rape does not compete well with weeds, so care must be taken to control weeds. In addition to cultural practices, herbicides can be used to control certain broadleaf weeds and grasses. When stems and pods turn straw-colored and the seeds are dark brown, the crop is ready for swathing. A combine with a pickup attachment can be used to thresh out the seed. Care must be taken to adjust the combine to minimize damage to the seeds. If the crop is free from weeds and ripens evenly, it can be combined directly without swathing. The seeds are small, round, and

Figure 15-7. A rapeseed plant. (*Walter H. Hodge*)

heavy, and they will leak through any small holes in trucks or bins. It is safe to store rapeseed that is down to 10.5 percent moisture.

Mustard

Another relatively minor oilseed crop that is important in local areas is mustard, *Brassica* spp. There are three types of mustard; yellow, brown, and oriental. Of these, yellow is the most commonly produced type in the United States. Most of the mustard in the United States is grown in the north central and northwestern states.

The plants resemble rapeseed plants, both being of the same genus. They vary in height from 30 to 45 in [75 to 112 cm].

The seedbed for mustard should be free from weeds and firm to facilitate shallow uniform planting of the small round seeds. The seed should be planted as early as possible while still avoiding frost. It is planted with a grain drill at about 10 to 14 lb/acre

Figure 15-8. Soybeans (left) are one source of oil, as are tree nuts such as pecans (upper right). Fruit-flesh oils can be extracted from olives (lower right). (*Grant Heilman*)

[11 to 16 kg/ha] with yellow types and at 6 to 8 lb/acre [7 to 9 kg/ha] for oriental and brown types. Nitrogen and phosphorus fertilizer often give a good response.

As with rapeseed, mustard is a poor competitor with weeds, so weed control is crucial to success in producing this crop. Both cultural and chemical control measures are often required. Insect and disease problems are similar to those with rapeseed.

Mustard is harvested by combine, either from the windrow or cut directly and combined. The seed can be stored safely at a moisture content of 10 percent or less. Marketing can be a problem, so growers should always have a market contract before seeding.

Other Oil Crops

Vegetable oil is extracted in the processing of corn and rice, which are significant sources of edible oil. In addition, there are a number of minor vegetable oil sources. Oils are extracted from oil-bearing agricultural wastes such as fruit pits, tomato pomace, wine pomace, and the seeds of citrus, apples, pears, pumpkins, squash, and pimento. In addition, oils come from tree nuts such as almond, walnut, and pecan; and fruit-flesh oils can be extracted from olive and avocado. Tung oil is produced from the seed of the tung tree for use in varnish. Castor beans and crambe are other potential oil crops (Figure 15-8).

Many plants once considered weeds or good only for forage for livestock have been developed into vegetable oil sources. It is likely that new oil crop species will be developed in the future to add to those described in this chapter.

SUMMARY

Crops from which oils and fats can be extracted are important, not only from the standpoint of meeting human food and animal feed needs, but also from the point of view of industry, which requires vegetable oils and fats for many manufactured products. The annual per capita consumption of fats and oils in the United States is rising, which indicates that these crops are destined to become more significant in importance economically. It is possible that vegetable oils could become a substitute for diesel fuel if it were to become short in supply.

Some of the major commercial sources of vegetable oils and fats are soybeans, cottonseed, peanuts, sunflower, safflower, flax, rapeseed, and mustard.

QUESTIONS FOR REVIEW

1. What are the five general sources of vegetable oils and fats?
2. What has been the trend in per capita consumption of vegetable oils in the United States?
3. What is the main method currently used by industry to separate oil from the meal in soybeans and other oilseed crops?
4. List the major oilseed crops grown in the United States.
5. Identify some of the major uses of each of the oilseed crops.
6. Pick one of the oilseed crops that is commonly grown in your area and describe the cultural practices applied by growers to produce it.

Chapter

16

Cotton

Cotton is one of the most important crops grown in the United States, ranking in dollar value behind only corn and wheat. It is one of nature's most valuable gifts to humanity. Its fluffy white fiber is the world's most important textile material. And—something less well known—the cotton plant produces even more human food and livestock feed than it does fiber. The seeds are a source of oil, which is used in the production of several types of food, and cottonseed meal and cottonseed hulls, which are important animal feeds.

Cotton is produced in the United States principally in the southern states. Texas produces approximately 35 percent of the nation's cotton annually. The western states of Arizona, New Mexico, and California produce about 30 percent, and the delta states of Mississippi, Arkansas, Louisiana, Tennessee, and Missouri about 30 percent. The other 5 percent is produced principally in Oklahoma, South Carolina, and Georgia (Figure 16-1).

Cotton is also an important crop worldwide. The leading producer of cotton is the U.S.S.R., followed by the United States, China, India, Brazil, Turkey, Pakistan, Egypt, and Mexico. The first five countries produce approximately 70 percent of the world crop.

Cotton has been grown for its fiber for several thousand years. One of the oldest records of cotton textiles, dating back about 5000 years, was found in the Indus River valley in what is now Pakistan. Cotton fabrics have also been found in the remains of some of the ancient civilizations of Egypt and in the ruins of the Indian pueblos of the southwestern United States, dating back hundreds of years before Christ.

Cotton was being grown in India in the third century B.C. and spread from there to western Asia, Europe, and North Africa. The Europeans who settled what is now the southeastern United States found that cotton was not generally grown in that area. It is reported that the colonists planted cotton in Florida in the sixteenth century and in Virginia and the Carolinas during the seventeenth and eighteenth centuries.

Two important events in the cotton industry may be credited for the increased development of cotton in the United States in the late eighteenth century. One occurred in 1790, when Samuel Slater

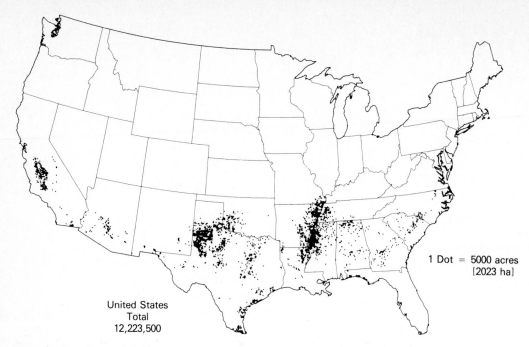

United States
Total
12,223,500

1 Dot = 5000 acres
[2023 ha]

Figure 16-1. Cotton production is primarily in the southern United States. The map indicates the distribution of cotton acreage by State.

established the first successful textile mill in the United States at Pawtucket, Rhode Island. The other was the invention of the cotton gin by Eli Whitney in 1793. Slater's mill showed the possibilities of the domestic market for cotton. Whitney's gin made possible the rapid separation of fiber and seed, thereby making greatly increased cotton production economically practicable.

The nineteenth century was marked by the steady expansion of American cotton production, interrupted only by the War Between the States. From a limited area along the Atlantic coast, cotton moved westward to the Mississippi River and on into the virgin lands of Arkansas, Louisiana, and Texas. During the twentieth century it has become a major crop in Oklahoma, southern Missouri, and the western states of Arizona, California, and New Mexico. In recent years, about four-fifths of the American cotton crop has been produced west of the Mississippi River.

CHAPTER GOALS

After studying this chapter you will be able to:

- Identify the areas of cotton production, both in the United States and worldwide.
- Describe the historical development of cotton.
- Describe the cotton plant and its various parts.
- Describe the types of cotton grown in the United States.
- Describe each of the practices used in the production of cotton.
- Identify the important cotton insects and diseases that affect cotton crops.
- Identify the methods used to harvest cotton.
- List the various products made of and uses of cotton.

THE COTTON PLANT

The cotton plant is *indeterminate* in its growth pattern, which means that both vegetative and reproductive growth can occur simultaneously. In tropical climates, the cotton plant is a *perennial,* that is, it will continue to grow year after year if permitted to do so; but in the temperate parts of the world, cotton acts as an *annual* plant. In the United States it is farmed as an annual.

The cotton plant has a prominent, erect main stem from which true leaves and branches arise. Two types of branches are usually produced—vegetative and fruiting. Few, if any, vegetative branches develop when plants are grown under crowded conditions, but several normally develop if plants are widely spaced and grown under conditions of adequate nitrogen and soil moisture. Up to three or four vegetative branches may be desirable, particularly for widely spaced plants, as they substantially increase the framework of the plant on which bolls may later develop.

Fruiting branches arise from both the main stem and vegetative branches. All the flowers and bolls produced on the plant originate from fruiting branches. Thus the main stem and vegetative branches must first branch to produce fruiting branches before flowers are produced by the plant. As the fruiting branch grows from the main stem or from the vegetative branch, it is terminated by a floral bud, or *square,* and a leaf (Figure 16-2).

The *ovary,* or young boll, develops rapidly after the ovules have been fertilized and reaches full size in about 21 days. When bolls mature, the carpel walls dry and split along the lines where the carpels meet. Under favorable conditions the boll opens rapidly, allowing the *locks* (tufts) of cotton to dry and fluff within a period of 3 to 4 days. The number of locks corresponds to the number of carpels. Bolls of most varieties of upland cotton will normally have four or five carpels each. The number of seeds per lock is usually about nine.

Each cotton fiber originates from an extension of a single cell on the outer layer of the seed coat. These cells begin to elongate about the time the flower opens. They continue to elongate as thin-walled tubular structures until maximum length is reached, usually about 18 to 20 days after blooming. The time required for maximum elongation is less for short-staple varieties than for long-staple varieties.

Vegetative growth and flowering usually decrease gradually and in the semideterminate cultivars practically cease after the

Figure 16-2. The fruiting process in cotton is illustrated by (*a*) the square, (*b*) a bloom, (*c*) a young boll, (*d*) a mature boll, and (*e*) an opened boll.

plants have become more or less heavily loaded with bolls late in the summer. Apparently the developing bolls have first claim to the plant's food supply, produced by photosynthesis and the nutrients absorbed by the root system.

The cotton plant usually has a comparatively deep root system. The depth of its penetration depends on the age and size of the plant and on the moisture, aeration, and structure of the soil. The primary root is a taproot which penetrates the soil rapidly and may reach a depth of 6 in [15 cm] or more by the time the cotyledons unfold. In the alluvial soils of the Mississippi valley, cotton roots have been traced to a depth of 6 ft [183 cm]. In certain irrigated soils of the Far West, roots have been observed at a depth of 3 to 5 ft [91 to 152 cm] by the time the plants were 8 to 10 in [20 to 25 cm] high and at a final depth of 10 ft [3 m] at maturity (Figure 16-3).

Numerous lateral roots grow outward from the taproot. They branch and rebranch to form a mat of roots which often extend outward from the base of a mature plant for several feet.

Cotton has a habit of shedding fruit, including squares, blooms, and bolls, which is in excess of the number of bolls it can finally mature. Shedding may be due to prolonged drought or excessive soil moisture; insects, disease, and other biological pests; an inadequate supply or an imbalance of nutrients; excessively high or low temperatures; strong winds; or even hereditary factors. The shedding will usually occur when the plant is under some type of stress from the conditions just mentioned.

SPECIES AND VARIETIES

There are three primary groups of cotton: (1) the upland varieties which vary in staple

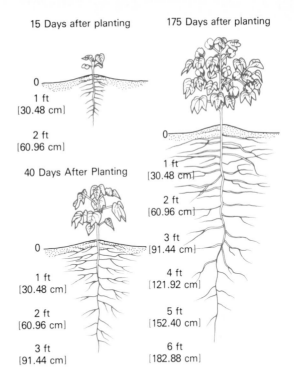

Figure 16-3. Taproot system of cotton showing top growth relative to root growth. Cotton has a taproot system with the length of root normally being approximately twice as long as the aboveground portion.

length (around 3/4 to 1 1/2 in [2 to 4 cm]); (2) a longer-staple cotton, commonly called *Egyptian* or *sea island* cotton; and (3) the shorter-staple cottons that are grown in southeastern Asia.

The upland and Egyptian cottons are the ones commonly grown in the United States. The short-staple varieties are usually classified as *Gossypium hirsutum*. The long-staple varieties are classified as *Gossypium barbadense*.

Cotton belongs to the genus *Gossypium* and is part of a family of plants called the Malvaceae. Several other common plants belong to the same family, including okra, hollyhocks, hibiscus, and althea; these plants are "cousins" of cotton.

The cultivars of upland cotton can be classified as long-staple or short-staple cottons, with a staple length of $1^1/_8$ in [3 cm] being the division between the long-staple and short-staple cottons. The long-staple upland cottons are normally grown in the Mississippi delta area as well as in California, Arizona, and New Mexico. The short-staple cottons are grown primarily in the northern portions of the cotton-producing areas such as Texas and Oklahoma. The Egyptian cottons are grown primarily in Arizona, New Mexico, and California and in the upper and lower Rio Grande valleys of Texas.

The Egyptian cotton is called *pima*. The pima cottons normally grow much taller and have longer, slender branches with leaves which generally resemble okra-plant leaves, in contrast to the upland varieties, which are shorter, with shorter branches and leaves more typical of cotton leaves. The flowers of pima cotton are yellow, while the upland cotton flowers are initially white, then pink, and subsequently red. The lint length for the pima cotton usually ranges from $1^1/_4$ to 2 in [3 to 5 cm], while the upland cottons range from $^3/_4$ to $1^1/_4$ in [2 to 3 cm] in length.

Considerable effort has been made by researchers and cotton breeders to improve cotton cultivars. They have attempted primarily to increase production of lint fiber and improve its quality. They also attempt to develop earlier-maturing varieties, more insect and disease resistance, and adaptation to mechanical harvesting.

An example of how cotton cultivars are developed for specific characteristics is the adaptation to mechanical harvesting. In areas where there is a relatively long growing season, cotton cultivars that set bolls early in the season and mature early are developed so that their characteristics are suitable for harvesting with spindle pickers.

This is the type of cultivar that is normally used in the Mississippi Delta (Figure 16-4). This can be contrasted to the types of cotton cultivars used in the Texas South Plains, for example, where the plants have been developed to be harvested with a stripper.

In this same area of stripper cotton, cultivars are developed which have storm resistance. Because of the strong winds that occasionally occur in that area, the cotton lint needs to be firmly attached to the boll, or production losses due to wind will be sustained. Cultivars used in stripper harvesting are normally considerably shorter. In the use of the spindle picker, the plants are disturbed as little as possible, so that additional bolls may mature and be picked at a later date, whereas in stripper harvesting, the harvest is conducted only once across the field, stripping the stalk of all the lint as well as the boll, leaves, and plant parts other than the stalk itself.

Many different cultivars have been developed by plant breeders in each major cotton-growing area which are peculiarly suited to that area. These varieties differ in maturity, staple length, fiber characteristics, and other plant characteristics.

PRODUCTION PRACTICES

Production practices for cotton vary widely, depending on areas of production. Cotton is grown principally in the southern United States, with climate and conditions ranging from the high-rainfall, humid areas of the South and Southeast, such as the Mississippi delta region of Arkansas, Louisiana, and Mississippi, to the arid and semiarid, low-rainfall environment of the Southwest in the states of Texas, New Mexico, and Arizona and in California. Much of the cotton in these areas is irrigated.

Figure 16-4. Mississippi delta cotton showing fiber in boll typically loosely held.

Tillage

Tillage practices start after the crop is harvested in the fall or winter and lead up to the preparation of the seedbed and planting of the next crop. Stalks are shredded and incorporated either by disking or chiseling. Early destruction of stalks may be desirable to aid in controlling insects. This is followed in some areas by breaking with a moldboard plow. A light disking or chiseling is used to destroy any weeds. In midwinter to early spring, the seedbed is prepared for planting. The bedding operation varies with soil type, physical condition, and soil moisture. Finer-textured soils (unlike sandy loams and sands) normally need to be bedded earlier so that a firm seedbed will be formed by planting time. Rainfall or the application of irrigation water will cause the seedbed to "settle" and become firm.

In some areas after the old cotton stalks are shredded, the middle of the old rows is plowed to reform the old beds with a plow called a *lister*. In some cases a lister plow is used to "bust-out" the middle of the old rows, and new beds are formed. This latter practice is usually followed in areas with fine-textured, heavy soils to allow plenty of time for the soil to "mellow" and for a firm seedbed to form. In case weeds become too numerous before planting time, the lister plow or other implement may be used to destroy the weeds. The seedbed normally is disturbed very little by this process.

In most areas, after the land has been initially tilled, subsequent tillage practices are kept to a minimum to cut moisture loss and save tillage costs.

Tillage operations often include an application of fertilizer or chemicals. These are discussed later in this chapter.

Seedbed Preparation

The shape of the seedbed depends on general climate conditions—amount of moisture, rainfall patterns, wind, and similar factors. In areas of high rainfall, listed beds are usually formed with seed to be planted on the top of the bed. If excessive moisture is not a problem after planting, the land may be planted flat without any beds being formed. In areas of low rainfall, where the top of the bed might tend to dry out, seed is often planted in the furrow. By being planted in the furrow, the young, emerging seedlings are somewhat protected from wind and sand damage. After planting, the beds may have a different shape or the land may be relatively level, with no bed structure.

The main point to remember on seedbed preparation and planting practices is that the planted seed should be in contact with moist soil and the row should be subjected to maximum moisture in the seed zone. In many cases it may be desirable to use a press-wheel or some attachment on the planter to press the soil firmly around the seed.

Planting and Stand Level

The date of planting is influenced by soil temperatures and soil moisture. In the Southwest, soil temperature at a 6 in [15 cm] depth should be around 60°F [15.6°C] as a 10-day average. In the Mississippi Delta, a soil temperature of about 68°F [20°C] at a 2 in [5 cm] depth is about right. When soil temperatures reach this level, the cottonseed is better able to withstand an attack by disease organisms than if planted in cold, damp soil. Often early plantings are less susceptible to insect damage, particularly weevil damage, but it may be more important to get good seedling vigor and to plant a little later at the appropriate soil temperature. Cotton planting begins in the lower Rio Grande valley of Texas in February and spreads northward and to the east and west as average temperatures increase.

The planting rate for cotton depends on anticipated weather and climatic conditions, varieties, soil conditions, size of seed or type of seed planted, and items of this nature. The amount of cottonseed normally planted varies by area. For most cotton-producing areas with a long growing season, planting rates are around 15 lb/acre [16.8 kg/ha] of delinted seed (seed which has had all the fiber removed) and 20 to 25 lb/acre [22.4 to 28 kg/ha] of nondelinted seed. For areas with a relatively short growing season, rates per acre are about 20 lb/acre [22.4 kg/ha] of delinted seed to 30 to 35 lb/acre [33.6 to 39.1 kg/ha] of seed which has not been delinted.

Cotton is usually planted at a much higher rate than would be needed if conditions were ideal. Since the cotton seedling is tender and soil conditions are often adverse, the seedling may not be able to emerge and withstand unfavorable climatic conditions during its early growth period. For example, a relatively wet spring in the Mississippi Delta might be conducive to a higher incidence of cotton seedling disease. Or in the Texas South Plains, winds and sand can cause a decreased stand level. Precision planting of cottonseed is becoming a practice in areas where weather and climatic conditions would normally have less of an effect on stand level.

Cotton is a plant which tends to compensate for stand level. When the plants are not too thickly spaced, the individual cotton plant tends to produce more fruiting stalks and more bolls, whereas if the cotton is planted more thickly this effect will not occur. The trend is toward higher plant population per acre. Research has generally shown that thicker stands produce higher yields.

Because the cotton seedling is highly susceptible to damage by disease, most cottonseed is treated with a chemical (usually a fungicide) before planting. This helps to improve the emergence of the plant, resulting in more uniform stands, and helps to prevent seedling diseases such as damping-off.

The depth at which cotton is planted is determined principally by soil moisture, since the cottonseed should be planted in soil which has adequate moisture. It is also determined by temperature and the seedling vigor of the particular cultivar and by whether the seed is delinted. The depth usually ranges from 1 to 3 in [2.5 to 7.5 cm], depending on the moisture level.

For row spacing, most cotton is planted in 38 or 40 in [97 to 102 cm] rows, but in some areas cotton has been planted in rows 20 to 30 in [50 to 75 cm] apart. In other areas it is planted in 10 in [25 cm] drill rows or even broadcast. The primary reason for planting in narrower rows or broadcasting is either to increase the plant population or to improve its spacing in order to take advantage of the additional light, moisture, and nutrients, as well as any weed control advantages. Closer planting may aid in weed control by shading more of the soil surface. Even though it has been shown that narrow rows usually result in a higher yield, most producers still plant in 38- to 40-in rows. The fact that machinery is adapted to plant, cultivate, irrigate, and harvest cotton in this conventional row spacing deters the trend toward different row spacing.

In areas where moisture is somewhat limited, it is common practice to plant two rows and leave one or two rows without any planting of cotton. This may be referred to as *skip-row* planting or as *two in—one out* or *two in—two out* spacing (Figure 16-5). The principal reason for this planting pattern is to permit the cotton plant to take advantage of the limited moisture. It also has the added advantage of providing additional light to the plants, since on one side there is no row to compete for the sunlight.

Pest Control

Weed Control. It is highly important that weeds be controlled for efficient cotton production. Weeds compete with cotton for

Figure 16-5. A "two in—one out" planting pattern for cotton.

sunlight, water, and nutrients, and this can result in decreased production. In addition, weeds may result in other costs, as when they impede the efficient harvest of the cotton and reduce the quality of the lint cotton. In addition, weeds often act as hosts for insects which are injurious to the cotton, decrease the efficient use of chemicals, and interfere with irrigation. Weeds in cotton can be controlled by mechanical cultivation, use of chemicals, and, in some areas, proper rotation of cotton with other crops (Figure 16-6).

Weed control starts prior to planting the cottonseed. The control of weeds in a crop planted one year will have an effect on the

(a)

(b)

Figure. 16-6. Fields of young cotton showing weed problems. (a) Volunteer corn. (b) Johnsongrass.

Figure 16-7. Cultivating young cotton to kill young weeds and break the crusted surface soil.

weed population the following year. One weed control method is to leave the land in a *fallow* state, i.e., a condition in which no crops are planted and either chemical or mechanical means are used to control weeds. This is a typical method for control of one of the worst weeds in cotton—johnsongrass. After seedbed preparation, the cultivator and rotary hoe are two of the principal mechanical means used to control weeds in cotton (Figure 16-7). The rotary hoe has become a popular tool for controlling weeds. To be most effective, it must be used as the weed seedlings are appearing at the soil surface. The use of the rotary hoe and other mechanical devices and chemical control have almost entirely eliminated the need for hand hoeing to control weeds.

Many of the tools used for controlling weeds, particularly the rotary hoe, will reduce the stand level of the cotton by a sig-

nificant percentage. This is one of the reasons that extra seed is often planted.

In addition to the use of mechanical means of control, herbicides for control of weeds are used on a large scale. The chemicals on the market today control most of the broadleaf weeds, as well as many of the grasses which compete with cotton. Some weeds are notable exceptions and cannot be controlled with chemicals—e.g., the weed commonly referred to as *nightshade* or *blueweed,* which persists in cotton in the Texas south plains. Herbicide rates and times of application vary by cotton-growing area. In the Southwest, for example, herbicides are usually applied as a preplanting application and incorporated into the top 1 to 2 in [2.5 to 5 cm]. In the delta, the herbicide is normally incorporated preplant as well as just prior to emergence of the cotton seedling (called *preemergence*) and also after the seedlings have emerged (called *post-emergence*). The time for the best application of chemicals will depend on the type of weed to be controlled and the particular conditions that exist in a field.

It is important that herbicides be used properly. For more information on the general handling of chemicals for herbicides as well as for fungicides, see Chapter 9, which discusses the use of chemicals for pest control.

Insect Control. Yield losses from insects have been estimated at about 15 to 20 percent per year. Insects can be controlled, but the producer must continuously battle the insect problem (Figure 16-8).

Types of insects that damage cotton vary from one area to another. Cotton growers must be thoroughly familiar with the various insect pests of cotton, their potential hazards, and the principles of control. Fields should be carefully checked at least once a week during the squaring, blooming, and fruiting periods to determine the prevailing infestation levels and their trends. Growers often employ a competent, well-trained, licensed person to perform this service for them. Promiscuous and excessive use of insecticides should be avoided.

The general philosophy of cotton insect control should be to use all methods of natural, biological, and cultural control and turn to insecticides when they are needed and are profitable to use. Cotton is no longer grown in some areas because of the excessive costs of controlling insects. Cultural practices can be effectively used to control insects. The following practices can be followed to reduce damage (although some may not be feasible in certain areas):

1. Destroy stub or volunteer cotton.
2. Plant recommended cultivars.
3. Try to maintain a proper stand level.

4. Use fertilizers properly.
5. Rotate cotton with other crops.
6. Irrigate when needed—but not excessively.
7. Harvest early and completely.
8. Destroy stalks early.

Control measures for specific insects should be obtained from the local agricultural extension advisor or commercial company representative.

Disease Control. Production losses of cotton from disease are estimated to be about 10 to 12 percent annually. The principal cotton diseases that contribute to these annual losses are seedling diseases, leaf blights, wilts, and boll rots. Most diseases can be controlled at costs that will be more than covered by increased yield.

With some diseases the losses are obvious, as in fusarium wilt or cotton root rot, where the plants are killed outright. With others, the losses are not as evident. For example, bacterial blight lesions on the leaves will cause a certain amount of defoliation, but later in the season plants may be relatively free of the disease. However, it has been demonstrated that in such cases defoliation alone may be responsible for up to 29 percent reduction in yield.

Usually a single control measure is not completely effective against any one disease. However, certain practices which contribute to the control of one disease usually help in the control of others. Practices which are beneficial in reducing losses, with maximum production of cotton, are as follows:

1. Plant disease-free seed.
2. Use disease-resistant cultivars.
3. Practice the rotation of cotton with other crops.
4. Destroy plant residues as early as possible.

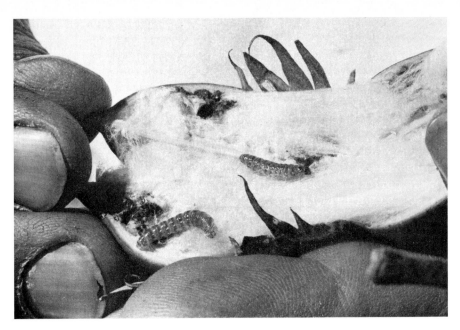

Figure 16-8. The pink bollworm is one of the most destructive insect pests for cotton. (*USDA*)

5. Use chemicals as recommended.
6. Fertilize and irrigate properly but not excessively.

Nematode Control. Nematodes are microscopic, wormlike organisms which attack cotton plant roots by using a swordlike stylet to pierce the root and feed on it. The root-knot nematode, one of the most prevalent, moves into the root, becomes stationary, and produces knots or galls on the roots preventing normal root development.

Symptoms of nematode damage include restricted growth of the root system; fewer large roots and coarse, stubby lateral roots; and, in the case of the root-knot nematode, the presence of galls. Above-ground symptoms include stunting, slower plant growth, variability of growth in different plants, wilting during the heat of the day, and reduced yields. Nematode damage is often accompanied by fusarium wilt. The injured root tips provide a means of entry for the fungus.

Nematodes can be controlled by cultural practices, some nematocides, resistant cultivars, and soil fumigation.

Fertilization

Cotton should be fertilized according to soil-test results. All states have soil-testing laboratories which are either state-managed or managed privately through industrial concerns. A cotton producer should test the soil at least once every 2 to 3 years and rely on the recommendations of those who have conducted the research in local areas and who have a knowledge of the local soil needs and growing conditions.

Another method of determining fertilizer need is through plant analysis. Programs have developed recently in which certain plant parts, specifically the petioles or the stems of the cotton leaf, are used for plant analysis and as a basis for applying additional amounts of nitrogen or other plant nutrients which may be needed to ensure better yields.

Even though plant analysis may be too late to provide information during the current crop season, it can often provide information as to whether there was an adequate fertilization program and such information may be of value as a guide for the following year.

Since the cotton plant has an indeterminate growth habit, the ideal fertilization program is to apply a type of fertilizer which will result in sufficient, but not excessive, vegetative growth. Sufficient phosphorus, potassium, and the other essential nutrients must be available to encourage proper reproductive growth and good set of bolls and their subsequent growth to maturity. Nitrogen, however, is the key element to vegetative growth and can be easily regulated during the season. In years when the temperature is relatively high, nitrogen levels should normally be somewhat higher than during a season when temperatures are lower than normal. Too much nitrogen during a cool growing season may result in too much vegetative growth and less boll set.

In areas where soil moisture is less than adequate and yield potential therefore less than normal, the nitrogen rate, as well as the rates of other nutrients, needs to be adjusted downward. If soil moisture is good and therefore yield potential is high, the rate should be adjusted to above normal. A good example of this is the Texas South Plains, where, if the soil profile has an available moisture level that is equivalent to 5 in [13 cm] of moisture, the cotton farmer is much more assured of a good cotton yield during the season than if the soil moisture is equivalent to less than 3 in [8 cm]. Because of this, many producers wait to apply nitrogen until after the stand has been established and after they can get some indication as to the type of season which appears to be developing. If the stand is established early and if it appears that the spring is warming up more rapidly than normal along with a better than average level of soil moisture, the producer might apply a rate of 40 to 50 lb/acre [45 to 56 kg/ha] of nitrogen, whereas if the conditions were less than optimum, a nitrogen rate of 20 to 30 lb/acre [22 to 34 kg/ha] might be used. In some seasons when moisture is minimal, no nitrogen should be used. In areas where the growing season is short, side-dressing of nitrogen should not be delayed too late in the season, since this could also delay the maturity of the cotton.

Irrigation

A significant percentage of the cotton produced in the United States receives supplemental water from irrigation. Even in some of the high-rainfall areas such as the Mississippi Delta, supplemental irrigation is used at critical times, such as the first weeks of blooming or when temperatures are high and dry periods occur.

The purpose of irrigation is to keep the soil supplied with readily available moisture for plant growth and development. The response to irrigation is influenced by the crop, the cultivar, the climate, and the soil on which the crop is grown. The benefit from irrigation varies from year to year because of differences in total rainfall and rainfall distribution.

Crop needs for water are not constant throughout the growing season. For cotton the needs steadily increase to a maximum in late summer and then decrease as the crop reaches maturity (Figure 16-9). Crop needs for water can be considered as the amount of water that is lost by transpiration through the plant and by evaporation from the soil.

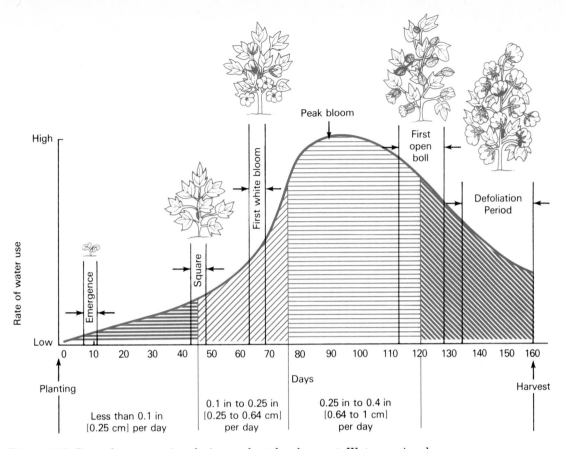

Figure 16-9. Rate of water use in relation to plant development. Water use is relatively low in the early days of plant growth to squaring, then increases rapidly with peak water use coming at peak bloom and then declining as the plant matures.

Fortunately, the soil acts as a buffer to climate. During periods of excessive rainfall, the soil can act as a reservoir and store water for use during those periods when crop requirements exceed rainfall.

Soil characteristics which affect intake, storage, availability, or removal of water have a direct influence on irrigation practices. Soils with low intake (infiltration) rates are not capable of absorbing much of the rainfall. Such soils are difficult to irrigate. Irrigation water must be applied at a low rate over a longer period of time, thereby increasing its cost. Sandy soils have a higher water-infiltration rate than silt or clay soils, but they hold less available water for use by plants.

The condition of the cotton plants forecasts the need for irrigation. After flowering begins, the plant should grow steadily but not luxuriantly. The squares at the top of the plants should be rather prominent. A few flowers should be readily observed among the top leaves. When only the leaves are noticeable, growth is too rapid. When only flowers are prominent and a flower-garden effect is noticed, vegetative growth has been restricted and irrigation is needed. Ordinarily, there should be 3 or 4 in [7.5 to 10 cm] of tender green stem between the terminal bud and the reddish coloring of the stalk. An extension of this reddish coloring toward the terminal bud also indicates a checking of growth and need of irrigation.

Careful observation will show a change in appearance of the plant before signs of wilting occur. The foliage will have a slightly bluish tinge, and in drier spots it will appear somewhat darker. These drier spots should be closely watched in the field, for they are the first to show a change in color. The producer should remember that if the beginning of irrigation is delayed, much of the acreage may be badly wilted before the irrigation is completed.

A soil tube or auger is useful for obtaining samples of the soil at various depths to determine the moisture condition of the soil at any time. It is a valuable aid in determining the need for irrigation and the depth of water penetration after irrigation.

HARVESTING

When the crop is sufficiently mature, it is desirable to initiate harvesting and attempt to harvest a large percentage of the crop with as high quality as possible. The time to begin harvesting will vary considerably in the various production areas of the nation. The crop's maturity, the acreage to be harvested, the number of pickers available, and the number of days of favorable weather that can be expected determine the time to begin picking cotton. In the Mississippi Delta, this may be when 65 to 75 percent of the cotton bolls are open and ready for picking, with the balance to be harvested at a later time. In other cotton-producing areas, such as parts of Texas, the harvesting procedure may consist of only a once-over operation called *stripping*.

Cotton-production practices prior to harvesting can influence the yield of cotton as well as the efficiency of the mechanical picker. Some of the accepted field preparation practices and conditions which will permit harvesting the highest quality cotton and preserving as much of the quality of the cotton as possible include the following: (1) drainage that leaves fields smooth, without potholes or poorly drained areas; (2) wide, smooth turning of rows; (3) a thick, uniform stand; (4) control of weeds and grasses; and (5) removal of any stumps, rocks, bricks, or old irrigation pipes or tubes. Clean fields not only improve the efficiency of the harvester but also improve the quality or grade of the cotton. The large mechanical cotton harvesters in use today operate with precision, and time can be lost if fields are not selected carefully and properly prepared for efficient harvesting (Figure 16-10).

In most cotton-producing areas it is desirable for the cotton to be defoliated prior to harvest. In some areas defoliation occurs naturally from frost, but in many areas it is desirable to apply a chemical to defoliate the cotton plant. Defoliation, or loss of the leaves, usually gives higher grades of cotton, because few leaves remain to add trash or to stain the fibers. It also helps lodged plants to straighten up, thus increasing plant exposure to sun and air and enabling the plant to dry more quickly and thoroughly and the mature bolls to open faster. It also reduces the population of damaging insects. Defoliation will permit earlier harvesting. It will usually pay to defoliate cotton unless an early frost does the job.

The best time to apply a defoliant varies with the cotton-producing area, but it should be remembered that the defoliant chemical normally kills the plant. Hence it should be applied only after judgment indicates that a large percentage of the mature bolls will open. It is normally better to apply it late in

Figure 16-10. A "rood" cotton harvester is designed especially to be used in fields where the initial harvesting operation was poor or where environmental conditions such as excess rainfall or wind caused cotton to be lost from the boll.

the season and ensure high-quality fiber as well as high yields. At least 75 percent of the bolls should be mature at the time of application.

A desiccant can also be used to defoliate cotton. The desiccant is a chemical which kills the leaves on contact; plants become dry as fast as weather permits. The leaves usually stay on the plant. The primary benefit of the desiccant is to reduce green leaf strains, but it usually also increases the trash content of the ginned lint.

A producer should remember that chemicals used for defoliation and desiccation are poisonous. All the precautions and safety recommendations given on the container label should be carefully followed. See Chapter 9 on the use of agricultural chemicals for more information.

The two principal types of cotton harvesters are the cotton picker and the cotton stripper (Figure 16-11). The five key parts of the cotton picker are the spindle, the moistener pad, the doffer, the stalk lifter, and the

pressure plate. The spindle is the most important part of the machine, because it is the one that will pick the mature cotton out of the bolls. In the cotton stripper, the teeth that strip the plant of all parts except the actual stalk and limbs are the key part.

An important key to an efficient cotton harvesting operation is the transporting of the harvested cotton to the gin. Until recent years, this has been done primarily by trailers which hold sufficient lint cotton to make approximately four to five 500-lb [227 kg] bales of cotton. Often the harvesting process is delayed by the unavailability of trailers to haul the cotton to the gin.

In recent years, cotton harvesting has been aided by the development of *module builders,* by means of which cotton can be stored in a rick in the field. The amount of cotton in a rick varies from the equivalent of about eight bales up to approximately sixteen bales. Using module builders and storing the cotton in a rick in the field, a producer can complete the harvest of the cotton

(a)

(b)

Figure 16-11. The two principle types of cotton harvesters. (a) Cotton picker. (b) Cotton stripper. (*American Textile Manufacturers' Institute, Inc.* [a]; *John Deere* [b])

Figure 16-12. (*a*) The cotton stripper unloads cotton into the module builder. (*b*) After compressing the cotton, the module is moved, leaving the rick of cotton. (*c*) A self-loading wagon or truck picks up the rick and transports it intact to the gin.

when it is ready and will therefore not be delayed by lack of equipment to haul the harvested cotton to the gin. The rick of cotton can then be picked up as one unit with an automatic loader, loaded onto a truck, and transported to the gin (Figure 16-12).

Two errors can affect modules. One is not taking sufficient time to build a solid module with a slightly rounded top. The second is harvesting the cotton too wet. The cotton's field moisture content should be 12 percent or less before harvesting is started.

COTTON QUALITY

Cotton quality depends primarily on staple length, grade, and *micronaire* (Figure 16-13).

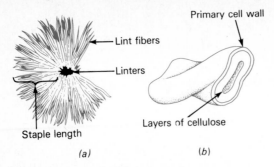

Figure 16-13. (*a*) Staple length of cottonseed (*b*) Micronaire is the strength or thickness of the cotton fiber.

These are the most important quality factors to the producer because they largely determine the price the cotton brings at the point of first sale. The three factors used to determine grade are color, leaf, and preparation. Color grades include white, light spotted, tinged, and similar classifications. The term *trashy* refers to leaves and foreign matter in the cotton. The term *preparation* is used to describe the degree of smoothness or roughness with which the cotton is ginned. Each of the three grade factors is taken into account in arriving at the grade of an individual sample. Staple length is measured in inches and fractions of an inch, generally in graduations of thirty-seconds of an inch. Staple length starts with less than $^{13}/_{16}$ and goes up to 2 in [2.1 to 5.1 cm]. Micronaire is an indication of fiber strength which is influenced by the fineness of the cotton fibers and is an indication of fiber maturity. Readings of 3.5 to 4.9 are desirable. Micronaire readings below and above those two values are discounted in the price received for the cotton.

UTILIZATION

Cotton is the principal crop used for clothing and other textiles. The cotton industry is one of the largest in the United States and one of the most important based on an agricultural commodity (Figure 16-14).

After cotton is harvested, the cotton gin separates the cotton fibers from the seed and other extraneous materials present. The fiber then moves into the market after being classified as to quality; the seed is transported to a processing plant; and the remainder, called *gin trash*, can be used as a source of organic matter to be applied on the soil or can be fed as roughage for livestock.

Fiber

Three principal operations are involved in cotton manufacturing. The fiber is processed into yarn and then into fabric. The cotton goods are then bleached, dyed, and finished. The third step occurs when the finished goods are manufactured into the end-use product. To make the wide range of cotton goods, the industry uses a wide variety of raw cottons. In a sense all cotton may look alike. However, there are differences in the degree of whiteness, amount and kind of trash, fiber fineness, staple length, cohesive ability, and other properties which greatly affect suitability for different types of end products.

Cotton has many uses in the final manufacturing state. In recent years cotton has been improved tremendously through research programs, to the point that cotton-fiber apparel is now very stylish and used in the finest of clothing as well as in home furnishings and many industrial products.

Seed

Where the stylish cotton fabrics and other uses leave off, the tough cottonseed is left for further processing and utilization. It is first delinted and then dehulled and the oil extracted, leaving a residue.

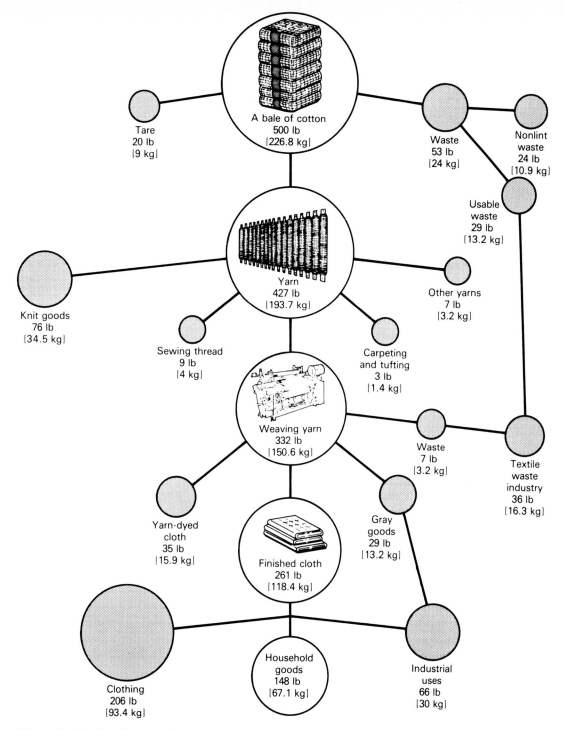

Figure 16-14. Distribution of an average bale of U.S. cotton. (*USDA*)

There is almost no waste in cottonseed processing. A mere 30 to 40 lb/ton [15 to 20 kg/t] is lost, most of which is dirt or trash which has been brought to the processing mill with the seed. Four basic marketable products, from which scores of by-products are derived, are produced from the versatile high-protein seed: oil, meal, linters, and hulls.

Oil manufactured from cottonseed is used to make salad and cooking oils, margarine, mayonnaise, shortening, and similar food items, as well as soap, glycerine, plastics, pesticides, and many other items. The greater part of cottonseed meal is marketed as protein, in the form of cottonseed cake, meal pellets, and similar products for livestock. Cottonseed offers a potentially rich and plentiful supply of protein for human consumption, provided an economic method is eventually developed for removing the toxic material (gossypol) which prevents its use for human consumption. Monogastric animals, such as humans, swine, and poultry, are unable to digest the substance. Processes have been developed in recent years by which this compound can be removed, providing an opportunity for the processing of the seed meal into human food products. When the process is further refined, protein from cottonseed meal could cost much less than an equivalent amount of protein from beef or other animals. Cottonseed meal is 35 to 45 percent protein, whereas soybean meal is around 45 percent protein. The hulls are used as livestock feed, as well as for soil conditioners, oil-drilling mud, pulp, synthetic rubber, and petroleum refining.

Another product coming from the cottonseed are the short fibers which cling to the seed, called *linters*. These linters are used in various products such as rayon, air hoses, industrial fabrics, films, smokeless powders, medical supplies, yarns, and many other types of products.

SUMMARY

Cotton is the principal source of fiber for clothing and other uses. It also provides human food and livestock feed. Cotton production and development have paralleled the development of humankind. The cotton plant grows both vegetatively and reproductively at the same time, producing both vegetative and fruiting branches. A cotton boll develops which contains the fiber and seed. The growth of the plant and production of the boll are quite sensitive to environmental conditions and management practices.

Producers need to be fully aware of the proper cultural practices to follow in a given area in order to maximize production. Harvesting needs to be properly timed to prevent harvest losses and deterioration in crop quality.

Cotton fiber is utilized in clothing as well as home furnishings and many industrial products. The seed provides linters, hulls, oil, and meal. The oil is used in food products such as margarine and salad dressings. The meal and the hulls are used as livestock feed, while the linters are used by industry in a variety of products.

QUESTIONS FOR REVIEW

1. What were the two events which resulted in the increased development of cotton in the United States?
2. What are the three primary groups of cottons and where are they grown?
3. What are possible reasons why one area (such as the southeastern United States) is no longer a major producer of cotton, while other areas have an increase in the acreage of cotton?
4. What are the reasons for the production of long-staple cotton in one area while another area produces primarily short-staple cotton?
5. What factors should be considered in determining the proper depth for planting cottonseed? What is the usual range of depth?
6. What are some practices that can be followed to reduce insect damage to cotton crops?
7. What are two conditions of cotton plants that indicate a need for irrigation?
8. What are five accepted field preparation practices and conditions which permit harvesting and preserving highest quality cotton?
9. What are the most important quality factors to a cotton producer?
10. What are the factors used to determine cotton grade?

Chapter
17

Fruits and Vegetables

Fruits, and to a lesser extent vegetables, have been a significant part of the earliest recorded history of humankind. There are references to fruits in Biblical accounts of the Creation, and wine—a product of grapes—is mentioned repeatedly in many early writings. Early travelers carried dried fruits, and most early explorations from Europe included a naturalist whose records frequently note exotic fruits in subtropical regions. In more recent times, the significance of citrus fruits as a source of vitamin C to prevent the development of scurvy was recognized by sailors.

In spite of extensive use by and obvious benefits to humans, fresh fruits and vegetables have often been viewed as a sort of luxury. Historically, this may in part be attributed to the fact that they were available only "in season," as fresh produce. Even today, fruits and vegetables are considered by many to be a luxury and are far more commonly included in the diets of people from more highly developed countries than in the diets of people of emerging nations. In many areas they are plentiful but seasonal, thus expensive.

Fruits and vegetables are valuable parts of a balanced diet. Green and yellow vegetables are rich sources of vitamin A, among other things. Citrus is an exceptional source of vitamin C; various types of nuts are sources of oils, and legumes are excellent protein sources. In addition, both fruits and vegetables are sources of roughage that aids in the total digestive process, and they add tasty diversity to the diet. Americans enjoy many fruits, but there are national differences in production, as you can see in Table 17-1.

The production of fruits and vegetables is big business. Although the average farm may be small, fruit and vegetable production requires significant capital investment, many risks, and exceptional management skills for profitable yields. Compared with many agronomic crops, such as wheat, most fruit and vegetable crops require far more critical management decisions. (This does not imply, of course, that rigorous management in producing cereal crops lacks importance.

CHAPTER GOALS

After studying this chapter you will be able to:

■ List the most important fruit and vegetable crops produced in the United States.

■ Identify the areas in which these valuable crops are produced and some of the factors that determine where they are grown.

■ Explain some of the basic production practices and understand the rigorous management required for successful production of fruit and vegetable crops.

TABLE 17-1 MAJOR AREAS OF PRODUCTION OF FRUITS AND NUTS, WORLDWIDE

Crop	Where Grown
Almond	United States, Spain, and Italy
Apple	United States, Italy, West Germany, France, Japan, and China
Apricot	United States, Spain, France, Italy, and Turkey
Cherry	West Germany, United States, Italy, France, and Yugoslavia
Chestnut	Italy, Spain, and Portugal
Filbert	Turkey, Italy, Spain, and United States
Peach	United States, Italy, France, Japan, Spain, Argentina, and Greece
Pear	Italy, China, United States, France, Japan, West Germany, and Spain
Pecan	United States and Mexico
Pistachio	Iran, Turkey, and Syria
Plum	Yugoslavia, United States, West Germany, France, and Italy
Walnut	United States, France, Italy, India, and Turkey

CLASSIFICATION OF FRUITS

Fruit crops can be classified in a multitude of ways. One common, convenient way is to generalize in terms of the environment in which the crop is commonly grown. This yields three groups of fruit crops: temperate crops, subtropical crops, and tropical crops. Within this broad grouping, fruits can be further classified according to the actual (botanical) type of fruit formed. Remember that a true fruit is a mature ovary, sometimes with stem or receptacle tissue included as part of the fruit. Seeds, the product of sexual fertilization or of apomictic processes, are formed in the fruit. The various types of fruits are pictured in Figure 17-1.

Temperate Crops

In the temperate regions of the world (Figure 17-2), pome fruits (such as the apple and pear) are produced, along with stone fruits (such as peaches and cherries) and a multitude of small fruits including grapes (which are discussed separately at the end of this chapter), and many berries. The pome fruits typically have a well-delineated core surrounded by fleshy fruit material, which is stem tissue. The core is the mature ovary.

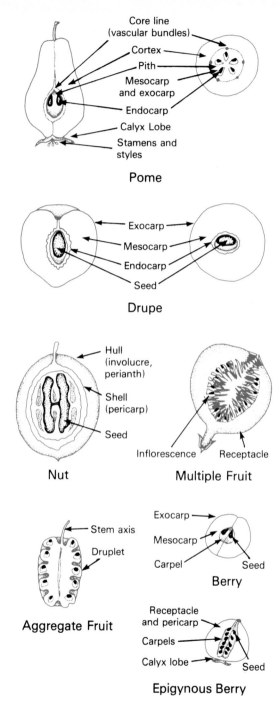

Core line
(vascular bundles)
Cortex
Pith
Mesocarp
and exocarp
Endocarp
Calyx Lobe
Stamens and
styles

Pome

Exocarp
Mesocarp
Endocarp
Seed

Drupe

Hull
(involucre,
perianth)
Shell
(pericarp)
Seed

Nut

Inflorescence   Receptacle

Multiple Fruit

Stem axis
Druplet

Aggregate Fruit

Exocarp
Mesocarp
Carpel
Seed

Berry

Receptacle
and pericarp
Carpels
Calyx lobe
Seed

Epigynous Berry

Figure 17-1. Examples of various types of fruit.

Stone fruits have the typical hard pit, the outer layer of which is the endoderm of the pericarp, and the fleshy material is the mesocarp. A true seed is found in the pit. Small fruits make up a confusing array of botanical fruit types. The grape is in fact a berry (so is the familiar tomato), and strawberries are in reality aggregate fruits composed of receptacle tissue with small achenes embedded in it.

Subtropical and Tropical Fruits

Florida, Texas, Arizona, California, and Hawaii produce the subtropical and tropical fruits grown in the United States. Citrus fruits are the most important, led by oranges and including grapefruits, limes, and tangerines. Avocados and olives are becoming more important, and dates and figs are significant. Pineapple production is limited to Hawaii and has declined. The leading tropical fruit is the banana. This group also includes papayas and some exotic species. The beverage crops, tea and coffee, may also be considered subtropical to tropical. Note that tea is grown for its leaf, not the fruit.

MAJOR FRUIT CROPS

Some of the major fruit crop species can be grown nearly throughout the continental United States in home gardens. However, commercial production of the major crops is concentrated in one or more specific areas. Temperature is a common factor that limits the production of fruit crops. Extremely low temperatures can kill trees even in the dormant stage. Citrus fruits can be devastated by temperatures only slightly below freezing. Although minimum temperatures are a concern, most fruit trees require winter chilling to produce flowers and, ultimately,

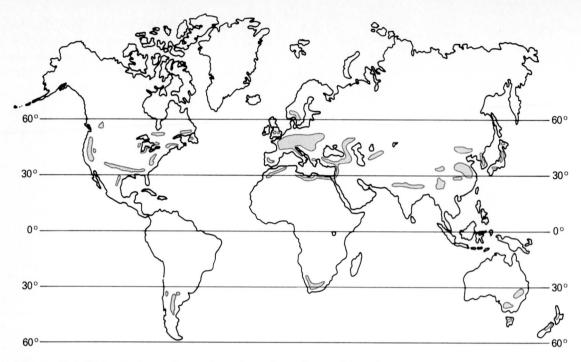

Figure 17-2. Major fruit growing regions throughout the world are in temperate regions (top) although altitude or proximity to large lakes or oceans may moderate temperature and allow fruit production in northern areas (bottom).

fruit. Thus tropiclike winter conditions limit the production of some fruits, such as peaches. Similarly, cool fall temperatures are needed to enhance fruit quality, such as the desired red coloring of certain varieties of apples.

Soil factors can also be a consideration, as can the prevalance of certain pests. Adequate moisture properly distributed through the growing season is a must, and the investment in orchards and value of the crop frequently justify irrigation.

In spite of recent advances in mechanization, the production of most fruit crops still requires a relatively large amount of hand labor. Thus labor supply can be important. Storage, transportation, and processing fa-

cilities are all of concern for most fruits, because they are very perishable and bulky to ship. Proximity to local markets can be a factor in deciding to produce fruits.

Fruit crop production is a detailed type of farming. It requires rigorous management at all stages of production and sizable investments of capital. Profits can be good, but weather and other factors make this a high-risk business.

Apples

Among the temperate crops, excluding grapes for now, apples are the leading fruit crop. Although commercial crops are produced in 35 states of the United States,

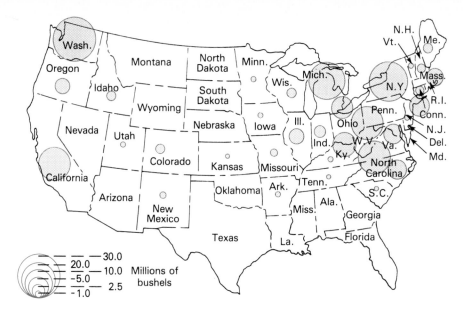

Figure 17-3. Major apple producing areas in the United States.

about 80 percent of the crop is produced in four states (Washington, New York, Michigan, and California) and in the Appalachian region from Pennsylvania to North Carolina (Figure 17-3). About half the annual crop is marketed as fresh fruit (although it may have been stored for several months), and half is processed. There are relatively few major cultivars: Delicious and Golden Delicious types account for about 60 percent of the annual crop. The northern limit of production is set by the minimum winter night temperature; apple trees do not tolerate temperatures below about −22 to −40°F [−30 to −40°C] even during winter months when trees are dormant. The southern limit is fixed by warm temperatures. During the winter, trees require from 1000 to 1500 hours of temperatures below 44°F [about 7°C] for normal reproductive growth. In addition, fruit coloring requires cool evening temperatures.

Pears

The annual pear crop is about one-third of the apple crop. The greater part of the crop, 95 percent, is produced in the three states of the Far West, California, Oregon, and Washington. Up to 60 percent of the crop is processed in some way.

Peaches

Peaches are the leading American stone fruit. California leads the nation's production, with over 60 percent of the crop, most of which are clingstone types grown for processing. South Carolina and Georgia are leading producers of freestone types for the fresh-fruit markets. The South Carolina crop is approximately one-third of the California crop. Like apples, the environmental limits for production are set by temperature. Peach trees require 650 to 1000 hours of chilling for normal development, although plant breeders are developing cultivars which may require as little as 250 hours of chilling, thereby moving production southward.

Plums

Plums rank second behind peaches in worldwide production. In the United States, approximately 90 percent of the crop is produced in California, and 80 percent of this crop is dried for prunes. Oregon, Washington, and Michigan produce significant amounts of plums.

Cherries

The cherry crop is about 25 percent of the peach crop. Two major types of cherries are recognized—sweet cherries and tart, or sour, cherries. Sweet cherries are produced in the same states as plums. In addition, there is a small band of production immediately adjacent to Flathead Lake in Montana. This represents a unique example of a local climatic situation. The heat-exchange capacity of the lake affects spring and fall temperatures to the extent that cherry trees thrive. Sour cherries are produced in the Great Lakes region, led by Michigan, as well as smaller plantings in New York. The development of suitable mechanical harvesters to replace diminishing and expensive hand labor has allowed the continued production and processing of cherries.

Citrus Fruits

The various types of citrus fruits are the most important subtropical and tropical fruits, and oranges are by far the most important type in the United States. Worldwide, citrus production only slightly exceeds banana production. Oranges are classified as common or navel and seedless or seeded. Florida and California are leading producers of oranges. Common types are grown in Florida and navel types in California. The overwhelming portion (over 90 percent) of the Florida crop is processed, while the California crop goes more to the fresh-fruit industry. Grapefruit is also important. Florida produces about three-quarters of the United States crop, followed by Texas, California, and Arizona. Lemons are produced mostly in California and Arizona.

Other Fruits

Avocados are limited to southern Florida and California. California produces the entire United States olive crop, 80 percent of which is canned. The same trees yield ripe (black) olives and green (immature) olives.

Other significant crops include bananas, cacao, tree nuts, and coffee.

MANAGEMENT OF FRUIT CROPS

Successful fruit culture requires intense management in all phases of production. Fruit trees are perennial; the establishment of an orchard is a long-term investment. In most instances, orchards yield little or nothing the first 2 to 4 years, yet land is occupied and trees must be managed along with the soil and potential problems arising from pests.

Management decisions require scientific, agricultural, and business skills. These decisions start with determining the type of tree to be planted. The grower must decide what species and cultivars are adapted to local soil and climatic conditions and what crops offer the best promise of returning a reasonable profit. The planting design of the orchard must consider the size of the trees at maturity, harvesting methods, the need to

Figure 17-4. Different methods are used to apply a variety of pesticides to orchards. (*USDA*)

move various types of equipment around the orchard, and possibly irrigation methods. Certain types of rootstocks for many crops yield dwarf trees. This may be desirable but must be included in the planting plan to achieve the best land use. With apples, dwarf trees bear fruit 2 to 3 years earlier than standard-size trees.

In many areas, provision must be made for frost protection with various types of heaters or fans. The orchard layout should include consideration of the pattern of air movement and cold-air drainage. Fertilizer practices are sometimes unusual compared with other crops. Most fruit trees require relatively large amounts of potassium and comparatively small amounts of phosphorus.

Compared with major field crops such as wheat, orchards are treated with many chemical pesticides (Figure 17-4). Extreme

care must be exercised when any pesticide is used. All safety precautions and label directions should be followed. Because of the large amount of hand labor involved in the production of many fruit crops and because much fruit is marketed fresh for direct consumption, extra care is needed to avoid toxic residues.

Fruit trees are frequently cross-pollinated. In designing and managing orchards, the proper pollinator trees must be included and insect pollinators must be available. Frequently this means keeping bees. Insect control programs must be designed to avoid harming essential insects.

Storage of the fresh crop and transportation and marketing are of major producer concern. Storage can be critical, depending on how the crop is to be utilized, and obviously, transportation is essential. Remember, fresh fruits are comparatively bulky and require special handling, which frequently includes refrigeration to minimize spoilage and retain the highest quality. In an energy-conscious world, these are major factors to consider. Cooling the crop and handling large quantities of fresh produce require large amounts of energy.

Pruning

Orchard crops, including grapes, require the same management considerations as other types of crops. In addition, unlike other crops, they require pruning. Pruning is far more than merely cutting a tree. Although all field labor need not be technically skilled, the manager who directs the pruning crew or work must have a basic understanding of tree growth, including how and where buds that yield fruit are formed and how pruning affects the yield, shape, and growth of trees. It is essential to understand that various trees differ in how, when, and why they should or must be pruned.

Pruning can be divided into two general operations: initial pruning and the seasonal pruning of producing orchards. Initial pruning is done during the first year or two of establishment of an orchard. One goal is to select the branch or branches that will become the main body of the tree. In addition, initial pruning is done to hold the top growth in balance with the young, frequently sparsely developed root system. The initial pruning must also serve to remove damaged, misshapen, or diseased parts of trees.

In many orchard crops, highest yields are obtained from trees that, after careful initial pruning, are not pruned for the next 3 to 5 years. Following this period a regular pruning schedule must be followed to sustain yields and protect trees. Pruning trees reduces the amount of top growth, encourages the development of floral buds that will ultimately yield fruit, removes weak or diseased branches, and serves to shape the tree for optimum growth (including access of all limbs to sunlight) and for ease of harvesting. Proper pruning demands understanding of the growth characteristics of the various types of fruit trees.

With apples and other pome fruits, fruit is formed from buds on spurs from wood that is 2 years old. Thus pruning is designed to encourage this type of growth (Figure 17-5). Excessive pruning reduces this wood (or short branches) and thus lowers yields for 1 or 2 years. Long, unfruitful spurs develop in response to excessive pruning. However, because the individual fruits are relatively heavy in crops like apples, compared with crabapples, these trees can be fairly heavily pruned even if the number of fruits is reduced, and the yield per tree will not suffer.

Sweet cherries require less pruning than any of the common fruit trees. The fruit is formed from buds that develop from short spurs; spur development is not stimulated by pruning, and improper pruning may remove too many potential fruits. The spurs are new growth; obviously, this growth forms on older branches. On older trees, trimming shoots to about 1 ft [30 cm] encourages branching and ultimately spur formation and fruit yield.

Peaches yield a relatively large fruit, and large size is encouraged by reducing the number of fruits formed. In fact, sometimes the crop may be thinned early during the season to encourage the production of large fruits. Little pruning of peach trees is required in the first 5 years after the initial training.

Four patterns of thinning are used by peach producers. The first is *corrective* —removal of diseased or broken limbs or pruning to establish or reestablish the desired shape of a tree. Second is *thinning,* a

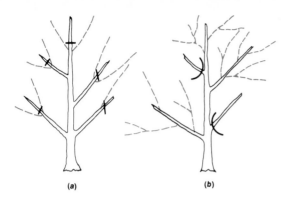

(a)   (b)

(c)

Figure 17-5. Pruning young apple trees is critical in determining their ultimate size and shape and affects their annual yield. The two major types of pruning cuts used in almost every type of pruning are (a) thinning, and (b) heading-back. (c) A woman in the process of pruning. (*Jane Hamilton-Merritt*[c])

process in which secondary and weak branches are removed and the tree is opened more. Third is *conventional thinning* and *heading back,* or trimming, of branches and shoots. The fourth is *severe pruning,* which means removing many mature branches and nearly restarting the tree. Severe pruning is done mostly in response to severe damage to trees from wind or other weather factors or insect or disease damage (Figure 17-6).

The pruning of citrus trees is designed to facilitate hand-harvesting. Virtually all pruning is to shape the tree; as a result, many trees take on a bushlike appearance. This pruning is made feasible by the fact that floral buds form on both new and older growth in citrus trees, thus either type of growth can be removed to shape the tree.

Harvesting

Fruit production requires relatively large amounts of hand labor for pruning crews and especially at harvest. For some crops, mechanical harvesting has become a reality (Figure 17-7). For cherries and other crops mechanical shakers have been developed, and fruits are either shaken to the ground and then picked up with a scoop or vacuum device or shaken into some type of cart or sled which is moved from tree to tree. This device is designed to minimize damage as the fruit falls, but there is still a significant loss. Obviously, the intended market affects how the fruit is harvested—fruit for the fresh-fruit market is still handpicked. Fruit to be processed into juices or other products may tolerate the damage associated with mechanical harvesting. Trees must be protected during harvest. Both human pickers and machines can damage trees and reduce yields in subsequent years. Although the giant apple tree may look sturdy, in terms of ability to produce high-quality fruit, it is delicate, like all other plants.

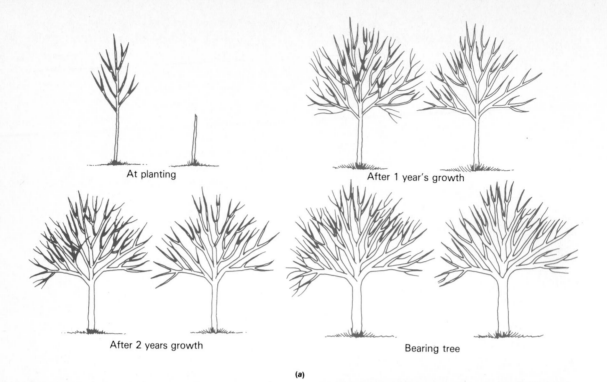

At planting

After 1 year's growth

After 2 years growth

Bearing tree

(a)

(b)

Figure 17-6. (a) Different patterns of pruning peach trees. (b) Mature, properly pruned peach orchard.

Figure 17-7. (*a*) Fruit harvesting requires large amounts of hand labor as shown by this laborer picking grapefruit in the Rio Grande Valley. Mechanical fruit harvesters have been developed for some crops such as (*b*) blueberries and (*c*) cherries. (*Texas State Department* [*a*], *USDA* [*b*], *USDA* [*c*])

The grower must know not only the proposed use of the crop but the required degree of ripeness. For this purpose, two concepts are clarified in the following section.

Maturity vs. Ripeness

For both fruit and vegetable production, the grower must know when the crop is ready to harvest. With field crops like wheat, the crop is said to be *ripe* when it is dry. With fruits and vegetables, first the fruit is *mature,* meaning that it is fully developed and will start to ripen. What constitutes a ripe crop of fruit depends on a variety of factors; a major one is the intended use of the fruit. For example, for the fresh-fruit market, a table tomato is ripe when it is fully red; there is little value in green tomatoes for this purpose, although they will "ripen" off the vine. At the opposite extreme, consider the banana. For consumption, a fresh, bright yellow fruit is wanted. Everyone knows how easily fresh bananas spoil. A banana will ripen easily off the tree. Thus, to the producer who ships bananas, the fruit is "ripe" when it is green. To the final consumer, it must be yellow. Citrus fruit, on the other hand, is more difficult to ripen off the tree, but for processed drinks this is not as critical.

VEGETABLES

Vegetables are perhaps the most diverse and widely grown crops in the United States, if not throughout the world. Home vegetable gardens have been a source of pleasure and pride to both city dwellers and country families for many years (Figure 17-8). Associated with the ever-popular vegetable garden are

Figure 17-8. Home vegetable gardens have been a source of pride and pleasure to both city and country families. (*National Garden Bureau, Inc.*)

pride in a source of fresh vegetables in season and the ever-popular home preserves. In spite of the larger number of vegetable gardens throughout the United States, home-grown vegetables are of minor economic significance, although commercial vegetable production is an important, complex element of our agricultural economy.

Commercial vegetable production can be traced virtually to the founding of the nation. Initially it involved trading produce for other goods and/or services and the sale of excess fresh produce from the home garden. From this sprang the local peddler who produced fresh vegetables in season for a local market—a town or village. Fresh vegetables were among the first and most common items for sale or trade in the traditional farmers' markets. These markets are experiencing something of a rebirth today, with a new twist—the roadside fresh fruit and vegetable stand that attracts both local citizens and tourists in many fruit- and vegetable-producing regions. Also, in some areas, the traditional farmers' market has become active again.

As transportation grew, the local peddler became the truck farmer who hauled fresh produce to more distant markets by truck. With the advent of refrigerated trucks and freight cars and field cooling and chilling facilities, the truck farmer serves markets from coast to coast: lettuce harvested in California's Salinas Valley on Monday is served in New York by the following Thursday.

Vegetable production can be described as either for the fresh market or for processing—canning or freezing. For the past 20 years, fresh-vegetable consumption has remained fairly constant, as have centers of production. In terms of percentage of crop acreage harvested, California leads the nation with over 30 percent, followed by Florida, Texas, Michigan, and New York. In terms of amount of production, again California is the leader with over 40 percent, followed by Florida, Texas, Arizona, and New York. For value, California leads with about 44 percent, again followed by Florida, Texas, Arizona, and New York. Note that Arizona is among the five leading states in percent of production and value but not acreage. This reflects the high yields of high-value crops.

Wisconsin leads the nation in acres of vegetables for processing, with over 20 percent, followed by California, Minnesota, Ohio, and Illinois. Wisconsin produces extensive acreages of peas and sweet corn for processing. California leads the nation in terms of percent of total production and value of vegetables produced for processing, with about 48 and 40 percent, respectively. Other leading states in production are, in order, Wisconsin, Ohio, Minnesota, and Oregon. For value, the leading states are Wisconsin, Oregon, Ohio, and Minnesota.

VEGETABLE PRODUCTION

Vegetable production is highly specialized, with specific practices for nearly every crop. Compared to most field crops, vegetables are grown on relatively small acreages, although a few large producers may account for a great portion of any crop nationwide. Commonly, vegetable producers are highly specialized and have specialized equipment to meet the needs of relatively few crops. Leasing land for the production of vegetable crops and producing contract crops are more common for vegetables than for other crops. Production is intense, and competition for markets is keen. Producers face the same challenges of seedbed preparation, seeding, pest management, fertilization, irrigation, and harvesting as other farmers. In some instances, such as tomato production for table

Fruits and Vegetables

or slicing tomatoes, fields are established by transplanting seedlings that have been started in greenhouses. At one time Ohio produced an early tomato crop entirely in greenhouses. With increasing fuel costs and demands for fuel economy, the cost of heating greenhouses for this apparent luxury is in question.

Shipping and processing are also major concerns to vegetable producers. Commonly, producers must schedule harvest dates to provide processors with a steady supply of raw materials for the processing facility. To do this the producer must know what quality of the produce is demanded by the processor, as well as the factors that influence the rate at which a crop matures. It may be necessary to stagger planting dates in the spring to satisfy the schedule at the processing factory in the fall. With each planting date, different management practices may be required for pest control, irrigation, and labor scheduling. All in all, production is complicated, requires relatively large amounts of labor, and is intense in terms or producing two or more crops on a field in a year.

MAJOR VEGETABLE CROPS

Vegetables are a diverse group of crops. For home-gardening purposes, most vegetables can be produced throughout the continental United States. Some vegetables are popular on a geographic basis, such as okra in the South and artichokes mainly in California. In part this is the result of the specific environmental conditions required to produce the crop. The central coastal region of California near Monterey, with its sandy soils and foggy but moderate climate, has the only significant artichoke production in the United States. In other instances, the crop may be popular by tradition or culture.

Commercial production of the major vegetable crops is confined to specific areas where the climate favors high yields. Usually temperature and length of growing season are major concerns. Because of the value of vegetable crops, intense management is practical, including irrigation. Sometimes rainfall during the growing season is not desirable because it may damage the crop in the field. Growers find it easier to provide all needed moisture through irrigation.

Vegetable crop production is highly dependent on labor. Also, access to processing facilities and transportation are major expenses in producing and marketing crops.

Potatoes

The common, or Irish, potato is the leading vegetable crop, in tons produced, in the United States. Sometimes potatoes are considered agronomic or field crops, an arbitrary distinction. This species is apparently native to the new world, but it was a disease that virtually wiped out the crop in Europe in the nineteenth century that led to the infamous Irish potato famine and illustrated the danger of nation's depending too heavily on a single crop for food. This significantly affected patterns of immigration to the United States and the culture of our nation today.

The United States potato crop averages about 13 million lb [5,889,000 kg] annually. Nearly 6 million lb [2,718,000 kg] are consumed fresh, although many have been stored. Over 7 million lb [3,182,000 kg] of potatoes are processed. Processed potatoes are among the most popular products; they range from chips and fries to instant mashed potatoes and potato flour.

Potatoes are grown in commercial quantities in 40 states, with Idaho and Maine the leaders (Figure 17-9). The potato plant is a

(a)

(b)

Figure 17-9. (a) Freshly dug potatoes lay on the ground on this Maine farm where they will be placed in barrels for shipment to the processing plant. (b) Potato harvester on west Texas high plains. (USDA [a], Burnett [b])

moderately tall annual member of the night-shade family. Production of the crop is unusual in that it is grown from vegetative parts, not seed; growers sow pieces of potato from which new plants grow and produce tubers, or potatoes. The tuber is the enlarged tip of an underground stem. Tuber formation is favored by relatively warm days and cool nights. As a result, some areas produce fall or winter crops and others spring crops. Idaho, Washington, and Maine produce a fall crop which is generally the largest.

Potatoes can be stored for prolonged periods of time under appropriate conditions. To prevent sprouting, they may be sprayed with maleic hydrazide. Tubers must be stored at low temperatures (but protected from freezing) and at relatively high humidity initially. Underground, or partially underground, cellars for storage, to provide natural insulation, are still a common sight in some potato-producing areas.

Tomatoes

Of the salad vegetables, tomatoes are the leading crop, with production above 8 million lb [3,624,000 kg] annually. The greater part of this is processed into juice, paste, and packed tomatoes; a small amount is consumed fresh.

California produces about 75 percent of the more than 6.6 million lb [2,989,800 kg] of processing tomatoes grown annually. This production is based on the development of efficient mechanical harvesters that replaced diminishing supplies of hand labor and the development of new cultivars of tomatoes that are adapted to mechanization (Figure 17-10). These new cultivars can be seeded directly, as opposed to transplanting. They mature uniformly, so they can be harvested at one time. (Older cultivars produced fruit throughout the season. Since the mechanical harvester destroys the plant, these cul-

Figure 17-10. A mechanical tomato harvester, such as the one shown, saves a great deal of labor, but plants with tough fruit that mature uniformly are essential. (*R. W. Reed/USDA-Soil Conservation Service*)

tivars were not suitable for processing because of excessive overripe and green fruit.) Finally, the new cultivars are firm and tolerate the rougher handling inherent in mechanical harvesting.

Florida and California each produce about one-third of the 950,000 lb [430,350 kg] of fresh tomatoes grown annually. Arizona, New Mexico, and Texas also produce some. In many areas, tomatoes are an extremely popular home-garden crop. Plants are not markedly frost-tolerant and seem to thrive best when started indoors and transplanted to the garden when the danger of frost has passed. Unfortunately, plants are also sensitive to early fall frost, which disrupts normal maturation. Fruit will mature off the vine, but many people feel that the quality of fruit so ripened is inferior to that of vine-ripened fruit.

Sweet Corn

About 75 percent of the sweet corn produced in the United States is processed; the other 25 percent is consumed as a fresh product. About 70 percent of the processed sweet corn is canned and 30 percent frozen, although the frozen food industry is becoming increasingly important. Sweet corn is commonly considered a fresh summer crop; indeed, commercial sweet corn as a summer crop is produced in some 20 states, led by New York, Ohio, Pennsylvania, and Michigan. A winter crop is produced in Florida and is in a sense a luxury fresh product. Spring crops are harvested in Florida, California, Texas, and Alabama, and a fall crop comes from Florida and California. Corn for processing is grown in the northern states, led by Wisconsin and Minnesota, which jointly account for about half the annual processing crop. Florida produces about 40 percent of all the fresh market corn.

The environmental requirements for sweet corn are comparable to those for field corn. Seeding rates are lower, and planting dates may be staggered to provide fresh produce to processing facilities for the maximum possible period.

Growers must estimate maturity for different cultivars on the basis of cultivar characteristics and the grower's knowledge of local climatic conditions. The concepts of degree days and photothermal units are important in estimating when a crop will reach an appropriate level of ripeness to harvest for processing (or optimum conditions for premium fresh produce). Excessive irrigation or heavy applications of nitrogen fertilizer may delay maturity, just as unexpected cool weather will.

Because the crop is "ripe" before the grain is fully dried, hand labor is frequently an important element in harvesting. Also, pest management in sweet corn is critical, especially for the crop that is marketed directly through the local grocery store or chain. Pesticide residues must be avoided, but the presence of insects or evidence of their presence can drastically lower the actual value of a crop.

Lettuce

Of all of the salad crops, lettuce is the leader. Unlike most other crops, all the lettuce produced is marketed as a fresh crop. Production is highly concentrated; 75 percent of the United States crop comes from the Salinas Valley of California, where sandy soils, fog, and moderate temperatures favor year-round production.

In spite of dramatic steps in mechanization, lettuce production, in terms of weed control, thinning, and particularly harvesting, depends heavily on hand labor. In addition, for nationwide marketing, the lettuce

Figure 17-11. Lettuce as it is grown and packed in the field. Boxes are chilled in portable refrigerators prior to shipping. (*BUD of California*)

industry depends on special refrigerated freight trains that receive priority scheduling on cross-country runs and on the availability of facilities to cool the fresh produce as it comes from the field and is packed for shipping (Figure 17-11).

Producers face the normal production problems of seedbed preparation, which includes forming beds, preirrigation, and possibly management of salty soils, thinning seedlings, and pest management. Lettuce growers must also be skilled in labor management and in transportation technology and management. The lettuce industry is a highly specialized and localized industry. In addition to the heavy production concentration in California, about 15 percent of the nation's crop is produced in Arizona.

Onions and Cabbage

Onions, a bulb crop, rank as the fourth leading vegetable (excluding potatoes), and cabbage, classified as a cole crop, ranks fifth. Onions are generally marketed fresh, or as dried onions. About 80 percent of the cabbage crop is marketed fresh and 20 percent is processed, a large portion as kraut.

Other Crops

The sixth through tenth leading vegetable crops are, in order, carrots, cucumbers, snap beans, celery, and green peas. Carrots are a biennial root crop, but they are farmed as an annual. Over 60 percent of the crop is consumed by the fresh market; the remainder is processed. California produces almost two-thirds of the United States crop, Texas and Michigan about 18 and 8 percent, respectively. Carrot production has become highly mechanized, although carrots are still popular items in home gardens. The roots can be stored for prolonged periods under suitable conditions, i.e., temperature at freezing and relative humidity from 90 to 95 percent. Carrots are produced nearly year-round in the major areas of production. Like most root and tuber crops, sandy-textured soils are favored for ease of harvesting.

Cucumbers are an important crop. About 28 percent are devoted to the fresh market, with the other 72 percent being processed, mainly as pickles. Pickle consumption jumped about 50 percent in the 15-year period from 1960 to 1974; the amount of cucumbers pickled increased proportionately. Florida produces nearly 40 percent of the fresh crop; California, Ohio, and Wisconsin produce cucumbers for processing. Although cucumbers need cool temperatures for optimum fruit development and a temperature of about 55.4°F [13°C] for germination, they do not tolerate freezing. Also, plants are cross-pollinated, and fruit yields may be low in the absence of appropriate pollinators (bees). This can be a problem in the small plantings that are typical of home gardens.

Snap beans, celery, and green peas rank eighth, ninth, and tenth in importance as vegetables. Both beans and peas are members of the legume family and are valuable sources of protein.

Celery is a popular relish and appetizer. It is the second leading salad crop in the United States. Like carrots and cabbage, the plants are biennials but are farmed as annuals for the vegetative stalks. California produces about two-thirds of the nation's crop and Florida about one-quarter. Other states producing small commercial crops are Michigan, New York, Ohio, and Washing-

ton. Fresh celery is available year-round. Very little celery is processed.

There are multitudes of other vegetable crops of local importance, and some are becoming of national significance. In addition to various greens (e.g., mustard greens), there are artichokes produced exclusively in California, broccoli, cauliflower, and Brussels sprouts, all of which are seen nationwide, and crops like okra, more typical of the South.

GRAPES

Grape culture may be older than the recorded history of humankind. Most certainly grapes were a significant crop in pre-Christian Greece and Rome. The grape, which may be classified as a small fruit, is consumed in one of three forms: in season as a fresh fruit, dried as raisins, and fermented as wine. By tonnage, grapes are the leading fruit crop in the United States, followed closely by apples.

Four types of grapes are grown in the United States: (1) *Vitis vinifera,* those cultivars with direct European ancestry; (2) American bunch grapes, *V. labrusca;* (3) muscadine grapes, *V. rotundifolia;* and (4) French hybrids of *V. vinifera* with American wild species.

Grape production requires rather special, temperate climatic conditions (Figure 17-12). The *cool* nights and warm, dry summer days and mild winters, along with fertile soils, make the Napa valley north of San Francisco an ideal and world-famous grape-producing area. Most grapes from this area are used for wine. Other similar areas inland and south of San Francisco are also recognized for superior grapes. Grapes produced in California are of the *V. vinifera* type. These include Tokay grapes, which

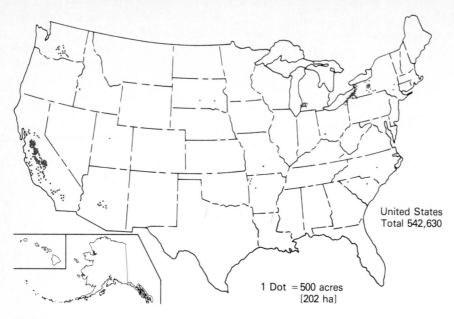

United States
Total 542,630

1 Dot = 500 acres
[202 ha]

Figure 17-12. Grape-producing regions in the United States.

are red and seedless; the Thompson seedless, perhaps the most common fresh table grape, which is light green or white and seedless (also a major source of raisins); and Emperor grapes, which are large, rose to purple, and a popular table grape. In addition, many cultivars of wine grapes are classified as *V. vinifera:* zinfandel is a well-known red grape yielding a premium red varietal wine in vintage years. About half the grape crop of the United States is used in making wines; in recent years, wine has become increasingly popular, and acreages of *V. vinifera* cultivars have increased. Plant breeders are working to expand the range of environmental conditions in which *V. vinifera* cultivars can be grown successfully.

The Great Lakes tend to moderate the climate of the land around them. In the eastern United States, major grape-producing regions are confined largely to the south shore of Lake Erie in Ohio, Pennsylvania, and New York. Generally, *V. labrusca* cultivars are produced in this area. They also are grown in Washington and Michigan.

Vineyard management is technical and complex. When grapes are grown for wine, management becomes in part an art to determine the precise stage of ripeness that will yield the desired wine. Establishing a vineyard involves the same considerations as establishing an orchard, including choice of rootstock. Pruning and training vines is critical, and hand labor for premium fruit is a must, although mechanization is coming to the grape harvest. Climatic conditions greatly affect the quality of grapes, hence that of the wine produced from them. As a result, the vineyard manager must as much as possible ensure consistent conditions year after year. Of course, weather cannot be controlled, but fertilizer practices, irrigation, pest management, and frost protection are all critical elements in controlling fruit quality.

SUMMARY

Compared to most field crops, fruit and vegetable crops may be grown on a much smaller scale. This does not mean they are unimportant. These crops are important in the human diet as a source of essential vitamins, minerals, and energy, and they add desired, if not needed, diversity to the diet. In the United States, fruit and vegetable crops are big business. Production and marketing require rigorous management of many diverse factors. Understanding basic principles is essential to effective management of fruit and vegetable production. This means growing, harvesting, storing, processing, and marketing the crop and, in perennial crops (fruit trees), tending plants, including pruning during nonbearing periods.

QUESTIONS FOR REVIEW

1. What do vegetables and fruits provide to the diet? Give examples.
2. Name the major fruit crops in the United States and describe where they are grown. Do the same thing for your state.
3. With examples, name four factors that limit where various kinds of fruit are grown and tell how these factors are managed by fruit producers.
4. Why is extra caution needed in applying any pesticide to a fruit or vegetable crop?
5. Why are trees pruned?
6. What factors must be considered in developing a pruning plan for an orchard?
7. What is the difference between a mature and a ripe fruit?
8. For what two markets are vegetables grown? How does this affect how and where they are produced?
9. What are the leading vegetables and vegetable-producing states? What limits where the major vegetable crops are grown?
10. What are the similarities and differences between vegetable production and marketing practices and practices for other crops?
11. What is unique about the production of potatoes?
12. Briefly describe the culture of grapes.
13. Describe the environmental or climatic conditions which favor grape production.
14. Where are the leading grape-producing areas in the United States? Be specific; use more than the name of a state, and explain why these areas are important.

Chapter
18

Forages

Forage is any plant material consumed by livestock. Among the most common forage crops are grasses and legumes grown primarily for their vegetation. They play an important role in the production of food. Livestock consume forage and convert it to meat, dairy products, poultry products, and wool which are of value to people.

Forage and livestock go together (Figure 18-1). Each is necessary for the other to be profitable. The most profitable use of forage occurs when livestock and forage are properly balanced with the right combination of management practices (Figure 18-2).

Over 11 billion acres [4.5 billion ha] of land are used worldwide in agricultural production—over 3.5 billion acres [1.4 billion ha] are cropped and 7.5 billion [3 billion ha] are in grasses and legumes. The United States, with about 2 billion acres [810 million ha] of land, has about one-fourth, almost 500 million acres [202.5 million ha], in cropland. Approximately one-half, or 1 billion acres [405 million ha], is in pasture and grazing land.

Introduced forage crops (those that are not native to an area) are grown throughout the United States but primarily in the eastern part of the country. The forages grown in the western United States (primarily west of the 100th meridian) are predominantly native species; these are discussed in Chapter 20.

Approximately one-half of the nutrients needed by livestock is provided by forage. These nutrients are derived from a diet for livestock estimated to average about 25 percent grain and 75 percent forage.

Forages are versatile crops. They can be harvested directly by grazing livestock or they can be harvested mechanically. Forage can be processed into hay by cutting it, letting it dry, and then storing it either in loose form or in some type of compacted bale or pellet. Forage which has been cut, chopped, and hauled directly to cattle for consumption while still in a green state is called *soilage*. Forage also can be stored for later consumption in a form called *silage*.

The use of forage crops is an important part of a soil management program and certain farming systems. Forage is important in maintaining soil productivity. Land that is unsuitable or poorly suited for crop production can be used effectively for forage production.

Forage is effective in decreasing erosion by providing a cover for the soil, so that movement of the soil by wind and water is kept at a minimum. The ground cover increases water infiltration, thereby holding more of the rainfall for use by the crop and also decreasing soil erosion. Forage also increases soil organic matter and thereby improves the physical condition of the soil.

Legumes add extra nitrogen to soil because of their unique ability to work "cooperatively" with bacteria in a process called *symbiotic nitrogen fixation*—"symbiotic" because the bacteria and the legume plant work together, and "fixation" because gaseous nitrogen is taken from the air and is fixed in a form which plants can use.

CHAPTER GOALS

After studying this chapter you will be able to:

■ Identify the factors that influence successful forage production.
■ Identify the types of forages and their uses.
■ Identify the methods used to harvest forage crops.
■ Explain how forage production is integrated with the production of livestock.

SELECTING A FORAGE

Selecting the correct forage is one of the keys to successful forage production and utilization. Several items need to be considered:

1. *Ability to Produce High Yields*   Production inputs such as fertilizer, weed control, insect control, and supplemental irrigation are several of the key factors that determine whether high yields can be obtained.
2. *Quality*   This can be described in terms of leafiness (leaf/stem ratio), palatability, and chemical composition.

3. *Resistance to Insects and Diseases*   Plant breeders are continuing to develop resistant varieties.
4. *Distribution of Production during the Year*   It is impossible in most of the United States to grow a single species of forage year round, hence the producer has to select a combination of species to provide nearly uniform grazing throughout the season. This may consist of two or more species grown together or several individual forage species grown in separate pastures.
5. *Compatibility of Species*   This is important in forage mixtures if it is necessary to assure a continual source of grazing

Figure 18-1. Forages are grown on all types of land and are consumed by livestock and converted to meat or dairy products. (*USDA*)

for livestock. Seldom will two or more species be completely compatible, but certain combinations offer a high degree of compatibility.

6. *Ease of Establishment* Early and rapid establishment will usually result in earlier production. Some grasses are established primarily by use of seeds, while some are established by the use of vegetative parts called *stolons* and *rhizomes*. Most legumes are established by using seed.

7. *Seed Production* This is important when a forage spreads by seed or when seed is to be harvested. The forage should be able to produce seed abundantly.

8. *Tolerance of Environmental Conditions* A forage may need to have cold tolerance, called *winter-hardiness,* in the northern United States or heat tolerance in the southern United States. Where it is desirable to plant wet-boggy areas to forage, tolerance of these conditions is desirable. Tolerance to conditions will vary by species.

9. *Longevity* Some forages are annuals and must be reseeded each year. If a forage crop is to be grown for several years, a species with appropriate longevity should be selected.

10. *Freedom from Toxicity* Some forages can be poisonous to livestock, but fortunately, very few are in this category. Prussic acid may accumulate in some species, especially sorghum, johnsongrass, and sudangrass when the plants are under stress from lack of moisture or recent frost. Nitrate can accumulate to undesirable levels in some of the grasses if unbalanced fertilization programs are used. Nitrates in forage can accumulate to levels high enough to reduce the rate of gain and, in a few instances, result in the death of animals.

TYPES OF FORAGE

There are many types of forage plants. The number of grass species in the world has been estimated to exceed 5000 and the number of legume species to exceed 10,000. Nearly all species are useful as forage for livestock consumption.

Grasses and legumes are two general categories of forage. Grasses belong to the family Gramineae and legumes to the family Leguminosae. The two families have several different characteristics, but both have one point in common—they provide feed for livestock and poultry.

Differing Characteristics

Characteristic differences occur in the germinating plant, leaves, flowers, root system, method of pollination, and nitrogen fixation. Figure 18-3 summarizes these differences between grasses and legumes.

The *cotyledons* (seed leaves) of most forage legumes are pushed above the surface of the soil at germination. They turn green and manufacture food for the seedling for a short time. Grass seedlings send a pointed shoot (*coleoptile*) directly to and through the soil surface at germination, while the cotyledon remains in the soil. The leaves of most forage legumes consist of three small leaflets; grasses have single leaf blades. Because of differences in location of growing points, grazing management is more critical with legumes than grasses.

The flowers of legumes usually are colorful, and pollination is mostly by insects. Grass flowers are inconspicuous and flowers are self-pollinated, or pollination is by air currents.

Most legumes have a deep, penetrating taproot with lateral branch roots. Grasses have a fibrous root system.

Another major difference between the two is that most legumes are readily able to fix nitrogen through a symbiotic relationship with *Rhizobium* bacteria.

Grasses have an advantage in that there is less risk of bloat in grazing cattle than with legumes. *Bloat* is a rapid formation of gas in the rumen, or stomach, of cattle, sheep, and other ruminant animals. If not properly treated, bloat can cause death. Antibloat agents are available to diminish the bloat problem. Not all legumes cause bloat.

Grasses can usually withstand weed competition better than legumes. Trampling by livestock will normally damage grasses less than legumes. Grasses also tend to tolerate

Figure 18-2. The utilization of forages in pastures for grazing is a complex subject which requires careful planning and coordination. Here, white-faced beef cattle are shown grazing on a well-managed pasture in southwestern Montana.

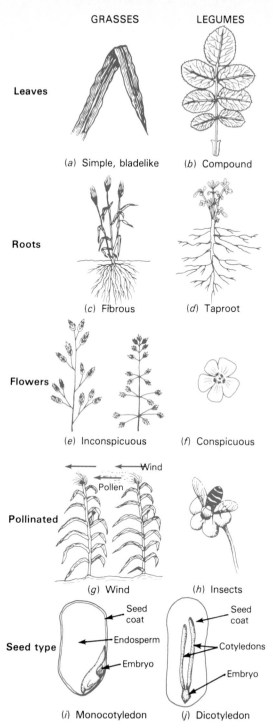

GRASSES   LEGUMES

Leaves

(a) Simple, bladelike   (b) Compound

Roots

(c) Fibrous   (d) Taproot

Flowers

(e) Inconspicuous   (f) Conspicuous

Pollinated

Wind

Pollen

(g) Wind   (h) Insects

Seed type

Seed coat

Endosperm

Embryo

Seed coat

Cotyledons

Embryo

(i) Monocotyledon   (j) Dicotyledon

Figure 18-3. Grasses and legumes have several different identifying characteristics.

spring flooding far better than most common forage legumes.

There are several principal advantages of legumes. One is nitrogen fixation. Significant quantities of nitrogen are added to the soil, and this benefits the crop following the legumes. A soybean crop may add 40 to 60 lb/acre [45 to 67 kg/ha] of nitrogen, while a high-yielding alfalfa crop may add up to 200 lb/acre [223 kg/ha]. Legumes usually have a higher protein content and less fiber, and they usually yield more tonnage than grasses provided comparable management practices are followed.

Some of the more commonly used forages of principal economic value in the United States are described in the next section.

Grasses

Grasses can be divided into warm-season and cool-season types. Warm-season grasses originated principally in Africa, eastern Asia, and South America, but some originated in North America. As the name implies, warm-season grasses require warm temperatures for optimum vegetative growth and short, warm days for formation of flowers and seed. There is little vegetative growth in early spring or late in the fall. These grasses are grown primarily in the southern United States and are also grown in parts of the Southwest and the Great Plains.

The cool-season grasses, grown primarily in the northern United States, originated primarily in northern Europe. Most are winter-hardy. They grow best under cool, moist conditions and usually slow down in growth during hot summer weather.

Warm-Season Grasses. Bermudagrass, *Cynodon dactylon,* is a moderately palatable, long-lived, perennial grass. Important forage types of bermudagrass are common, coastal, giant, midland, and Suwanee. It spreads by rhizomes, stolons, and seeds. Bermudagrass is one of the most salt-tolerant of the grasses.

Blue panicgrass, *Panicum antidotole,* is a vigorous perennial. This warm-season grass is most productive from May to October. It is productive during midsummer, when legumes may be dormant. Blue panicgrass will grow on moderately saline soils.

Dallisgrass, *Paspalum dilatatum,* is a tall grass that is found primarily in the South, with most production in the humid areas. It is a leafy grass produced from a clump—a bunch-forming type of grass. It should not be overgrazed or it will die out, but it should be grazed frequently enough to keep it palatable.

Johnsongrass, *Sorghum halapense,* can be a highly productive forage for both grazing and hay. It is, however, a serious pest in many crops throughout the South, especially cotton, corn, and grain sorghum. It has broad leaves and may reach 6 to 8 ft [183 to 244 cm] in height. It is similar to forage sorghums such as sudan. It is used extensively as a hay and pasture forage in the southeastern United States.

Pangolagrass, *Digitaria decumbens,* is grown on moist, fertile soils along the Gulf Coast. It is a sod-forming grass and spreads by stolons. It ranges in height from 2 to 4 ft [61 to 122 cm].

Carpetgrass, *Axonopus affinis,* is adapted to the coastal plains of the South. It spreads by stolons and seed. Its normal height is 9 to 10 in [23 to 25 cm], and it will grow on low-fertility soils if adequate moisture is available.

A large number of other warm-season grasses are tropical or semitropical forages and grow primarily in warm, humid areas such as southern Florida and southern Texas. They include species such as bahiagrass, vaseygrass, centipedegrass, paragrass, digitgrass, limpograss, and stargrass.

Most of these grasses are able to reproduce vegetatively through stolons.

Cool-Season Grasses. Orchardgrass, *Dactylis glomerata,* is a palatable, long-lived, perennial bunch grass best adapted to sandy loam soils. It tolerates considerable shade. This drought-tolerant grass can be highly productive, especially early in the season.

Perennial ryegrass, *Lolium perenne,* is a short-lived grass, usually persisting for 3 or 4 years. There is an annual as well as a perennial ryegrass. Perennial ryegrass is a poor sod former. This grass is sometimes used in mixtures with fescue because of its rapid emergence. It produces some forage before fescue is established. Most growth occurs in the spring and fall.

Italian ryegrass, *Lolium multiflorum,* is one of the annual types of this species. It is similar to perennial ryegrass but usually lives for 1 year only. Stands of both perennial and annual ryegrass are established with seed.

Smooth bromegrass, *Bromus inermis,* is resistant to extremes of temperature and will survive periods of drought. Smooth bromegrass develops a large, fibrous root system. It is established by seed but also spreads vegetatively by vigorous rhizomes which develop near the soil surface. Smooth bromegrass has an especially high leaf/stem ratio. Smooth bromegrass is palatable even when mature.

Tall fescue, *Festuca arundinacea,* is a vigorous, long-lived perennial bunch grass. It has a massive root structure, is an excellent soil-improvement crop, and is tolerant of saline soil. It is best adapted to clay or clay-loam soils. In some areas a disorder known as fescue foot occurs occasionally in cattle eating tall fescue. It is well to be aware of this possibility and to observe cattle consuming fescue forage for symptoms of this disease.

Figure 18-4. Reed canarygrass which has been cut for hay. (*USDA*)

Timothy, *Phleum pratense,* is a short-lived perennial plant. It reproduces both vegetatively and from seed. It is easy to establish and produces forage for harvesting each year. Timothy hay is especially popular for horses. It grows well in mixtures, is winter-hardy, and is relatively shallow-rooted.

Reed canarygrass, *Phalaris arundinacea,* is especially well adapted to wet areas. It is also well suited for grassed waterways and gully control. It needs to be grazed early for a highly nutritious forage (Figure 18-4). The grass can spread vegetatively from rhizomes.

Kentucky bluegrass, *Poa pratensis,* grows extensively across the United States, particularly in the northern and middle states. It is relatively low-yielding compared to other forages but produces high-quality forage for grazing under cool, moist conditions.

Legumes

The legume forages can be subdivided into three groups—alfalfa, the true clovers, and the miscellaneous legumes.

Alfalfa. Alfalfa, *Medicago sativa,* is grown on more acres in the United States than any other forage legume (Figure 18-5).

Alfalfa is a high-yielding, drought-resistant, palatable perennial legume. It is well adapted to loams and clay loams but is grown on all soil types. It is relatively easy to establish and is moderately salt-tolerant.

The roots of established stands of alfalfa penetrate the soil to a depth of 6 ft (183 cm) or more when there are no compacted soil layers to impede their development.

Alfalfa grows well in association with grasses. It recovers more rapidly after harvest than most grasses, especially during the spring and fall. The root system of alfalfa is usually deeper than that of forages so the roots remove moisture and nutrients from different zones or depths.

Alfalfa is seldom used for pasture for ruminants because of the danger of bloat.

Yields of alfalfa will normally range from 3 to 10 tons/acre [6.7 to 22.4 t/ha], depending on soil fertility and moisture. Winter-hardy varieties are available for the north, while nonhardy types are used in the South.

Figure 18-5. An alfalfa plant.

TABLE 18-1 COMMON AND SCIENTIFIC NAMES OF SEVERAL TRUE CLOVERS

Common Name	Scientific Name
Red clover	*Trifolium pratense*
Alsike clover	*T. hybridum*
White clover	*T. repens*
Crimson clover	*T. incarnatum*
Subterranean clover	*T. subterraneum*
Hop clover	*T. dubium*
Persian clover	*T. resupinatum*
Strawberry clover	*T. fragiferum*
Berseem clover	*T. alexandrinum*

True Clovers. The genus *Trifolium* contains the true clovers. A relatively large number of species in this group are of economic importance. Table 18-1 lists some of the important true clovers with their scientific names. Two well-known species—sweetclover and bur clover—are not true clovers and will be described later.

Other true clovers include kura, zigzag, lappa, cluster and rose. These are not widespread but are of importance in localized areas. Only the first four clovers listed in Table 18-1 are described here.

Red clover (Figure 18-6) is grown primarily in the northern part of the United States as a hay crop. It is a short-lived perennial. The two common types in the United States are mammoth and medium, with the principal difference being that the mammoth produces only one hay crop while the medium can produce two crops. Red clover grown with a grass makes an excellent pasture.

Alsike is grown primarily in New England and the upper Midwest as a hay crop, but it is also used with timothy or other grasses for pasture. It is managed similarly to mammoth red clover.

White clover is a short-lived perennial of three main types—large, intermediate, and small. Ladino clover is the most common of the large types, normally ranging from 10 to 12 in [25 to 30 cm] in height. White Dutch, an intermediate type, ranges from 4 to 6 in [10 to 15 cm] in height. The small white clover is seldom cultivated and grows voluntarily in pastures. Ladino, because of its normally higher yields, is used more extensively than the other white clovers.

Both flowers and leaves of white clover are produced on creeping stems or stolons, which develop at or near the ground level. The plants spread vegetatively by these creeping stems: when the stem comes in contact with the soil, a new plant develops if growing conditions are satisfactory.

Figure 18-6. Red clover—a forage legume. (*Grant Heilman*)

This legume, when grown where adapted, is about as long-lived as alfalfa or birdsfoot trefoil but usually produces less dry matter per acre than these legumes. The problem of bloat on ladino clover pasture is as great as for alfalfa or greater. When ladino clover is grown in combination with grasses or other legumes, competition from other plants can be reduced by grazing or cutting them.

Crimson clover, a winter annual, is grown primarily in the Southeast and the coastal areas of the western states. It is the most widely grown winter-forage legume in the South. It is used as a hay and pasture crop and also for soil building and erosion control.

Other Legumes. Four principal legumes of economic importance not described previously are sweetclover, bur clover, birdsfoot trefoil, and lespedeza.

Yellow- and white-blossomed sweetclover, *Melilotus officinalis* and *Melilotus alba,* are tall-growing, deep-rooted, biennial legumes adapted to most soil types. These legumes are sometimes included in pasture mixtures. Sweetclover is not a true clover but is closely related to alfalfa. It is easy to establish and is relatively drought-tolerant even in the seedling stage.

Sweetclover is one of the best forages to use as a soil builder. It is seldom used, however, for hay. An anti-blood-clotting compound may be formed in moldy sweetclover hay which can be toxic to livestock that consume it.

Bur clover, *Medicago hispida,* is a winter annual. This clover usually reseeds naturally and, when used in mixtures, extends the pasture season in southern regions, since it is most productive during the late fall, winter, and early spring. Growth tends to be prostrate, and seedpods usually have small, stiff-hooked spines.

Black medic, *Medicago lupulina,* is closely related to bur clover but is slightly less cold-tolerant.

Birdsfoot trefoil, *Lotus corniculatus,* is a long-lived perennial legume. While it resembles alfalfa, it has a finer stem and is relatively low-growing. The branching taproot extends to depths of 5 ft [152 cm] or more. The leaves of birdsfoot trefoil have five leaflets instead of the usual three for most legumes. Birdsfoot trefoil has not been found to cause bloat in cattle. It is somewhat salt-tolerant and grows well where drainage is poor.

Birdsfoot trefoil may be used in a simple mixture with one grass. The grass and legume seed should normally be planted in alternate rows. The growth of trefoil during the first year after planting is slow. If it is not seeded in alternate rows when used in a grass-legume mixture, it may be crowded out by the grass.

Trefoil produces very little pasture during the year following planting. The trefoil plants must be managed carefully or they will not persist. Birdsfoot trefoil withstands relatively high summer temperatures as well as freezing temperatures. It is best adapted to the Northeast and Midwest and, when irrigated, to the West.

Lespedeza is best adapted to the Southeast and is used mainly as a pasture crop. Two common annual lespedezas are Korean (*Lespedeza stipulacea*) and Kobe (*L. striata*). The principal perennial lespedeza is sericea (*L. cuneata*). Yields are relatively low compared to other legumes, but it grows relatively well on acid, low-fertility soils.

A number of other forage legumes are used principally for soil building or as cover crops. They include crotalaria, vetch, sainfoin, black medic, Alyce clover, and hairy indigo.

Field Crops

In addition to the forages normally classed as warm-season and cool-season grasses, some crops normally grouped in the field-crop category are grown for forage. These crops also belong to the grass family, Gramineae. They include corn, sorghum, and the small grains, principally wheat, oats, and rye. Corn and sorghum are used as silage crops, while sorghum/sudan types are used for both pasture and hay. About 10 percent of the corn acreage and 5 percent of the sorghum acreage is used as forage. The small grains are used principally for temporary pastures but occasionally for hay. Since these crops have been described in Chapter 13, they are not further described in this section.

HARVESTING FORAGE

Forage is harvested either by grazing livestock or mechanically. If properly managed, harvesting of forage by livestock can be more profitable than using mechanical harvesters (Figure 18-7).

Harvesting by Livestock

The key to efficient utilization of forage by livestock is coordinated management of the forage system and the livestock. The initial pasture establishment, if other than native grass, is important. Several steps can be taken to ensure a good forage system for livestock:

1. Seed pastures on land that has been fertilized and, if necessary, limed according to soil-test recommendations.
2. Select the best forage or combination of forages for your area and climate and for the livestock system that will utilize the forage.
3. Follow good production practices: (*a*) start with a properly prepared seedbed and correct planting procedures (be sure to inoculate legumes), (*b*) practice weed control, (*c*) apply needed maintenance amounts of fertilizer and lime, and (*d*) follow good irrigation practices if water is available.
4. Fit the grazing patterns and stocking rate to the production of the forage.
5. Use rotational grazing patterns in order to utilize forage more efficiently and get greater productivity. Do not overgraze the pasture, because regrowth will be slow and desirable species might be reduced.
6. Clip pastures periodically to reduce unpalatable growth of certain grasses and to reduce weeds.

Pastures can be classed as permanent or temporary. Permanent pastures consist of native vegetation that has never been plowed up for cropland or fields that have been seeded to remain indefinitely in pastures. Permanent pastures range in quality

Figure 18-7. Forage harvesting by Charolais cattle shown grazing on a mixture of ball clover, small grain, and bahia grass.

from brushy woodland pastures, which normally produce very little usable forage, to improved, highly productive, well-managed pastures. Permanent grass pastures are often improved by overseeding with a legume.

Temporary pastures are usually planted as part of a cropping sequence and may be used for 2 to 3 years or longer or may be used for only one winter or summer. Those lasting more than 1 year are normally seeded to perennial or self-seeding annual grasses and/or legumes. Temporary pastures to be used for only 1 year are normally planted for either the fall-winter-early spring period or the spring-summer-early fall period. Grasses for the cooler period will usually be the cereal grains, such as wheat, rye, or oats, or ryegrass. For the warmer periods, crops such as sudangrass, forage sorghums, and millet are normally used.

Planning a forage production system and a livestock unit which will efficiently utilize all the forage produced requires careful management. The type of livestock system—beef cow-calf system, dairy herd, stocker calves, or hogs—and the type of forage system needs to be properly planned and coordinated for best results.

A beef cow-calf herd requires either year-round grazing or grazing plus supplemental hay or feed. Year-round grazing systems are easier to describe than to produce. In the South, where winters are usually mild, a year-round program of forage could start with annual legumes in the late fall and winter followed by seeded cereal grains such as rye for late-winter and spring grazing. Summer and fall grazing could be from the permanent or long-term perennial grasses or from temporary pastures such as sudangrass. Climatic problems—high or low temperatures and/or lack of rainfall—can cause gaps in any forage program. Consequently hay or silage may often be part of the forage system to ensure year-round forage availability.

In the north, where forage is available for grazing only 5 to 8 months of the year, year-round grazing is impossible. In this case, cattle raisers use hay and silage (or in some cases grain) for the period when forage is not available for grazing.

As a result of the attempt to provide year-round grazing, excess forage is often available. A producer will often buy additional livestock—for example, stocker calves weighing 300 to 500 lb [136 to 227 kg]—to consume the extra forage. The excess may be harvested as hay in some cases.

Excess forage often is stemmy and may be low in quality and unpalatable. In such cases the pastures are often cut or chopped or even burned to get rid of the excess forage so that regrowth can occur.

Utilization of forages in pastures for grazing is a complex subject about which there are volumes of information. Many different types of forage and many different soil and climatic conditions influence the system to be used. Local farm advisers should be contacted for specific recommendations for local areas.

Harvesting by Machines

Forages can be harvested mechanically, processed, and stored for consumption by livestock as needed. The two principal methods used to preserve forage are as hay or as silage. Hay is the green foliage which is cut, dried, and stored in that form. Silage is the above-ground portion of the forage plant that has been cut and immediately chopped and stored in an airtight structure.

The principal advantages of harvesting forage mechanically compared to harvesting by livestock are (1) use of excess forage and (2) production of higher yields as well as better quality. The principal disadvantage is the cost. Investment in equipment and costs of operation are relatively high. Mechani-

cally harvested forage also has to be stored, resulting in additional costs. Silage normally costs more than hay to harvest, handle, and store.

Hay Harvesting. Harvesting forage at the right time is the key to obtaining the best combination of quality and yield of hay. Each species of forage has an optimum stage at which it should be cut for hay. In legumes, the time to cut is usually best determined by bloom stage. For example, alfalfa should be cut when it is at one-tenth bloom, while crimson clover should be cut at full bloom. The best time to cut grasses such as fescue, orchardgrass, bromegrass, timothy, johnsongrass, and dallisgrass (grasses which produce a seed head) is at the boot stage (when the seed head is emerging). For grasses such as bermudagrass, which may not produce a seed head, it is more difficult to describe the best time to cut. The grass should be cut while quite leafy, but stem should also be apparent. A good estimate would be an appearance of at least 50 percent leaf area compared to stems.

Quality is important in forage, regardless of method of harvest, but it is an especially important consideration in hay, because quality can decrease during the haymaking process. Hay quality is measured principally by (1) leafiness, (2) color, (3) nutrient content, and (4) amount of weeds and foreign material. Leaves contain most of the nutrients (about three-fourths of the protein and almost all the carotene), hence a high proportion of leaves to stems is desirable. Carotene, or vitamin A, content is closely related to the greenness of the leaves, hence a bright-green color is important. Nutrient content, which includes proteins, vitamins, and minerals, can be determined only by chemical analysis. A significant percentage of weeds or foreign material reduces the value and quality of hay. Most weeds are less palatable than grasses or legumes. Weed seeds can be spread from farm to farm through hay sales.

Most hay is cut and allowed to cure in the field under natural conditions. Some hay is dried in storage, but high energy costs have reduced "artificial" drying to a minimum.

The amount of time required for drying in the field depends on local weather conditions—e.g., temperature, humidity, and cloudiness—but most forage should dry to the required 12 to 18 percent moisture within 36 hours after cutting, provided there is no rain.

The hay-curing process consists primarily of losing moisture (from 70 to 90 percent moisture at cutting down to 12 to 18 percent moisture at time of baling). During the curing process, nutrients are also lost. Exposure to sunlight reduces carotene content.

Rain on cut hay can be damaging to the quality and quantity of the hay taken to storage. Dry-matter losses can amount to one-fourth of the original yield, and as much as 40 percent of the nutrients may be lost.

The curing process is one of the keys to the management and efficient use of forage harvested for hay. It is important to dry the forage as soon as possible after cutting to maintain the nutrient content and to preserve the overall quality of the hay.

Hay Storage. Hay can be stored in a number of ways: loose hay, in bales, or as cubed pellets or wafers (Figure 18-8).

Loose hay can be stored in the same form as when it was cut, it can be chopped, or it can be dehydrated and stored.

Long loose hay can be stored at a higher moisture content than it can in bales. This is the form in which nearly all hay was handled before modern harvesting machinery was perfected. This long, loose hay is often stored in the field, and cattle eat the forage from the haystack (Figure 18-9).

Hay is sometimes chopped into lengths of 1 to 2 in [2.5 to 5 cm] and stored in the loose form, but it is usually compacted in some manner with a stack wagon or module builder (similar to those used to make ricks of cotton for field storage). Cattle are usually permitted to self-feed from stacks of chopped hay. If properly compacted, ricks of chopped hay can be moved with the same equipment used to transport ricks of cotton. Chopped hay should seldom exceed 20 percent moisture.

Forage can be chopped even finer and dehydrated and stored in this form. Alfalfa is the principal forage handled in this manner. Called *alfalfa meal,* it is used primarily as a feed additive (Figure 18-10). By going directly from cutting to the dehydrator, a larger percentage of leaves are saved, resulting in higher levels of protein, vitamins, and minerals.

Hay is usually marketed within a relatively short distance of where it is produced, partly because of the cost of transporting such a bulky product.

If hay is to be transported any significant distance from the point of production, it is usually baled. The forage is forced into a tight package, reducing the amount of space needed for storage. Most bales are either rectangular (often called *square* bales) or cylindrical (often called *round* bales). The rectangular bales vary considerably in size and weight but approximate $16 \times 18 \times 40$ in [$41 \times 46 \times 102$ cm] and weigh around 75 lb [34 kg]. The cylindrical bales likewise vary in size and weight but are approximately 4 ft [122 cm] in diameter and 5 ft [152 cm] long. Weight ranges from 400 lb [181 kg] to as much as 2000 lb [906 kg] each.

The rectangular bales are better for storage and transporting long distances. Cylindrical bales are usually transported only short distances and are mechanically loaded, unloaded, and stored.

Figure 18-8. Forages can be stored in several forms. (*a*) Loose stacks of hay. (*b*) Round bales. (*c*) Cubed and piled in field.

Figure 18-9. A loose-cut stack of hay which has been unloaded from the harvesting wagon.

Figure 18-10. Alfalfa hay can be processed into pellets for livestock feeding.

Cubed hay is becoming popular in areas where complete mechanization of handling is desirable, as in cattle feedlots. A high capital investment is required for a cubing operation. The cubes are small in size (approximately a $1\frac{1}{2}$- to 2-in [3.8- to 5-cm] cube). In this form the hay can be handled mechanically and efficiently stored and fed to cattle. The process and equipment for cubing has been perfected only recently. The process requires hay that is normally drier, in the 8 to 12 percent moisture range, than when other forms of handling are used. Because of this, most of the cubing of hay is done in the arid, western states.

Silage. Machine-harvested forage can also be stored as a product called *silage*. Forage is chopped while still green and stored in this form in an airtight structure. Two basic types of structure are the upright silo and the horizontal silo. The upright silo is a tall cylindrical structure normally made of concrete or steel. The horizontal silo, often called *trench* silo, is usually built by simply digging a trench in the ground, usually on the side of a slope. The bottom and three sides are sometimes soil, but concrete or plastic liners are also used. The fourth side is usually left open for access when the silage is used. The top is often left exposed to the air or covered with plastic or similar material to reduce spoilage (Figure 18-11).

Silage is formed when the green chopped material ferments in the absence of air. The fermentation process results from bacteria breaking down sugars and producing organic acids which further help preserve the forage in a condition palatable to livestock.

Advantages of silage are (1) preservation of the quality and quantity of the forage with no losses due to field curing; (2) earlier harvest of forage with higher quality (but perhaps lower yield); (3) destruction of weed seeds; (4) fully mechanized harvesting and handling; and (5) the fact that, if properly ensiled initially, silage will keep for several years. The principal disadvantage is the high cost of mechanization for handling.

Nearly all forage can be used for silage, but some types are better suited than others. Corn is the principal crop used, followed by the forage sorghums, sorghum/sudan grasses, and small grains. Special varieties of both corn and grain sorghum have been developed for silage. Corn, because of its higher grain/stover ratio, is usually of higher quality, but yield may be less than for sorghum. Corn produces from 22 to 30 tons/acre [49 to 67 t/ha] of forage with good management, while the forage sorghum varieties commonly produce over 30 tons. Some sorghum varieties used primarily for grain production are also used for silage. They produce a higher-quality silage, because it contains more grain, but yield is lower than with the standard forage sorghums.

For the silage-forming process to work, moisture in the forage should be in the 60 to 75 percent range, depending on the crop to be ensiled. Corn and sorghum are usually in the correct moisture range when the grains are in the medium to hard dough stage. Other forage should be harvested at about the stage of growth best suited for hay–in the early boot stage or early to medium bloom stage.

Soilage. Soilage is another form in which forage can be harvested mechanically. The term describes forage that is cut green, chopped, and fed directly to livestock. Since forage for grazing is usually available at the same time that soilage can be harvested, this method of utilizing forage is minimally used.

(a)

(b)

Figure 18-11. (a) Three tractors pushing silage. (b) Tractors driving on silage.

MARKETING

Much of the forage produced is marketed through livestock, by way of markets that are quite well defined and available. The market for hay, however, is much less structured than the markets for most agricultural commodities.

Hay is usually purchased on the basis of appearance, considering aroma, proportion and size of stems, moisture content, and presence of weeds and foreign materials. It is occasionally bought on the basis of protein and total digestible nutrients. Quality and types of hay are usually discussed between buyer and seller.

Efforts to establish classes and grades of hay have been made in the past, but the precise classes and grades of hay as established by the USDA are seldom used. Chemical composition is often mentioned as one of the criteria that should be used in grading hay. Owing to the variability in type of hay, mixture, quality, chemical composition, and other important characteristics, it is difficult to establish precise standards and even more difficult to get the standards used by those who trade in the hay market.

SUMMARY

Forage plays an important role in the production of food. Livestock consume forage which is normally not suitable for human consumption and convert it to food for humans. A significant acreage—two-thirds of the world's agricultural lands—is used for forage.

Forage is versatile. It can be consumed directly by livestock or harvested mechanically and stored as hay, pellets, or silage for later consumption by livestock. It can be grown in cold or hot climates, under arid or humid conditions, in deep or shallow soils, and under almost any condition.

The keys to profitable production of forage are (1) selecting the right forage or combination of forages; (2) using proper cultural practices; (3) coordinating production with utilization; and (4) using proper harvesting procedures.

There are many different species of forage, and at least one is well suited to almost any environment.

Forage is classified in two groups—grasses and legumes. Grasses have simple, leaves and inconspicuous flowers, are self-pollinated or are pollinated by wind, and have a fibrous root system. Legumes have compound leaves and a conspicuous flower, many are pollinated by insects, and have a taproot system. Grasses cause less bloat and can withstand weed competition better than legumes. Legumes have a higher protein content and usually yield more tonnage. Legumes also have the symbiotic nitrogen-fixation characteristic.

QUESTIONS FOR REVIEW

1. What is the unique role of forage in providing food for human consumption?
2. What are the principal differences between grasses and legumes?
3. Why is forage described as being a versatile crop?
4. What are the two major categories of grasses and where are they grown?
5. What is the most important factor in efficient utilization of forage by livestock?
6. Name four problems related to the mechanical harvesting of forages.
7. What are the principal means used for storage of hay?
8. Describe silage and explain the advantages of producing it from machine-harvested forage.

Chapter

19

Sugar Crops

The major sources of sugar in the United States are sugar beets and sugarcane (Figure 19-1). Another important source is corn. Small quantities of sugar come from maple syrup, sorghums, and honey. Of the total produced in the United States from cane and beets, approximately 55 percent is from sugar beets and 45 percent from sugarcane.

Sugar beets are grown principally in the temperate areas of the world, while most of the sugarcane is grown in tropical or subtropical climates. Principal producing areas in the United States for sugarcane are in Hawaii, Florida, and Louisiana. Hawaii and Florida each produce slightly less than 40 percent of the cane sugar in the United States. Louisiana produces about 20 percent of the total, and Texas produces about 2 percent. Sugar beets are grown in California, Minnesota, North Dakota, Idaho, and twelve other western and midwestern states. California produces almost 20 percent of the beet sugar, with Minnesota producing about 15 percent. Idaho and North Dakota each account for slightly over 10 percent of the nation's production. Table 19-1 and Figure 19-2 show the distribution of the sugarcane and sugar beet acreage in the United States.

All domestic sugar production combined still falls short of consumption, and in most years, slightly less than half of the sugar consumed in the United States is imported. Imported sugar normally exceeds $1 billion in value annually, topped only by coffee, meat, fruits and nuts, and cocoa.

Twenty-five percent of the sugar consumed by Americans is in the form of packaged sugar, 22 percent in beverages, 21 percent in bakery and confectionery products, and 32 percent in processed foods and other products (Figure 19-3). Yearly per capita consumption of sugar in the United States is about 90 lb [41 kg].

Cane sugar and beet sugar—sucrose—are identical in quality and have the same chemical analysis and configuration, therefore they have the same sweetening value. Corn sugar, however, is fructose and dextrose, which is twice as sweet as sucrose. It is being used in increasing quantities in the soft-drink industry.

Figure 19-1. (*a*) Sugar beets grow up to 3 ft tall with the root being used for sugar. (*b*) Sugarcane is a tall, upright plant with the stalk being used for sugar.

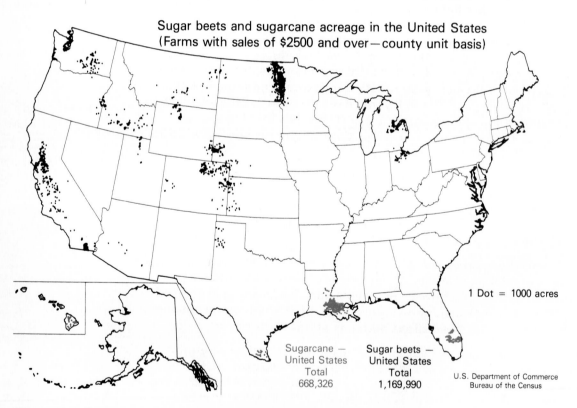

Sugar beets and sugarcane acreage in the United States
(Farms with sales of $2500 and over — county unit basis)

1 Dot = 1000 acres

Sugarcane —
United States
Total
668,326

Sugar beets —
United States
Total
1,169,990

U.S. Department of Commerce
Bureau of the Census

Figure 19-2. Sugarcane and sugar beet producing regions of the United States.

TABLE 19-1 SUGARCANE AND SUGAR BEET ACREAGE, PRODUCTION, AND VALUE, BY STATE

State	Acres (thousands)	Production (tons)	Dollar Value of Sugar (thousands)
Sugarcane			
All states	692	26,587	484,109
Florida	315	10,647	187,780
Hawaii	102	10,045	182,904
Louisiana	243	4,981	102,986
Texas	32	914	10,439
*Sugar Beets**			
All states	1,120	21,996	648,221
California	214	5,731	120,796
Minnesota	244	3,782	108,368
Idaho	126	2,804	76,591
North Dakota	143	2,304	69,937

*Other states growing sugar beets with less than $50,000 value include Colorado, Michigan, Nebraska, Montana, Wyoming, Arizona, Kansas, New Mexico, Ohio, Oregon, Texas, and Utah.
Source: Data are from the USDA Agricultural Statistics Handbook, 1980.

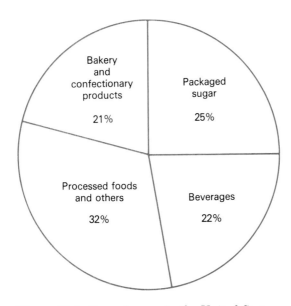

Figure 19-3. Uses of sugar in the United States.

CHAPTER GOALS

After studying this chapter you will be able to:

- Identify the major sources of sugar and where the crops are grown.
- Describe the consumption and uses of sugar in the United States.
- Identify the production practices used in production of sugar crops.
- Describe the harvesting and processing procedures for sugar crops.

SUGARCANE

Sugarcane is a giant, robust, tropical grass native to Asia, where it has been grown in gardens for over 4000 years. The methods for manufacturing sugar from sugarcane were developed in India about 400 B.C. It is believed that sugarcane originated in northeastern India. Christopher Columbus brought the plant to the West Indies, and today sugarcane is cultivated in tropical and subtropical regions throughout the world.

Sugarcane was introduced into Louisiana in the mid-1700s by Jesuit priests. It is now grown in several states, with Florida and Louisiana each producing approximately 15 percent of the total sugar (both beet and cane) produced in the United States. The production of sugarcane in the continental United States increased significantly in the early 1960s because of the Cuban sugar embargo. Prior to the embargo, only 50,000 acres [20,250 ha] of sugarcane were grown in Florida. Today, about 300,000 acres [121,500 ha] are grown.

Sugarcane is a perennial and is a member of the grass family Gramineae. The scientific name is *Saccharum officinarum*. Two other species used in plant-breeding programs are *S. spontaneum* and *S. barberi*.

Sugarcane is a tall plant, with a normal full-grown height of 15 to 20 ft [4.5 to 6 m]. Stems range from 1 to 3 in [2.5 to 7.5 cm] in diameter. A stalk averages about 3 lb [1.35 kg] if growth is normal. One 3-lb stalk will yield about $1/3$ lb [0.15 kg] of sugar.

Production Practices

Soil preparation for planting sugarcane is similar to that for most other crops. Vegetation is first destroyed and plant residue is incorporated into the soil for decomposition. Since sugarcane is propagated vegetatively, the stalks, which would ordinarily be cut and hauled to the mill for processing, are harvested from mature stands and cut into 20- to 36-in [51- to 91-cm] sections, which are laid end-to-end in furrows and covered with 2 to 3 in [5 to 8 cm] of soil (Figure 19-4). A new cane plant sprouts from each node (approximately 2 to 4 in [5 to 10 cm] apart) along the stalk. Within 2 to 3 weeks, if soil moisture and temperature are favorable, the shoots emerge and secondary shoots are produced which provide a stand of cane. Thinning may be necessary to ensure optimum growth and highest yield of sugar. The plants in a stand should normally be 9 to 12 in [23 to 30.5 cm] apart. Thinning is done by machines.

The time of planting will vary with location. In the continental United States, planting is normally done in the spring. In Hawaii, sugarcane can be planted any time based on temperature during the year.

Figure 19-4. Sugarcane is planted by laying 2 to 3 ft lengths of the cane in rows and then covering with soil.

Figure 19-5. Sugarcane borers can damage stalks. Note the holes at the internodes and the borer excrement.

In Florida and Texas, sugarcane produces white or reddish plumes during the winter and spring. The plumes are flowers and seed heads of the cane. Each plume consists of several thousand tiny flowers, each capable of producing one seed.

Seeds are so small (1000 per gram, or 500,000 per pound) that they cannot be planted directly in the field. Sugarcane does not breed true, and every seedling is a new genotype; hence seeds are not used for commercial planting. Varietal uniformity within fields is important for commercial production, hence stalks are used to vegetatively propagate new plants to ensure uniformity.

Since the vegetative portion of the sugarcane plant is the part that is harvested and utilized, nitrogen use is high. Since it is also important that sugars be formed in the plant, other nutrients should be supplied, if needed, to provide a balance of nutrients in order to optimize tonnage and sugar content of the plant and maximize the total yield of sugar.

Like most other plants, sugarcane will produce the highest yields on soils that are fertile, of moderate texture and exchange capacity, and with good physical condition.

An estimated 95 percent of the sugarcane crop is treated with herbicides to control weeds. Rotation of crops and mechanical practices can be used in conjunction with chemicals to reduce weeds. Since sugarcane is a grass, various grass species are among the most difficult to control with chemicals.

The sugarcane borer (Figure 19-5) and sugarcane beetle are the most important insect pests for cane. Certain corn pests—specifically, the corn borers—also damage sugarcane if not controlled. The sugarcane borer attacks the top of the plant and causes it to die. Chemicals are the principal means of controlling the insect.

Diseases also cause problems and loss of yield in sugarcane. Virus diseases are often carried by infected seed pieces. Treatment of the seed pieces provides satisfactory control. Various fungi also cause diseases in sugarcane. The principal fungal disease is red rot, which causes the seed piece to rot after it is planted. Disease-resistant varieties provide the best control, but adequate soil moisture and favorable temperature of about 70°F [21°C] or above at planting also reduce disease losses.

Harvesting

In Louisiana, cane fields from a single planting are usually harvested each year for 3 years only, then planted to a different crop. Florida growers normally harvest a crop each year for 4 to 5 years. In Hawaii, growers harvest twice or three times in 6 to 8 years before replanting.

After a field has been harvested for the first time, it is maintained free of weeds and a second crop of stalks is produced by the old plants. This and successive crops are called *ratoon* crops. When production declines to an unacceptable level, the cane is replanted as described earlier.

Sugarcane is harvested from November through April in the continental United States, but it can be harvested in any month in Hawaii. If there is no freeze, sugar yields in the United States are highest after January 1, but some fields must be harvested before they have reached maximum yield to allow time for processing.

Fields are burned the day before harvest. The fires are rather spectacular but of short duration (a 40-acre [16 ha] field burns in about 15 minutes). They burn off dead leaves which would otherwise impede harvest and milling of the stalks. Burning is now normally done in the daytime, when dispersal of the smoke by air currents causes minimum nuisance. Some fields are harvested by hand with the use of cane knives, while others are cut by machines. In recent years there has been a considerable increase in mechanical harvesting.

The cane, if harvested by hand, is loaded into tractor-drawn wagons by machines called *continuous loaders*. Machine-cut cane is deposited directly into wagons by the harvester (Figure 19-6). At special ramps near the field, the cane is dumped from the wagon into rail cars or highway trailers for transport to the mills.

Processing

At the mill, the cane is crushed between heavy rollers to squeeze out the juice. Water is added, and the fiber is again squeezed to remove as much sugar as possible. The water is then evaporated, leaving a crystallized product called *raw sugar*. The coarse, brownish raw sugar contains impurities which must be removed in a separate refining process. The raw sugar is stored in bulk in large warehouses to await shipment to a refinery, where it is further processed into the form in which it is sold for consumption.

The fibrous plant material remaining after removal of the juice is called *bagasse*. It consists mostly of stalk fibers but also contains leaves and other trash brought in with the cane. Bagasse is about half water and half dry matter, a small fraction of which is sugar that was not removed. Bagasse may be burned as fuel or used to produce an industrial alcohol called *furfural,* fed as roughage to cattle, or made into particle board.

Figure 19-6. After being cut and gathered into a bundle, cane is being picked up.

Figure 19-7. The sugar beet root is converted into a food for human consumption—sugar.

The sugar beet is one of nature's most bountiful creations. It is a miniature factory and warehouse. Through the process of photosynthesis, it combines elements of air and water into sucrose, with the power (or energy) being supplied by the sun. The manufacturing takes place in the leaf, and the sugar is translocated into the root for storage (Figure 19-7).

The fact that sugar beets manufacture and store sugar and that it can be extracted in crystalline form was proved in 1747 by a German chemist, Andreas Marggraf. The sugar-beet industry subsequently developed in Europe in the early 1800s. The first commercial beet-sugar factory in the United States was built in 1870 at Alvarado, California. This led to the construction of other plants, and sugar beets became established as an important sugar source in the late 1800s. Establishment of sugar-beet plants was usually followed by increased livestock feeding, because the by-products beet pulp and molasses, as well as beet tops, were good sources of feed for livestock.

Owing to the international problems of the last several decades, the United States has placed increasing reliance on beet sugar to assure consumers of adequate supplies at all times.

The sugar-beet plant (Figure 19-8) is a biennial and belongs to the Chenopodiaceae family. Its scientific name is *Beta vulgaris*. The sugar beet has a taproot system. It is large at the top and tapers into a relatively small root. The roots range from less than 1 lb [0.5 kg] to over 5 lb [2.2 kg] in extreme cases. The average mature beet size will be around 2 to 3 lb [0.9 to 1.4 kg]. From 10 to 20 percent of the root is sugar, with the average sugar content ranging from about 13 to 18 percent in normal years. The percentage of sugar is greatly influenced by the

Figure 19-8. An individual sugar beet plant showing top growth and enlarged root protruding above the ground surface.

growing season and by management practices such as irrigation and nitrogen levels.

A desirable growing year for highest sugar production is one in which the spring warms up rapidly and early vegetative growth occurs. Warm days, with an average mean temperature around 70 to 75°F [21 to 24°C], and cool nights tend to maximize sugar content, provided moisture and nitrogen are being depleted normally. Rainfall or irrigation water should be plentiful, with good distribution during the season. The sugar content of beets is at a maximum at the end of the first year.

Production Practices

Land preparation for eventual planting of beets is similar to most other crops. The

Figure 19-9. A typical stand of young sugar beets in 30-in rows in a field when an application of irrigation water has been applied in alternate furrows.

seedbed should be firm and moist, with no compacted layers. Beets can germinate in relatively cool soil and, after emergence, can withstand light frosts. The seeds should normally be planted about $^3/_4$ to 1 in [1.9 to 2.5 cm] deep. Sugar-beet seeds are precision planted, with an attempt to get an exact spacing of plants. Because beet seedlings, like other plants, are subject to the ravages of wind, sand, excessive temperatures, and weed control practices, it is often necessary to increase the seeding rate. After germination and early seedling growth, plants are thinned to the desired stand. Final stand levels will vary, but there should normally be about 2 plants per foot [30 cm] of row, for 30,000 to 35,000 plants per acre [about 74,000 to 86,500 per hectare]. Rows are usually 30 in [76 cm] apart but may be 36 to 40 in [91 to 102 cm] apart in some beet-producing areas (Figure 19-9).

Monogerm seed, a relatively recent development, permits precision planting. Prior to the development of a monogerm seed, which produces a single plant, the multigerm seed was used. It produces a cluster of plants which makes thinning more difficult. Thinning is done mostly by machine but occasionally by hand hoeing.

Cultivars used in sugar-beet production are normally recommended and frequently developed by the contracting company. The company normally also supplies the seed. Most sugar-beet cultivars in use today are disease-resistant and will not bolt (produce a seed head in the first year of growth). Research scientists have improved the beet in many respects, including higher sugar percentage, root storability, the monogerm seed, and better seedling vigor.

Fertilization programs for sugar beets are variable and should be based on soil tests. If the soil cannot supply needed plant nutrients, they must be applied. Since most beets are grown in the western United States, the principal nutrient needs are nitrogen and phosphorus.

Management of nitrogen is the key to proper fertilization of beets. Adequate nitrogen is needed for early vegetative growth and good continued growth during the season, but nitrogen level should start declining in the last 30 days of the season, so that vegetative growth is slowed considerably. This permits the plant to use its energy in producing sugar rather than vegetative growth. The ideal situation is for soil nitrogen to be near depletion about 1 month before harvesting.

Available nitrogen and irrigation practices need to be coordinated during the season. Plant-tissue or petiole analysis can be used to monitor the nitrogen status during this period. If the nitrogen content of the plant is high within a month of harvest, the producer can decrease the application of irrigation water to make nitrogen less available. If nitrogen is low in the plant, soil moisture levels should be kept adequate and normal.

In determining nitrogen needs for beets, the subsoil nitrogen needs to be considered. The taproot of the beet plant will often grow 24 to 48 in [61 to 122 cm] or more deep (Figure 19-10), and if nitrogen is higher than normal at that depth, the excess nitrogen can keep the plant growing vegetatively too late in the season and result in lowered sugar content.

In areas where irrigation is practiced, light and frequent applications of water are usually recommended. The soil profile should have a good level of moisture prior to

Figure 19-10. Sugar beets are a deep rooted crop—they can penetrate to 6 to 7 ft if soil is sufficiently deep.

planting. The crop should normally be permitted to undergo moisture stress before harvest.

Pest control is a must with sugar beets as with any other crop. Weeds compete for moisture, nutrients, and sunlight. They may also interfere with harvesting. Mechanical control as well as the use of chemicals are principal means of controlling weeds.

Beets are also subject to damage from numerous insects and diseases. The usual methods of control are use of chemicals and cultural practices. The leafhopper is one of the principal insect vectors of diseases that damage beets; it carries and transmits the

Figure 19-11. Sugar beets are harvested mechanically and transferred to a truck for transporting to the plant or a central collecting point.

virus which causes curly top. Aphids can transmit the yellows virus. Fungus diseases affecting sugar beets include cercospora leaf spot and black rot. Beets should be grown on a field only once every 4 to 5 years, because of carry-over of certain diseases and nematodes.

Harvesting

Sugar beets for sugar are normally harvested at the end of the first season of growth. If grown for seed, they are harvested at the end of the second year, since the crop is a biennial. Harvesting usually starts ap-proximately 5 to 7 months after planting. The process consists of first topping the beets with a defoliator followed by a scalping blade. After harvest, livestock are often per-mitted to "graze" the tops of the beets left in the field, or the tops are often baled or en-siled for use as a livestock feed. Second, a wedge-type lift wheel pops the beet onto a conveyor and into a truck (Figure 19-11). For the highest sugar percentage, nitrogen should be depleted about 1 month before harvest; moisture should be depleted as much as possible; and cool nighttime tem-peratures should prevail. However, most sugar-processing plants need to start the refining process early in the season so that processing can be finished before the sugar beets start deteriorating in storage. Harvesting is therefore usually initiated by the processing company near the end of the growing season. A harvest schedule is established by which all growers harvest a portion of their crop prior to the most desirable time. Most areas complete the harvest by December, but in some areas, such as California, the beets are left to overwinter in the field. As a result of overwintering, tonnage is reduced as the plants use stored sugars, but the following spring the percentage of sugar is increased as complex carbohydrates are converted to sugar. Thus, pounds of sugar per acre would be reduced only slightly. Also, overwintering allows processing plants to extend their period of operation.

Processing

The beets are transported to the plant and stored in huge piles to await processing (Figure 19-12). Processing consists of washing, slicing, and cooking to extract the sugar. The liquid, containing the sugar, is separated from the solids, purified, and evaporated,

Figure 19-12. After delivery to plant, beets are stored outside in huge piles and subsequently processed for the sugar.

leaving the sugar. The sugar is further cleaned, purified, dried, packaged, and sent to the consumer.

Molasses is a byproduct of processing and is used as livestock feed and in the production of alcohol. The beet pulp remaining after the extraction of the sugar is usually fed to livestock.

SUGAR FROM CORN

Corn has been used as a source of sugar and syrup for many years. New processing techniques have now resulted in increased use of corn as a sugar source.

Sugar produced from the grain of corn is called *fructose* and *dextrose,* while the sugar from cane and beets is *sucrose*. The nature of the sugars is such that fructose and dextrose have twice the sweetening value of sucrose.

Since the fructose/dextrose combination is more concentrated, only half as much needs to be transported, stored, and handled, resulting in reduced costs for volume users of sugar. Fructose/dextrose is handled in the liquid form and is being used in increasing quantities by the soft-drink industry. An estimated 25 percent of the sugar used in that industry reportedly is in the form of corn sugar.

Corn production practices and the harvesting and storing of corn are presented in Chapter 13, "Cereal Grains." Corn-sugar processors buy corn from producers or grain marketers.

Starch is extracted from the corn grain and is enzymatically converted to dextrose. Approximately half the dextrose is further converted to fructose to produce a liquid fructose/dextrose combination.

By-products of corn processing include the steep water, gluten, and meal, which all contain significant quantities of protein and are fed to livestock. The germ of the grain, separated initially from the endosperm (or starch), is extracted for oil. This corn oil is used for cooking oil, margarine, and similar products. The bran, or fiber, that is left as a residue in the processing of corn is also fed to cattle.

SUGAR MARKETING

In the United States, as in most of the rest of the world, the production and marketing of sugar are regulated by law. Approximately 90 percent of all sugar produced and marketed in the world is under some form of market control, such as the British Commonwealth Agreement and France's program with its former colonies. Principles of the United States sugar program were established in 1934 to eliminate the disastrous price fluctuations which characterized the industry. However, there is no current system of regulated acreage, marketing allotments, and country-by-country import quotas, so United States consumers cannot be assured of adequate supplies of sugar at reasonable prices.

SUMMARY

Sugar beets and sugarcane are the principal sugar-producing crops in the United States, but corn is being used in increasing quantities as a source of sugar. Beets are grown principally in the temperate area, while cane is grown in the tropical and subtropical parts of the nation. The United States produces only about one-half of its sugar needs.

Sugarcane, a member of the grass family, is a perennial which will continue to produce as long as conditions are favorable; most cane fields, however, are replanted every 3 to 6 years. The stalk is the portion of the plant which is harvested. It usually contains 10 to 14 percent sugar.

The sugar beet is a biennial, even though it is harvested at the end of the first season of growth. The root of the beet—the portion harvested—contains 13 to 18 percent sugar in a normal season.

Both sugar crops have to be properly managed (Figure 19-13) to ensure optimum sugar yield. Harvesting and the initial processing are unique to each crop. The subsequent processing into the final refined sugar for the consumer, however, are quite similar. Sugar configuration and quality are the same for both sugar cane and sugar beets.

(a)

(b)

Figure 19-13. Research is important for all crops. (a) Scientists take cores from beets to check for disease that causes rotting during storage. (b) A scientist checks the top/root ratio of young beets in an effort to increase root weight relative to the vegetative weight.

QUESTIONS FOR REVIEW

1. What are principal producing areas for sugar cane and sugar beets in the United States?
2. What are the major crop pests and diseases that effect sugarcane and sugar beet crops?
3. Describe how sugarcane is harvested.
4. How is sugarcane processed?
5. How are sugar beets harvested?
6. How are sugar beets processed?
7. Why are sugar beets, a biennial plant, harvested at the end of the first season of growth instead of at the end of the second season?
8. What are the differences between fructose/dextrose and sucrose?

Chapter
20

Managing Our Rangelands

Rangeland is land on which the existing native vegetation is predominantly grasses, grasslike plants, forbs, or shrubs suitable for grazing or browsing by domestic or native animals. This land is generally not well adapted to field-crop production because of erratic or low rainfall, rough topography, poor soil, or cold or hot temperatures.

The term *rangeland* refers to those areas of land that are unsuitable for farming and more intensive agricultural production and are used primarily for livestock grazing. They are also used for wildlife habitat, recreation, and timber production. In addition, rangelands serve as huge watersheds, which need proper management to conserve their water and soil resources. There are over 700 million acres [284 million ha] of rangeland in the United States.

CHAPTER GOALS

After studying this chapter, you will be able to:

- Understand why rangeland resource management is needed.
- Define common characteristics of some of the major range plant species.
- Explain concepts related to managing our rangelands, such as range ecosystems, plant succession, carbohydrate storage cycle, range condition, and climax vegetation.
- Describe practices used by managers of range resources.

RANGE MANAGEMENT

Range management refers to the care and use of rangelands to obtain the highest continuing production of forage without damage to the resource. It deals with the growth of vegetation and the response of the grazing animals. It also involves decision making based on climatic, topographic, and hydrologic factors as related to the kind of use and the degree to which the range will be used. In addition, range managers must be aware of the needs of society as they consider alternative uses of our range resource.

IMPORTANCE OF OUR RANGE RESOURCE

When considering land use on a worldwide basis, rangelands rank as a major component. Over 30 percent of the world's land area is grassland, 27 percent is forested, and only 10 percent is classed as farmland for intensive cropping. Much of the area classified as forest is grazed, and many areas classified as deserts provide significant amounts of forage in years when adequate moisture is available. When all these areas providing forage for grazing animals are combined, it becomes apparent that our range resource is of great significance to the well-being of humankind.

Most of the rangelands of the United States are found in the seventeen continental states west of an irregular north and south line running from North Dakota to Texas. They are interspersed with irrigated valleys and dry-land farming areas. In the southern part of this region, year-round grazing may be possible. The northern climate limits winter grazing. About half the rangeland in the eleven western states is privately owned; the other half is federally owned. Grazing permits are issued to ranchers by the Bureau of Land Manage-

ment and the Forest Service for grazing of livestock on federal lands they administer.

In the western United States, more than half the forage produced comes from rangeland. Management of the huge rangeland area is an important job. In addition to livestock production, proper management of the range results in favorable wildlife habitat, efficient use of water, soil conservation, and recreational benefits.

The range-livestock industry played an important role in American history. As the early pioneers migrated westward, a new pattern of living was established on western rangelands from Texas on west.

The introduction of domestic livestock on the range did not introduce an entirely new influence on the ecosystem. Wild mammals in large numbers already had been grazing on these lands for centuries. The introduction of domestic animals, however, increased the intensity of grazing pressure, and in much of the area the range was misused or overused. These abuses were recognized, and people began to realize that the grazing capacity of the land was declining and that research should be initiated to develop management practices to improve and sustain the once-productive rangelands. From this realization was born a new scientific field-range science. Hundreds of people are now employed in the field of range science and as managers of our vast rangelands.

RANGE PLANTS

The native and introduced plants of the range are often called grass. Besides true grasses, there are shrubs, sedges, and forbs.

Managers can distinguish among these plants by observing the structure, size, and shape of their parts, i.e., stems, leaves, and flowers. The common range plants can be classified into four kinds: grasses, grasslike plants, forbs, and shrubs (Figure 20-1).

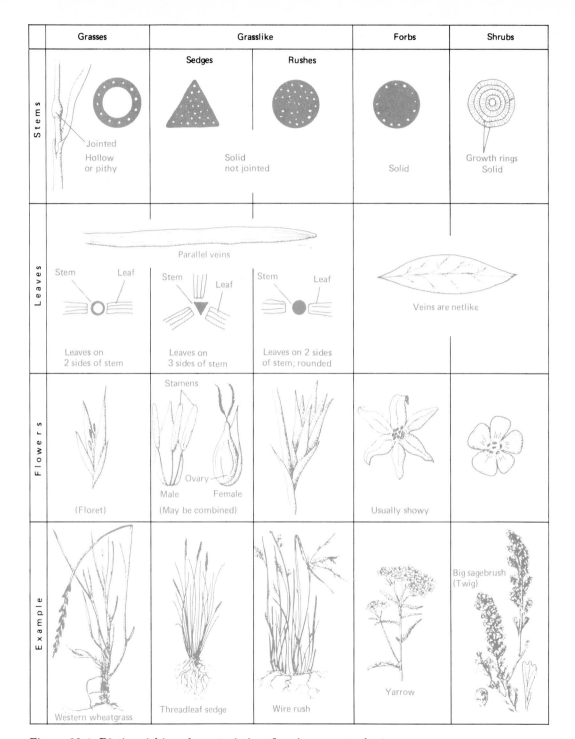

Figure 20-1. Distinguishing characteristics of various range plants.

Grasses. The stems, or culms, of grasses have nodes and internodes and usually are hollow. The leaves are in two rows on the stem. The veins of the leaves are parallel. Flowers are not prominent.

Grasslike Plants. These are the sedges and rushes that resemble grass but have solid stems and no nodes. There are three rows of leaves on the stem. The veins of the leaves are parallel, as with the grasses. Flowers are not prominent.

Forbs. Forbs are herbaceous plants with solid stems that die back to the ground annually. Often the roots of forbs are perennial or biennial. The leaves have netted veins and are broad compared to most grasses. Flowers are usually showy.

Shrubs. Shrubs are woody perennial plants with stems that live through the winter. Their growth is relatively low, distinguishing them from taller-growing trees.

Managers of rangeland should know which plants are adapted to their particular area. Adapted species differ significantly from region to region as soil types, rainfall, growing season, temperature, and other environmental factors change. Some common range plants are listed, along with some of their distinguishing characteristics, in Table 20-1.

THE RANGE ECOSYSTEM

Ecology is a science that deals with the response of organisms to their environment. This response can be dealt with in terms of the response of an individual organism or, more broadly, the response of plants, animals, and other organisms interacting and responding as a group to their environment.

TABLE 20-1   COMMON RANGE PLANTS

Plant Name	Type of Plant			Life Span		Origin of Plant		Season of Growth		Grazing Response			Forage Value			
	Grass	Shrub	Forb	Annual	Perennial	Native	Introduced	Cool	Warm	Decreases	Increases	Invades	Good	Fair	Poor	Poisonous
Antelope bitterbrush		X			X	X		X		X			X			
Arrowleaf balsamroot			X		X	X		X			X			X		
Arrow grass			X		X	X		X			X					X
Baltic rush	Rush				X	X		X			X				X	
Basin wild rye	X			X	X	X		X		X				X		
Big bluegrass	X				X	X		X		X			X			
Big bluestem	X				X	X			X	X			X			
Big sagebrush		X			X	X		X			X		X	to	X	
Black grama	X				X	X			X	X			X			
Bluebunch wheatgrass	X				X	X		X		X			X			
Blue grama	X				X	X			X			X	X			

TABLE 20-1 COMMON RANGE PLANTS (Continued)

Plant Name	Type of Plant			Life Span		Origin of Plant		Season of Growth		Grazing Response			Forage Value			Poisonous
	Grass	Shrub	Forb	Annual	Perennial	Native	Introduced	Cool	Warm	Decreases	Increases	Invades	Good	Fair	Poor	Poisonous
Broom snakeweed		X			X	X			X			X			X	
Buffalo grass	X				X	X			X		X		X			
Cactus		X			X	X					X				X	
Canada bluegrass	X				X		X	X								
Canada wild rye	X				X	X		X		X			X			
Canby bluegrass	X				X	X		X		X			X			
Cheatgrass brome	X			X			X	X				X		X to X		
Common chokecherry		X			X	X			X		X			X		X
Creosotebush		X			X	X			X			X			X	
Cudweed sagewort			X		X	X					X	X				X
Curlycup gumweed		X		Biennial		X			X		X				X	
Curly mesquite grass	X				X	X			X		X			X		
Dandelion			X		X		X	X			X			X		
Death camas			X		X	X		X			X					X
Fringed sagewort		X			X	X		X			X			X		
Foxtail barley	X				X	X		X				X		X to X		
Geranium (sticky)			X		X	X		X		X				X		
Greasewood		X			X	X			X		X			X		X
Green needlegrass	X				X	X		X		X			X			
Idaho fescue	X				X	X		X		X			X			
Indian ricegrass	X				X	X		X		X			X			
June grass	X				X	X		X			X		X			
Kentucky bluegrass	X				X		X	X				X	X			
Larkspur			X		X	X		X			X				X	X
Little barley	X			X		X			X		X				X	
Little bluestem	X				X	X			X	X			X to X			
Loco			X		X	X		X			X				X	X
Lupine			X		X	X		X			X				X	X
Meadow barley	X				X	X		X			X			X		
Mesquite		X			X	X			X		X				X	
Mountain brome	X				X	X		X		X			X			

TABLE 20-1   COMMON RANGE PLANTS (Continued)

Plant Name	Type of Plant			Life Span		Origin of Plant		Season of Growth		Grazing Response			Forage Value			
	Grass	Shrub	Forb	Annual	Perennial	Native	Introduced	Cool	Warm	Decreases	Increases	Invades	Good	Fair	Poor	Poisonous
Narrow-leaf milk vetch			X		X	X		X			X				X	
Needle and thread	X				X	X		X		X or X			X			
Nuttall's saltbush		X			X	X			X	X			X			
Phlox			X		X	X		X			X				X	
Plains muhly	X				X	X		X			X			X to X		
Prairie sand reed grass	X				X	X			X	X			X to X			
Red three-awn	X				X	X			X			X			X	
Rough fescue	X				X	X		X		X			X			
Rubber rabbit brush		X			X	X		X				X			X	
Sandberg bluegrass	X				X	X		X			X		X			
Sand dropseed	X				X	X			X		X		X			
Serviceberry		X			X	X		X			X			X		
Shad scale		X			X	X			X	X			X			
Sheep fescue	X				X	X		X			X		X			
Side oats grama	X				X	X			X	X			X			
Silver sagebrush		X			X	X			X		X			X		
Slender wheatgrass	X				X	X		X		X			X			
Snakeweed		X			X	X			X			X			X	
Snowberry		X			X	X		X			X				X	
Stream bank wheatgrass	X		X	X		X		X		X			X			
Thick-spike wheatgrass	X				X	X		X		X			X			
Threadleaf sedge	Sedge (Grasslike)				X	X		X			X		X			
Timber oat grass	X				X	X		X		X			X			
Tumble grass	X				X	X			X			X			X	
Two-grooved loco			X		X	X		X			X					X
Western wheatgrass	X				X	X		X			X		X			
Wild onion			X		X	X		X			X			X to X		
Winter fat		X			X	X		X		X			X			
Yarrow			X		X	X		X			X				X	

Source: Parker, Karl G., Montana Cooperative Extension Service Bulletin 1028, Bozeman, MT. 1965.

The term *ecosystem* refers to the entire complex of organisms and how they interact within their environment.

In a range ecosystem, carbon, hydrogen, oxygen, nitrogen, and many other elements are continuously cycled among the plants, animals, microorganisms, soil, and atmosphere.

The amount of each of these elements remains relatively constant in a stable system, but there can be changes within each division of the system. For instance, flowing water and water-soluble nutrients can move out of a division of the system, or nitrogen can be released into the atmosphere by combustion during a range fire. Eventually, however, these components reenter the ecosystem from reservoirs such as the atmosphere, parent rock, and underground water systems.

Energy enters the ecosystem as radiation from the sun. Organisms containing chlorophyll are able to capture and convert this energy into compounds essential to the sustenance of living organisms. For example, range plants are harvested by wildlife and domestic animals. The energy stored in the plants is used to sustain the animals. The energy stored in the animals may be released and used by humans who eat the meat or by other animals and microorganisms that consume the carcass. And so the cycle continues, with numerous possible paths. The flow of energy through various levels of the ecosystem is illustrated in Figure 20-2. Note the energy loss from the sys-

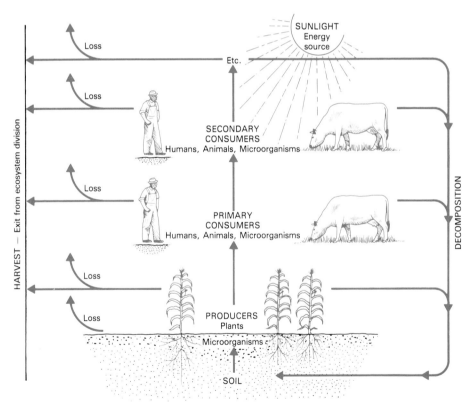

Figure 20-2. The flow of energy through successive levels of an ecosystem.

Managing Our Rangelands

tem at each level. The energy stored in the biomass is greater at the producer level and diminishes with each successive consumer level.

The range manager needs to be aware of the interrelationship of the various components of the ecosystem. The producer level of the system is where the range manager is able to exert the greatest influence.

Plant Succession

The range as an ecosystem is a complex natural environment in which organisms such as plants, animals, and microorganisms interact with the soil, the climate, the topography, and other physical factors of the environment. Range managers and range scientists are concerned with all these factors as they relate to the plant community on the range. The term *plant community* refers to a group of plants in a given environment that have similar living requirements.

Plant communities originally develop on bare rock or parent material in an orderly process of change called *plant succession,* a gradual process involving a series of changes that follow a somewhat regular course. It starts with the weathering and degradation of the surface of the parent material to form soil. This results in a change in the habitat and the growth of new species of plants. A series of these changes occurs on a given site until the vegetation and soil reach an *equilibrium,* or are in balance with the local environment. This state of equilibrium is called *climax,* and the plants that exist are the *climax community* (Figure 20-3). Climax is not an absolutely stable state: if a plant community is disturbed by grazing, fire, or other disruptions, it may revert to one of its earlier stages. With more favorable conditions, it will again progress toward climax.

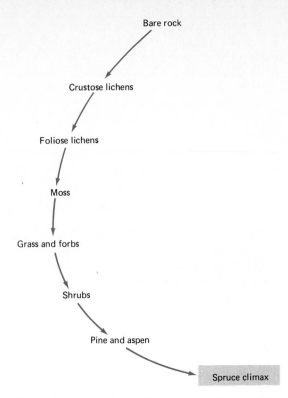

Figure 20-3. One example of the stages of plant succession from bare rock through final development of a spruce forest.

Carbohydrate Storage Cycle

Plants manufacture food (carbohydrates) by the process of photosynthesis. These carbohydrates are used for growth and respiration or may be stored for later use in the base of stems, in roots, and in seeds of plants. They serve as a food reserve that can be used later to sustain the plant during dormancy, for growth following dormancy, or for reproductive growth and development (Figure 20-4). Because these functions are so vital to survival and optimum production of range plants, it is important for range managers to allow plants to manufacture and store sufficient carbohydrates during the season of active growth to meet these needs. For ex-

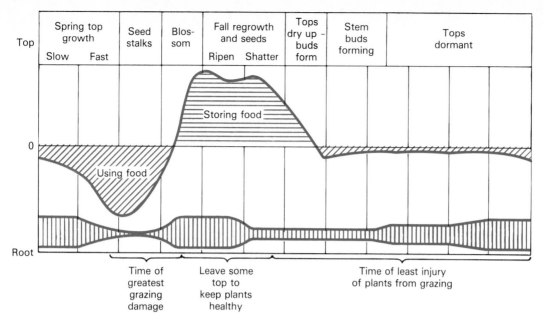

Top	Spring top growth		Seed stalks	Blos-som	Fall regrowth and seeds		Tops dry up - buds form	Stem buds forming	Tops dormant
	Slow	Fast			Ripen	Shatter			

Storing food

0

Using food

Root

Time of greatest grazing damage

Leave some top to keep plants healthy

Time of least injury of plants from grazing

Figure 20-4. Characteristic pattern of carbohydrate storage in range grasses. (*Source: Jefferies, 1977, Montana Cooperative Extension Service Bulletin 1018*)

ample, spring grazing of range plants should be delayed until the plants have adequate leaf area to carry out photosynthesis and rebuild the carbohydrate supply losses due to winter use when little or no photosynthesis took place. This is important in *all* forage management for perennials.

INFLUENCE OF SITE ON RANGE CONDITION

There are significant differences in productivity capacity within given ecosystem units or divisions. The areas of differing potential productivity are called *range sites*. It can be assumed that a good site is more productive than a poor site. Range sites are defined on the basis of their potential to produce.

In evaluating the potential productivity of a given area, soil type, climatic factors, and topographical differences must all be considered. For example, the kind and amount of vegetation can vary greatly within a given rainfall zone, as shown in Figure 20-5.

The illustration shows a cross section of a typical range site. Precipitation falls relatively uniformly over the entire area. Less water is available for plant growth on the hilltops, because part of the precipitation runs off before it can infiltrate into the soil. On the terraces, benchlands, and rolling hills, an average amount of moisture will penetrate the soil. The valley bottoms receive extra run-in water from the sides and from draws and upper reaches of the drainage. In the run-in areas, therefore, more moisture is available for forage production than the average precipitation would indicate.

Within each of these three situations, differences such as soil texture, depth, perme-

Managing Our Rangelands

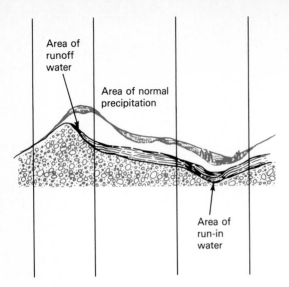

Area of
runoff
water

Area of normal
precipitation

Area of
run-in
water

Figure 20-5. Cross section of a typical range showing the three conditions that determine kind and amount of vegetation.

ability, salinity, or degree of wetness may exist. An area in which the factors are different enough to result in a significant difference in either the kind or amount of climax vegetation is called a *range site*. Only natural grasslands are classified as range sites. The Soil Conservation Service (SCS) has classified range sites by soil group name, precipitation zone, and geographic location. Table 20-2 lists SCS range site designations in the general order of their productivity, with descriptive criteria.

Range condition is classified in relation to what is considered to be its *climax vegetation,* defined as the stage of development in which plants, animals, and other organisms come into balance with their environment. As a range approaches climax, it receives a higher condition rating; as the species present deteriorate from climax, the site receives a lower rating. To describe range condition, the preferred terms are *excellent, good, fair,* and *poor*. On excellent-condition range, 76 to 100 percent of the vegetation is

made up of climax species. Good-condition range has 51 to 75 percent, fair has 26 to 50 percent, and poor has 0 to 25 percent climax species (Figure 20-6).

If too much grazing pressure is applied to a given range, the taller, higher-producing, most palatable species tend to disappear from the plant community because livestock prefer them. The disappearing plants are called *decreaser species*. As these species decline in number, shorter grasses and less palatable species increase in number to fill the void; these are called *increaser species*. If the range continues to be abused, the increaser species may diminish in size and amount, and weedy plants that were not a part of the original plant community move in. These plants are called *invaders*.

Range managers should keep a close watch on their ranges to observe any change in vigor or number of plants and to determine the direction of these changes. Climax vegetation, especially the decreasers, include the most desirable and, when properly managed, the most productive species. If there is a high proportion of climax species present, the range is in healthy condition. A growing number of increaser species is a warning that the range condition is deteriorating. When invader species come into the plant community in large numbers, this is a definite signal that management practices should be improved.

MANAGEMENT PRACTICES

When ranchers sell their products, they are paid for pounds (or kilograms) of meat, not for the number of animals on their rangeland. The amount of meat produced per unit of area and the proper care of the range resource are criteria for measuring the success of range livestock producers.

Range management practices must result in a profit for the producer. The manager

TABLE 20-2 RANGE SITES

1. Soil groups that can produce more herbage than ordinary range uplands because of superior soil moisture availability

Wetland: Lands where seepage raises the water table to above the surface during only a *part* of the growing season. Too wet for cultivated crops but too dry for common reed, cattails, or true aquatics.

Subirrigated: Lands with an effective subsurface groundwater table and water rarely over the surface during the growing season.

Saline lowland: Subirrigated and overflow lands where salt and/or alkali accumulations are apparent and salt-tolerant plants occur over a major part of the area.

Overflow: Areas regularly receiving more than normal soil moisture because of run-in or stream overflow.

2. Soil groups with no obvious soil or moisture limiting factors

Sands: Sands and loamy sands more than 20 in [50.8 cm] deep.

Sandy: Coarse to fine sandy loams more than 20 in deep.

Silty: Soils more than 20 in deep of very fine sandy loam, loam, or silt loam. This includes soils with 2 in [5.1 cm] or more of loam or silt loam over clayey subsoils.

Clayey: Granular clay loam, silty clay loam, silty clay, sandy clay, or clay more than 20 in deep.

3. Soil groups with characteristics that limit moisture-holding capacity or affect infiltration rates

Thin hilly: Loamy or clay soils on steep or hilly landscapes with a thin A horizon and weak or no structure in the subsoil, but with significant root penetration deeper than 20 in.

Stony: Soils more than 20 in deep with cobbles or stones occupying 40 to 80 percent of the surface.

Limy: Soils more than 20 in deep that are nearly white and very limy within 4 in [10.2 cm] of the surface.

Shallow clay: Shallow, granular clay soils that are 10 to 20 in [25.4 to 50.8 cm] deep.

Shallow to gravel: Soils that are 10 to 20 in deep to sandy gravel. Few roots penetrate deeper than 20 in.

Shallow: Soils 10 to 20 in deep to hard rock or soft beds of decomposed granite, siltstone, or sandstone. Few roots penetrate deeper than 20 in.

Panspots: Areas of silty, clayey, or sandy soils with shallow depressions of hard clays or other materials at or near the surface. The shallow depressions occupy 20 to 50 percent of the site.

Dense clay: Deep, nongranular clays. The layer is very hard to extremely hard when dry and very sticky when wet.

Thin breaks: Mixed soils of various depths with hard rock or other resistant bed outcroppings at different levels on steep, irregular slopes. Trees may occur locally above outcrops.

Gravel: Coarse-textured soils with more than half gravel and cobbles with loose sand and gravel at less than 20 in.

Very shallow: Areas where few roots can penetrate deeper than 10 in. Outcropping of gravel or bedrock is characteristic. Deep-soil pockets usually marked by tall grasses, shrubs, or stunted trees.

Saline upland: Soils more than 20 in deep with salt and/or alkali accumulations. Salt-tolerant plants occur over a major part of the area.

Shale: Readily puddled uplands where some raw shale fragments are exposed at the surface and little, if any, soil profile development is evident.

Badlands: Nearly barren lands broken by drainage, with small grazable areas.

Source: Montana Soil Conservation Service Technical Guide, Section II-E-3, May 1971.

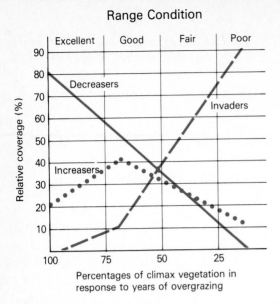

Figure 20-6. How increasers, decreasers, and invaders affect range condition. (*Source: USDA Soil Conservation Service*)

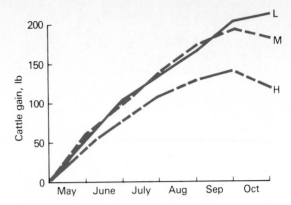

Figure 20-7. Pounds of grain with steers grazing under light (L), moderate (M), and heavy (H) stocking rates. Note weight loss at end of season under heavy stocking rates.

must strike a balance between the desire to improve range condition through deferred grazing or other management practices and the desire to maximize immediate income at the expense of long-term deterioration of the range resource.

Proper grazing pressure on the range can significantly influence the success of the ranching business.

Animal Units

The grazing capacity or carrying capacity of a given range is the maximum stocking rate or number of animal units that can graze in a given area for a given time period without damaging the vegetation or related resources. An animal unit is usually considered to be a 1000-lb [450-kg] animal or a mature cow and her calf or the equivalent. A horse is considered to be 1.25 animal units; a sheep, 0.2 animal units. The amount of forage required by an animal unit for 1 month is called an *animal unit month. Grazing pressure* refers to the relationship between the available forage and the number of animals grazing the area.

Stocking rates are extremely important, and managers must consider several factors—the capacity of the range to produce new growth, the amount of growth already present, the seasonal difference in production of various species, and other considerations that affect the amount of forage available throughout the grazing season. If the stocking rates are too high, the forage is grazed closer, less feed is available per animal, and gains per animal are reduced. If the stocking rate is too low, the animals have plenty of forage and gain well, but more usable forage is available than is consumed, and the animal produced per acre (pounds of meat) is less than optimum. So on a given range, as the number of animals per acre per unit of time increases from zero, the quantity of animal produced per acre increases. However, it reaches a peak as the number of animal units approaches the optimum grazing capacity of that site and falls off rapidly as the stocking rate increases beyond the optimum (Figure 20-7).

Moderate grazing usually gives the greatest return to the rancher in the long run. Table 20-3 describes the various degrees of grazing use.

Grazing Practices

Range management involves making decisions that take into consideration such factors as climate, soil types, plants, geology, topography, and drainage as they relate to plant growth, reproduction, and adaptation to a particular environment.

Rangelands vary widely in their productive capacity. Managers must consider the proper time during the season to graze, the most advantageous distribution of livestock on the range, the proper stocking rate (number of animals per unit of area), and the proper species of livestock for a particu-

TABLE 20-3 GUIDE TO DEGREE OF GRAZING USE

Degree of use	Description
Unused 0%	No livestock use.
Slight 0–20%	Practically undisturbed. Only choice areas and choice forage are grazed.
Moderate 20–40%	Most of the accessible range shows grazing. There is little or no use of poor forage and little trailing to grazing.
Full (This or less use is proper use.) 40–50%	All fully accessible areas are grazed. Major sites have key forage species properly utilized. Overused areas are less than 10 percent of pasture area.
Close 50–60%	All accessible range plainly shows use, and major sections are closely cropped. Livestock is forced to use much poor forage.
Severe 60–80%	Key forage species are used almost completely. There is a low-value forage-carrying grazing load. Trampling damage is widespread in accessible areas.
Extreme 80–100%	Range appears stripped of vegetation. Key forage species are weak from continual grazing of regrowth. Poor-quality forage is closely grazed.

1. Determine the degree of use at or near the end of the grazing period.
2. Proper use determination is based on key species on major sites, not total vegetation.
3. When properly grazed, the vegetation left will supply adequate cover for soil protection and will maintain or improve the quantity and quality of desirable vegetation.

Proper use of annual growth depends on the *season of use:*
Spring use   Moderate
Summer and early fall use   Full
Late fall and winter use   Dormant season (close)

Source: USDA—Soil Conservation Service.

lar environment. If wildlife grazing is involved, managers must be aware of the animal's grazing habits and needs.

Overgrazing can cause serious range deterioration, but grazing with good management can be beneficial to rangelands. Proper grazing use is the basic objective of any range management program. To achieve this goal, livestock numbers must be kept in balance with the annual forage supply. Most grass plants cannot stand having more than about half their year's growth grazed off. Grazing closer than that lowers the vigor of the plant, reduces root growth, and reduces the plant's ability to reach moisture and plant nutrients.

It should be recognized, however, that some range plants are well adapted to heavy grazing and browsing. These species are able to survive under severe grazing stress. Grazing is a means of harvesting the products of the range and should not be considered bad; it is harmful only if it results in significant deterioration of the range condition. In fact, ungrazed grasslands may stagnate and become less productive as a result of much buildup, cooler soil, increased incidence of mildew, and other factors. Continuous grazing can, however, reduce the vigor and stand of desired species. Moving the animals from one area to another will prevent them from grazing the same palatable plants repeatedly.

Sometimes the best management scheme is to combine several practices. A grazing system can be planned to defer grazing or rest some areas on a systematic basis. The animals will use the forage more fully in the smaller units. Because the animals are on a unit for only a short time, they will not graze the same plants repeatedly. The plants will then have part of the growing season to regrow and to store food reserves (Figures 20-8 and 20-9).

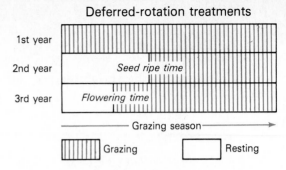

Deferred-rotation treatments

Figure 20-8. Grazing formula for deferred-rotation system utilizing three pastures, one of which would be receiving one of the treatments each year. The deferred-rotation system does not include the year of full rest in the formula, as does rest-rotation grazing.

A deferred or rest rotation system of grazing will regulate the season of use and, to a large degree, the intensity of use of some of the major species. The grazing system used must be flexible and specifically tailored to the range involved to be effective.

Livestock prefer some kinds of plants to others. These preferences result in some species being grazed more readily during certain seasons and at different stages of plant development. Some plants are grazed off at ground level; on others, only the seed heads are eaten; and on still others, it is the basal leaves that are taken. Some plants are overused every time livestock graze a range.

Even though overuse of a plant for one season may not kill it, overuse will weaken the plant and curtail its production, allowing less desirable plants to increase. A systematic deferment and change in season of use prevents livestock from overusing the same plants every year.

To supply needed forage early in the grazing season, managers may want to consider seeding part of their grazing land to an early, cool-season species such as crested

Rest-rotation treatments

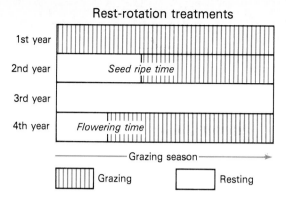

Figure 20-9. Grazing formula for rest-rotation system utilizing four pastures, one of which would be receiving one of the indicated treatments each year. The rest-rotation systems differ from deferred-rotation grazing by including the full year of rest in the formula.

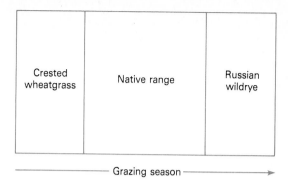

Figure 20-10. A practical grazing sequence for range livestock operations in the semi-arid to arid regions of Montana. This scheme utilizes the concept of complementary pastures.

wheatgrass (Figure 20-10). In some cases, ranges are in such a depleted condition that seeding of native species or adapted introduced species might be justified. Some examples of species that might be used include green needlegrass, meadow bromegrass, western wheatgrass, thickspike wheatgrass, and Russian wild rye.

In some areas, interseeding, pitting, or chiseling may help improve range condition and productivity. Interseeding is a practice in which narrow bands of soil are cultivated and seeded while leaving the remaining vegetation undisturbed. Pitting and chiseling cause irregular depressions which trap runoff water and provide a more favorable environment for plant growth.

Control of undesirable plants is a problem for the range manager as well as the crop producer. Range weeds are vigorous competitors but are generally low in palatability. Weeds invade overgrazed areas and dominate the plant community. Deferred grazing may permit the desired species to come back

as the dominant plant. Sometimes it is necessary to use herbicides, burning, mowing, or plowing and reseeding to control the weeds. Poisonous plants may also be a problem requiring control measures.

Fencing may help to use the range more effectively by making a rotation grazing system possible. Fencing should be carefully planned to avoid bunching the livestock in corners or in certain areas of the field. Because fencing is expensive, consideration must be given to the best possible layout to assure a reasonable return on the investment.

Developing water in appropriate areas and placing salt to assure proper distribution of livestock over the entire range are other important management considerations.

Livestock should be distributed as evenly as possible over the entire range. Water for livestock must be accessible at all times; a rule of thumb is to have it available within 1 mile [1.609 km] on relatively level range and within $\frac{1}{2}$ mile [0.8 km] in rough terrain. When water is available, the animals distribute themselves more evenly over the range and the forage is more efficiently consumed with less damage to vigor and stand. Sometimes it is necessary to fence the range into smaller units to get proper distribution (Figure 20-11).

"Take half and leave half" is not the whole answer to range management, but when used in combination with a sound, practical grazing system it can allow the range to improve in condition and absorb more moisture when it rains. It brings erosion to a minimum. It produces more grass, puts more dollars in the rancher's pocket, and preserves the range resource for future generations.

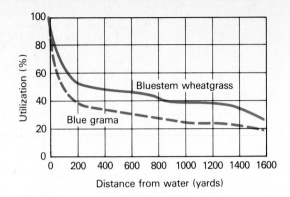

Figure 20-11. Relationship between percentage utilization of major forage species and distance from water on winter range.

SUMMARY

Rangelands cover vast areas of America, mostly concentrated in the western states. A primary objective of managing our rangeland is to obtain maximum livestock productivity on a sustained basis, while at the same time conserving the water and soil resources of these regions.

The range can be considered an ecosystem, which is a complex of organisms, both plants and animals, that interact within a given environment. Wise management of the range resource depends on a knowledge and understanding of the range ecosystem. Management practices must take into consideration the many factors of the range environment and how and when these practices can be applied to achieve optimum range productivity.

QUESTIONS FOR REVIEW

1. Describe what is meant by the term *rangeland* and how rangeland differs from cultivated or other types of land.
2. What are four classifications of range plants?
3. What are the characteristics that illustrate the differences between the four classes?
4. What are the components of a range ecosystem? How do they interact?

5. What is the range site and how is its potential productivity evaluated?
6. What is climax vegetation?
7. What are increasers, decreasers, and invaders?
8. What are the major factors that a rangeland manager must consider in determining the productive capacity of a range site?

Chapter
21

Food and the Challenges of the Future

Food, clothing, and shelter have been considered the three essentials for human life. The struggle for food is older than the written record of the human race. It is a tragic fact that starvation seems to have been part of the history of the human race. In spite of amazing strides in technology, an inadequate diet remains a serious concern to half or more of the world's population; 60 to 80 percent of the deaths in the world today can be attributed to dietary problems (an inadequate or an unbalanced diet). The threat of starvation is indeed tragic, but there are marked successes in the past and hope for the future. We must accelerate the rate of progress to further reduce the human suffering associated with undernutrition and malnutrition.

The tragic problems we face today are not new; they have existed for generations. Research and education have, in the past, helped overcome these problems. Research and education are our hope for the future.

CHAPTER GOALS

After studying this chapter you will be able to:

- Discuss the role agriculture has played in the growth of the United States as the world's leading industrialized nation.
- Discuss the growth and success of American agriculture in terms of labor requirements and increased productivity.
- Identify critical issues, such as erosion control and energy conservation, and possible solutions as they relate to crop production and human welfare.
- Identify career opportunity areas in agriculture and education requirements needed to seize these opportunities.

DEVELOPMENT OF AGRICULTURE

In the more highly developed nations, as is particularly shown in the United States, agricultural productivity has exploded in about the past 125 to 150 years. At the end of the first quarter of the nineteenth century (about 1825), farmers each produced enough to feed themselves and four others. As was true of most of the world, the economy of the United States depended nearly exclusively on agricultural production. To survive, most people were, to some extent, farmers. The first concern was to ensure an adequate supply of food. As a result of abundant natural resources, plentiful, fertile soil, and a favorable climate, including adequate rainfall, the potential for high crop yields existed. American ingenuity, in the form of a wide range of equipment, allowed this potential to be realized (Figure 21-1). Producing more than was needed on a single farm made it possible for men and women to turn to pursuits other than farming. In part, efforts went to further mechanize farming. Human resources were also released to the benefit of the professions—medicine, law, education, and even politics. At the same time, human resources were devoted to comprehensive exploration of the nation, from the Atlantic to the Pacific oceans, and to participating in the westward migration of families and in the California gold rush of 1849.

The success of the mechanization of agriculture can be measured in two ways. First, today a single farmer can produce an abundant diet for over sixty-five persons (Figure 21-2). Second, and as a result of the high productivity of the individual farmer/rancher, the professions have flourished in the United States. Although certain rural communities still need skilled local professionals (especially physicians), as a nation we have an extraordinarily high level of health care available. Law is a vigorous and critical profession in our free democratic society, and education for all is a keystone of our freedom and our national productivity. Our role as worldwide technological leader must thus be attributed directly to the high level of agricultural productivity that has released human resources for other pursuits, such as space exploration, or for solving critical problems such as energy issues.

The successes of American agriculture can be measured in far more than the ratio of farmers to consumers. In addition to reducing the number of people required to till an acre of soil, the number of people fed per acre has increased, as a result of increasing the amount produced (yield) per acre. Yields of traditional crops have increased dramatically in the past 50 to 80 years: corn yield has more than doubled, as has cotton. Soybean yields have far more than doubled, and the use of the crop has changed from horse feed to a valuable, diverse food source. These increases in productivity may be attributed to at least three factors: superior new cultivars, better total field management, and new uses of old crops (Table 21-1).

The dramatic increase in crop yield reflects the first two of these. First, the development of well-adapted hybrid cultivars with high yield potential has been partly responsible. Second, proper management practices have been used to help ensure realization of this potential. Management included much heavier seeding rates, greatly increased fertilizer rates, and improved methods of pest management.

Soybeans clearly illustrate the third factor. Initially, in the late nineteenth century, soybeans were introduced as a potential feed crop for livestock, particularly horses. As a forage or feed crop, they were less than successful. In the mid-twentieth century, soybeans were grown on the West Coast as a

(a)

(b)

(c)

Figure 21-1. The development of modern farming equipment has contributed to the high crop yields possible on today's American farms. Examples shown here are *(a)* a multipurpose planter, *(b)* a wheat harvester, and *(c)* a corn head. (*John Deere* [*a*], *Tim McCabe/USDA* [*b*], *International Harvester* [*c*])

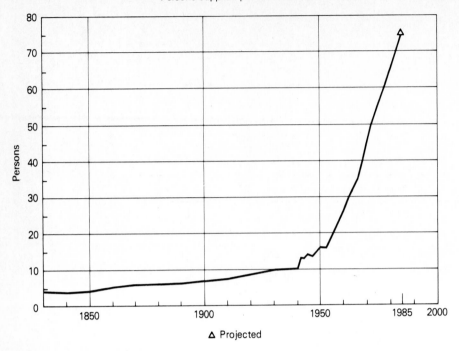

Persons supplied per farm worker

Persons

△ Projected

Figure 21-2. Increase in the number of persons supported per farmer.

seed-oil crop. They continue to be a valuable crop, but in more recent years, as a result of new technology, soybean meal is recognized as a prime source of proteins essential in the human diet. Many nutritious meat substitutes and so-called meat extenders are produced from soybean meal.

We have made great progress toward alleviating human suffering due to inadequate diets, but we still face many major challenges. The fundamental dietary needs in terms of quality and quantity of food required will not change. The way we meet these needs will become critical. The population of the world will continue to grow. Given the finite size of the earth, the ultimate solution of the world's food problems must include some type of long-term population control. The resources required for food

production are not limitless; over varying periods of time, many are renewable, but some, such as available space, are not (Table 21-2).

CHALLENGES FOR THE FUTURE

In terms of a meaningful goal for human well-being, none seems more appropriate than ensuring an adequate and nutritious diet for everyone. In spite of periodic reports of extensive crop surpluses in some areas, at no time in the recorded history of the human race has this goal been attained. Part of the problem seems to lie in socioeconomic factors that inhibit the distribution of food or crops from areas of so-called surpluses to deficiency areas. But even with essentially perfect distribution in the short term, some

TABLE 21-1 INCREASED YIELDS PER ACRE AND IN-CREASED FERTILIZER USE

	1962–64	1972–74	Percent Change
Acreage (thousands)			
Feed grains	101,377	99,001	−2
Food grains	49,896	58,728	18
Soybean	29,005	51,318	77
Cotton	14,613	12,541	−14
Production (thousands)			
Feed grains, tons	143,235	190,009	33
Food grains, tons	39,690	56,003	41
Soybean, bu	689,757	1,350,407	96
Cotton, bales	15,127	12,793	−15
Tons of Plant Nutrients Used (thousands)			
N	3,884.0	8,478.3	118
P_2O_5	3,085.9	5,006.5	62
K_2O	2,501.2	4,687.1	87
Total	9,471.1	18,150.0	92

Source: U.S. Department of Agriculture

dietary problems would continue to exist. On a long-term basis, redistribution of food seems not to be the ultimate answer, although the world must continue to strive for equitable distribution of foods and production of maximum yields.

Even with the most optimistic projections and analysis of natural resources, we must admit that the capacity of the earth to produce food is not limitless. Although the oceans of the world represent a significant resource that has not been fully exploited, they are of limited ultimate value. One possible solution to the long-term problems of malnutrition and undernutrition is control of population growth.

The question of birth control continues to involve serious religious, social, and moral issues. These issues seem not to be resolvable by simple yes-or-no answers. In addition, in spite of improved technology, widespread adoption of nearly any of the most effective methods of birth control, particularly in the less developed countries, will depend on extensive educational programs (Figure 21-3). Also, users need to be protected from the potential dangers of various methods through regular health checks.

TABLE 21-2 SPACE AVAILABILITY IN NORTH AMERICA COMPARED WITH THE WORLD AND SELECTED OTHER AREAS

Geographic Area	Estimated Total Area of Arable Land (million hectares*)			Projected Area per Person (hectares*)	
	1950	1970	1985	1970	1985
World	1418	1461	1610	0.40	0.33
North America	313	355	400	1.04	0.90
Oceania	24	47	95	3.03	3.52
Asia	463	467	484	0.23	0.17

*1 hectare = approximately 2.5 acres.
Source: H.O. Carter et al., "Perspectives on Prime Lands," 1975, USDA.

Figure 21-3. Family planning educational programs are an important resource for citizens of developing countries. (*F. Botts/FAO*)

LAND USE

Agricultural productivity draws heavily on both renewable and nonrenewable resources. Space now seems to be a factor that ultimately will limit agricultural productivity. Even considering the strides that have been made in "farming the sea," humans will continue to depend on traditional farm and ranch products to fulfill their dietary needs. Obviously, this means land for crop and livestock production. Of the nearly 30 percent of the earth's surface that is land, about 10 percent is ideally suited for crop production in terms of both soil and climate. Nearly all this land is now in crop production. Worldwide, the greater part of the land area that can readily be cultivated is currently under cultivation.

Human crowding is becoming a serious issue in many areas; Japan is a good example. In the United States, unfortunately, thousands of acres of high-quality cropland are diverted annually to nonfarm uses—housing, roads, and other urban needs. In some areas, laws have been passed to encourage reserving this valuable, productive land for food production, but questions of personal legal rights have been raised. Other issues concern pollution resulting from some production enterprises as urban areas move to previously rural settings. This has become a serious issue in poultry and dairy production and now involves also crops and the use of pesticides. On the other hand, pollutants from urban areas also are known to damage crops: in parts of southern California, citrus crops have been significantly harmed by automobile-produced air pollutants. Thus the problem becomes circular: the effects of urbanization on crops and livestock production and the effects of these, in turn, on urban growth (Figure 21-4).

Ultimately, the question of the best land use must be asked and answered. Seemingly, there are many areas where houses can be built and people can live but where crop or livestock production would be inef-

Figure 21-4. The use of land poses difficult questions. The irrigated orchards needed for food production and the housing development needed for shelter juxtaposed here illustrates the need for careful planning. *(D.J. Weir/H.U.D.)*

ficient and difficult, if not impossible. Such areas include the foothill areas adjacent to fertile valleys. When these areas are developed, care must obviously be taken to protect watersheds. Part of the land-use issue must include a deep concern for soil conservation.

Although more is written and broadcast about various forms of air and other types of pollution, soil erosion continues to be the nation's leading ecological problem (Figure 21-5). Unfortunately, far too few citizens realize that thousands of acres are lost annually to the forces of wind and water erosion. Land for which crop production is not the best use may be ideally suited to livestock grazing and have multiple-use potential as livestock habitat. Detailed analyses are critical to determine the best land use.

Until the mid-twentieth century in the United States, some increase in productivity could be expected merely by cultivating more land. This no longer is feasible. The high mountain ranges of the Sierra Nevadas and Rocky Mountains are not suitable for crop production, although some alpine or subalpine meadows are important summer grazing areas for livestock.

The greatest potential for increasing cultivated land appears to be major desert areas in the United States and worldwide. In these regions water is a major limiting factor, but where conditions otherwise are favorable, the high cost of developing water-distribution systems may well be justified. Several outstanding examples of the impact of water development are recognized in the western United States. In the Southern tip of California, for example, a veritable garden blooms in what had been barren desert (Figure 21-6), as a result of the development of distribution systems for water from the Colorado River. Crop production in Arizona relies heavily on irrigation and on water development. In some high-plateau desert regions, however, even if water were made available, other conditions would preclude significant crop production. Unfavorable conditions in parts of Nevada and Utah include extremely salty soil and, in some places, growing seasons too short to justify the expenses of water development and irrigation. One of the most outstanding examples of the impact of water development on crop production can be seen in the San Joaquin Valley in California. Thousands of acres that are naturally suitable primarily for dryland grain production now produce world-famous crops of grapes, vegetables, fruits, melons, beans, and cereals.

Worldwide, the greatest potential for bringing new land under cultivation seems to be the extensive jungle areas of South America and Africa. Climatic conditions in terms of length of growing season and available moisture indicate real potential for crop production. Three major problems would have to be overcome before such areas could be successfully cultivated. First, the extremely dense native vegetation would have to be removed and reinvasion restricted. Removing this natural ground cover could lead to severe soil-erosion problems and to flooding. Careful planning based on facts, not speculation, is critical. Second, most jungle soils are highly leached (from heavy rainfalls) and hence fairly to markedly infertile and acid. These problems could be overcome by the use of fertilizers and lime, but costs would be high. The third problem is one of economics: systems of transportation to the potentially most productive jungle areas are minimal if not absent, and people with adequate technical skills are scarce. Thus transportation systems would have to be developed and qualified technicians encouraged to relocate to these areas before their potential for food production could be realized.

Figure 21-5. Soil erosion remains the number one conservation problem in the United States. (*a*) Unprotected cropland erodes, washing away topsoil and polluting streams with sediment. (*b*) Crops planted up and down a slope accelerate the erosion of sandy soil. (*Tom Pozarnsky/USDA-Soil Conservation Service* [*a*], *Gene Alexander/USDA-Soil Conservation Service* [*b*])

Figure 21-6. Lettuce flourishes in this irrigated farm land in California. *(E.E. Hertzog/USDI-Bureau of Reclamation)*

INCREASED YIELD

Although events associated with the American space projects and advances in medical technology may seem more visible, the increase in crop (and livestock) yields illustrates the results of some of the most successful research efforts. During the most intense period of the United States space efforts, increases in the yield of major crops, such as soybeans, wheat, and corn, averaged 350 percent. To attain the goal of a satisfactory diet for every human, we must harvest more from each acre. This in itself is a major scientific challenge. Plant breeders have developed cultivars with greater yield potential and resistance to

disease and other pests. Other scientists have determined optimum planting rates and fertilizer management programs, while agricultural engineers have provided the equipment to handle large acreages and move and process large quantities of crop products. It is absolutely critical to realize that when greater demands are made on the soil in terms of higher yields, greater care must be given to the soil.

With higher yields per acre, more plant nutrients per acre are removed from the soil. This means that more fertilizer must be used to replace those elements removed with the crop. To justify the increased inputs, more extensive pest control must be practiced. This requires extra caution to avoid

Figure 21-7. Educational programs are an important factor in reaching the goal of a satisfactory diet for people throughout the world. (*Murray Berman/USDA*)

pollution and contamination from pesticides and to the soil from excessive cultivation arising from attempts at weed control or from multiple cropping. Remember, higher yields mean greater demand on the soil. The soil is the fundamental resource on which the crop product rests; this resource must be protected.

IMPROVED NUTRITION

High yields alone are not the whole answer in reaching the goal of a satisfactory diet for all. Agriculture must produce certain kinds of food to provide needed calories and to ensure good nutrition. We must know dietary needs and how they can be met, and land must be devoted, in the best way, to helping meet these needs. For example, corn and soybeans require essentially the same environment for good yields. In the corn belt, pig raising has flourished because appropriate feed is locally available. Now the question arises whether growing corn to produce pork can be justified when soybeans yield needed protein more efficiently. Some balance must be reached. Obviously, the same question could be asked about growing alfalfa for beef animals where beans or vegetables could be produced. Agriculturists should strive to cooperate as team members in producing the right amount and *kind* of food.

A major educational program is necessary if the goal of a satisfactory diet for all is to be reached (Figure 21-7). Advanced education is needed to train scientists to solve the problems of increasing the yield per acre.

Food and the Challenges of the Future

Equally important are educational programs to bring new technology to farmers and ranchers. Looking to the future, in the United States, more farmers and ranchers and more of the people serving them will hold college degrees. Various types of extension programs will increase in importance, and vocational programs will assume an upgraded and increasingly important practical role.

In addition to the expansion of these traditional educational programs on a worldwide scale, a major new program must be initiated. In some, if not all, parts of the world, people's eating habits and other cultural factors must be changed. In the interest of human welfare, the use of feed for protected, sacred animals cannot be justified. By the same logic, we must clearly distinguish luxury from necessity and commit ourselves to providing the necessities for all before a privileged few enjoy culinary luxuries. This does not automatically mean giving up meat for many in the United States; the country is blessed with abundant resources, the best use of which may well be livestock production. However, we may be forced to consider grass-fed beef as opposed to feedlot-fattened, grain-fed beef.

The United States cannot feed the world, and we need not be ashamed of the high-quality diet with which we are blessed. At the same time, we must recognize international factors that must be included in our decisions as to what constitutes the best land use. This and other related issues involved with the international marketing of agricultural products are very complex. It is essential that coming generations of United States agriculturists be educated in crop and livestock production and advanced technology, but they must also understand and become involved in economic decision making and politics. The agriculturist is becoming an international business agent.

ENERGY

The decade of the 1970s made a frightening fact clear to most of us in the United States: our supplies of energy, particularly petroleum products, are not limitless. Limited supplies and skyrocketing prices have forced agriculturists to examine the amount of energy used in crop production and to seek ways of reducing this consumption. At the same time, alternative energy sources must be developed.

As noted in Chapter 12, agriculture is not an *excessive* user of energy, but it is a major consumer. Current production, processing, and distribution practices require energy, much of it from fossil fuel. A drop in available supplies could lead to reduced production. Increased costs drive food prices upward in the absence of regulations. Farmers are in business and must make a profit.

Energy is required at all levels of the food chain. Production practices, from seedbed preparation through planting and cultivating to harvest, require energy. Some energy savings are expected with the development of more efficient engines and farm equipment. In many instances, several field operations can be combined, e.g., final seedbed preparation and actual seeding. This has reached its peak with the adoption in some areas for some crops of "no-till," or minimum-tillage methods (Figure 21-8). Weeds and other vegetation are controlled by specialized chemicals. In a single operation, a narrow seedbed band is cultivated, seed is placed in this banded area along with fertilizers, and sometimes pesticides are applied. There is no cultivation for weed control: the next operation is harvesting.

Methods of utilizing solar energy are being developed for drying grain (Figure 21-9) and hay—activities that have traditionally used fossil fuels. Solar methods are also being explored for curing tobacco.

(a)

(b)

Figure 21-8. "No-till" methods help save energy and reduce soil erosion. (a) A farmer uses the "no-till" method to plant corn. (b) "No-till" corn after first cutting. *(Tim McCabe/USDA-Soil Conservation Service* [a], *Gene Alexander/USDA-Soil Conservation Service* [b])

Figure 21-9. Methods of using solar energy for grain-drying are being developed. Three types of solar collection systems are being tested at this U.S. Grain Marketing Research Center in Manhattan, Kansas. (*USDA*)

Making nitrogen fertilizer requires fossil fuels; the need for nitrogen will increase as yields go up. More efforts must be made to take advantage of legumes in crop rotations to add nitrogen to the soil. Scientists are attempting to breed into cereals the symbiotic nitrogen-fixing capabilities found in plants of the legume (pea) family and *Rhizobium* bacteria. To date little success has been realized, but this is the type of major step we must continue to seek.

Energy is used in the production and application of agricultural chemicals, from fertilizers to weed killers and insecticides. The concept of integrated pest management must include energy conservation: do not spray needlessly or excessively; use the most effective material at the lowest rate; and where possible combine efforts to solve problems in an integrated manner. For example, effective weed control which is part of seedbed preparation may destroy the over-

wintering habitat of a serious insect pest, thus minimizing damage from the pest. A little extra effort at this stage of production may eliminate the need for special operations for insect control. There is a continuing challenge to determine how various activities can be better linked to achieve the most efficient production system.

Looking to the future, transportation and processing represent challenges for reducing energy consumption. Obviously, the crop is of little value until it reaches the consumer. Unfortunately, some crops are inefficient in terms of transportation—e.g., sugar beets. A field may yield 20 to 30 tons/acre [44.8 to 67.2 t/ha] of sugar beets, but the beets contain only 16 to 20 percent sugar. Thus the field yields 6000 to 10,000 lb/acre, or 3 to 5 tons [6700 to 11,000 kg/ha] of sugar. A great mass of relatively valueless material is therefore transported. Two points are important: (1) location of processing plants relatively close to major centers of production (this has been done in many areas) and (2) the finding of good uses for the by-products (this also has been done; beet pulp is a good feed). If the by-product requires extensive shipping, the energy issue arises again. The same phenomenon is seen in cotton. Generally, to minimize shipping, cotton gins are in close proximity to producing areas. The energy cost of moving fresh produce from the production site to the consumer is a problem that has no easy solution: for certain areas, many types of fresh produce will remain a luxury (see Chapter 17 for a review of where leading fruit and vegetable crops are produced).

To meet the goal of an adequate diet, food wastage must be minimized. Loss of crops growing in the field, during harvest, in storage, and during and after processing must all be greatly reduced. The technology to accomplish much of this is available, but energy input is essential. Certain factors must be carefully considered. On the one hand, more efficient processing and packaging methods are being developed; the microwave oven is a good example of a new, energy-efficient cooker (Figure 21-10). The return to reuseable containers, although at times a minor inconvenience, is a step in conservation, as is increasing recycling of aluminum containers. On the other hand, in terms of energy efficiency, how many convenience foods can be justified? How much energy are we willing to spend to have precooked, packaged meals that require only reheating? Note the term *reheating:* Energy has already been used once in processing.

CONSERVATION

If we are to meet the challenges of the future, conservation of all natural resources will become even more important. We must conserve and protect all the resources on which food production depends, including air, water, and soil.

Conservation of air may seem unnecessary, but remember that crops are damaged by air pollution. Crop plants need clean air as much as people need it.

Water conservation involves many factors. We must not waste water, although it appears to be replenished through rain and snow. Underground water supplies in some areas are being depleted as irrigation water is pumped out faster than water returns to them. We must exercise real caution to avoid polluting lakes and rivers. This demands care with chemicals and with all types of waste, both urban and rural. Livestock feedlots have been recognized as having significant pollution potential. Finally, water conservation may include the development of dams, reservoirs, and systems of canals and ditches to distribute irrigation water. This type of development requires intensive planning.

Soil must be considered the basic resource on which crop production, hence human welfare, depends; conserving the soil must be a top national priority. The importance of proper or optimum land use has been discussed. In terms of soil conservation, one consideration has not been adequately emphasized—control of erosion. In spite of all out technology and sincere efforts, soil erosion today remains our nation's most serious conservation problem. Wind and water move and destroy thousands of tons of irreplaceable topsoil every day. Not only is the soil lost to crop production, but it silts and contaminates rivers and streams. For example, estimates suggest that every year the Mississippi River carries in excess of 500 million tons [454 million t] of silt from valuable farmlands into the Gulf of Mexico. Soil losses can be as high as 8 tons/acre [17.9 t/ha]; a goal of reducing this loss to 2 to 3

Food and the Challenges of the Future

Figure 21-10. Using the microwave oven is one way in which we can reduce energy consumption while cooking. (*Jane Hamilton-Merritt*)

tons/acre [4.5 to 6.7 t/ha] would still leave serious damage. Erosion control requires thorough understanding of soils and of cropping practices. Careful management with awareness of the causes of and preventive measures for erosion must be the responsibility of every farmer (Figure 21-11).

CAREER OPPORTUNITIES

The challenges are great but surely not insurmountable. Most certainly the need for highly trained people will increase. Producers must be better prepared to recognize and solve problems and work with highly skilled technical representatives who deal in agricultural chemicals and related problems. Farm machinery and equipment development, modification, maintenance, and operation will require far more sophistication than today. The food-processing and packaging industry will increase in importance and in demands for technically competent professionals. Costs and prices will continue to swell, and business management

Figure 21-11. The control of erosion is everyone's responsibility. Farmers must adopt practices to protect the soil and others must learn not to abuse it. (a) Grass-backed, tile outlet terraces conserve soil and moisture. (b) White-pines provide protection from winds. (c) Contour strip cropping protects the soil. *(Tim Mc-Cabe/USDA-Soil Conservation Service [a], Erwin W. Cole/USDA-Soil Conservation Service [b], Erwin W. Cole/USDA-Soil Conservation Service [c])*

skills will become increasingly critical not only for farmers but for the galaxy of industries that serve and support farmers.

Truly exciting career opportunities exist today in all phases of agriculture, and opportunities will expand in the future. To take advantage of these opportunities requires mastery of basic principles and application of these principles to real problems. You must develop the ability to recognize problems and apply these principles to them.

Vocational agricultural programs give you a fine start. You should have clearly fixed in your mind that even when we consider fundamental scientific principles, we consider them in the light of real, practical problems, usually technical problems. You should realize that farming or ranching is a demanding technical career that takes practical experience. You cannot learn to grow crops or produce livestock from a book. (You should also realize, however, that the majority of career openings are not in direct production, and for them farm or ranch experience is of less importance.) You can get experience; your vocational agricultural program is an outstanding way. Then extend your education in agriculture to the college level. Even without a farm background you can find an outstanding, rewarding career. Discuss these opportunities with your teacher, your counselor, and your parents. In meeting the challenge, you can make a difference.

SUMMARY

Human welfare depends directly on successful crop production. The growth of human society is reflected in its agriculture. As the population of the world grows, more must be produced from every acre in order to feed people. There are many challenges to be faced if we are to improve human nutrition. These include determining the best use of lands. Increasing yields will make greater demands on the soil, which in turn will require more fertilizers and better pest management. Crop production requires energy. We must conserve energy and seek efficient ways to produce crops. Conservation of all natural resources is essential.

If agriculture is to succeed, the demand for well-educated technical specialists will increase. Although fewer people may be required directly in growing crops, the demand for people to provide farmers with technical advice and help is rapidly expanding. Career opportunities in agriculture are excellent. To take advantage of these, education is essential. The vocational agriculture program is an excellent starting place.

QUESTIONS FOR REVIEW

1. How much has the productivity of the United States farmer increased in about the last century? What factors might be responsible for this? Can we expect this type of increase in the immediate future? Why?

2. What are the major factors limiting increased food production in the United States? Worldwide?
3. What proportion of the earth's land surface is ideally suited for crop production? What factors make the rest of the earth's surface less well suited for crop production?
4. Briefly discuss the concept of "best land use." Why is an education in agriculture important in determining "best land use"?
5. What is the most serious conservation problem facing the United States today? Approximately what is its magnitude in terms of losses?
6. Where might new lands be opened to increase food production? What would be required in order to do this?
7. If plant breeders develop cultivars of major crop species with higher yield potential, what will growers have to do to use them?
8. Can American agriculture feed the world?
9. How might energy shortages affect crop production? Use examples.
10. What aspects of the agricultural enterprise consume energy in the greatest and in the least amounts?
11. Discuss why education in agriculture will become increasingly important to human welfare.
12. List ten career areas in agriculture other than farming.

Glossary

Adenosine diphosphate or monophosphate (ADP, AMP) Low-energy molecules produced by splitting off one phosphate group (ADP) or two phosphate groups (AMP) from the ATP molecule with subsequent release of energy for cell work. They are built back up to ATP using energy from the breakdown of glucose. (See also Glycolysis.)

Adenosine triphosphate (ATP) A special energy-storing molecule with three phosphate groups, two of which are attached by high-energy bonds. When a phosphate group is split from the molecule, energy is released.

Aerobic respiration A process by which oxygen is used to break down molecules, resulting in a release of energy, carbon dioxide, and water.

Amino acid An organic acid containing nitrogen, which serves as a building block for protein.

Anaerobic respiration A process used to break down molecules when there is an inadequate supply of oxygen. This process results in production of less ATP than occurs with aerobic respiration and the end products include acids (in bacteria) or alcohols (in yeast), rather than CO_2 and water.

Animal unit The amount of forage required by a 1000-lb (453.6-kg) animal or by a mature cow with or without calf.

Annual A plant that completes its entire life cycle in one year or in one growing season.

Antioxidant A chemical that will slow fat deterioration.

Aquatic Term applied to plants or animals that live in water.

Arthropoda A phylum of animals that have segmented bodies, jointed legs, and exoskeletons.

Asexual reproduction Propagation of plants from vegetative parts, such as stems, leaves, or roots, or from modified stems such as bulbs, tubers, rhizomes, and stolons. This is accomplished without union of gametes.

Available moisture The water in a soil that is available for plant use. The amount is between the field capacity and permanent wilting percentage.

Bacteria A group of one-celled, microscopic organisms.

Bacterium A single-celled microorganism.

Bagasse Crushed sugarcane or beet refuse resulting from sugar processing.

Bedding A method of preparing the seedbed for cotton.

Best (or optimum) land use Use of a land area that yields the greatest good to humans with the least damage to natural resources (after consideration also of possible benefits of alternative uses of the land).

Biennial A plant that requires two years or two growing seasons to complete its life cycle.

Biological control Any of a wide variety of substances and/or methods of pest control that emphasize the use of living organisms or products derived directly from them.

Biotic component of soil The living component of the soil, composed of microorganisms, plants, and animals. (See also Mineral component of soil.)

Bloat The rapid buildup of excessive gasses in the stomach of an animal.

Bolt To produce a seedhead prior to the time one is normally produced; frequently applied to a biennial that produces a flower in the first year of growth.

Boot stage Period of growth when the inflorescence of the plant expands in the upper leaf.

Budding The implanting of a bud from one plant into the stem of another plant.

Bulb A specialized, short stem with densely grouped, fleshy leaves. The common onion is a bulb.

Bunchgrass Grasses with a nonspreading growth habit; they have no rhizomes or stolons.

Calorie The unit of heat energy needed to raise the temperature of 1 g of water 1 ° C.

Cambium The layer of cells in the stems and roots of plants that give rise to phloem and xylem tissue.

Carrying capacity The maximum stocking rate or number of animal units that can graze in a given area for a given time period without damaging vegetation or related resources.

Carbohydrate An organic compound composed of carbon (C), hydrogen (H), and oxygen (O); these include sugars, starches, and celluloses, energy-producing foods.

Carpel A simple pistil or a single member of a compound pistil, e.g., the outer covering of a cotton boll.

Ceres The name of an ancient Italian goddess of agriculture from which the word cereal was derived.

Certified seed Seed that meets rigid standards of purity and of ability to germinate and is so designated by an authorized agency.

Chemical pesticide A chemical or other material that will destroy or control a crop pest.

Chilling requirement The requirements of cultivars of many crop plants, commonly fruit trees, for specific periods of cool temperatures needed to cause the plants to break dormance and initiate floral development.

Chloroplast A part of certain cells in green plants. It contains chlorophyll, which is necessary to photosynthesis.

Chlorosis A lack of chlorophyll; yellowness of normally green plant tissue due to partial failure of chlorophyll to develop. Many diseases cause chlorosis.

Citric acid cycle The complex series of biochemical reactions in which the products from glycolysis are further oxidized to yield ATP; CO_2 is given off.

Climate Long-term meteorological conditions of temperature, precipitation, wind, etc.

Climax vegetation The stage of development in which the plants, animals, and other organisms in an ecosystem come into balance, or equilibrium, with their environment.

Coarse-textured soil Soil with a relatively high percentage of sand; includes loamy sands and sandy loams, except very fine sand-loams. (See also Fine-textured soil; Medium-textured soil, and Light soil.)

Coleoptile   The first leaf above the ground forming a sheath around the stem tip.

Combine   A machine for cutting and threshing grain; used to gather and thresh seed crops directly from the field in which they are grown.

Configuration   The relative position in space of the atoms of a molecule.

Contamination   The introduction of physical, chemical, and/or biological agents that render food unfit to consume or reduce its nutritional value. (See also Spoilage and Wastage.)

Contour plowing   Plowing around a hill at a common elevation, rather than over it, to limit water or wind erosion; a very important soil conservation practice.

Convenience foods   Any of a wide variety of products that are processed and packaged for easy and rapid home preparation.

Cooperative   A business organization owned and operated for the benefit of those using its services.

Corm   A specialized part of a stem; a short enlarged base of a stem where food is stored. Timothy grass reproduces by means of corms.

Cotyledons   Leaves at the first node of the primary stem. In dicots they are the food-storage part of the seed.

Cross pollination   The process by which pollen from a flower on one plant fertilizes a flower on another plant.

Cultivar (of a crop)   An identifiable strain of a species, usually developed by plant breeders, that has specific characteristics of crop quality, adaptation, or resistance to pests. It is somewhat comparable to a breed of animals.

Cutter bar   The part of the combine that separates the heads from the standing stalk.

Cutting   A part of a plant that is used for propagation, such as the root, leaf, or stem.

Decreaser species   The palatable, nutritious climax species desired on a range site. When the range is overgrazed, they start to disappear.

Deficiency symptom   The specific characteristics displayed by plants when they have an inadequate supply of a specific essential element.

Defoliant   A chemical product that causes the leaves of plants to dry more quickly.

Defoliating agent   A type of chemical applied shortly before harvest to cause the plant to shed its leaves.

Dehisce (or dehiscence)   The natural splitting open of either fruits (usually dry fruits) or anthers. Also refers to the breaking off of plant parts.

Dehydrating   The controlled reduction of water content in food.

Desiccant   A drying agent commonly used in the food production process.

Determinate   Having the primary and each secondary axis ending in a flower or bud, thus preventing further elongation.

Diastase   An enzyme produced by the plant embryo which is instrumental in converting starch to sugar.

Dicotyledon (or dicot)   A subclass of flowering plants that have two cotyledons at the first node of the primary stem.

Digestion   A series of chemical reactions in which food is partially broken down, chiefly through the action of enzymes.

Dormancy of seeds and or plants   An arrested or slowed stage of metabolism, growth, or development. In plants this may be in response to decreasing temperatures to insure protection against cold.

Duration   The time a given temperature exists.

Ecology   A science that deals with the response of organisms to their environment.

Ecosystem   A community of living things, such as plants or animals, as it interacts with the soil, water, sunlight, and other parts of the environment.

Elongate   To increase in length.

Embryo   The miniature plant that results from sexual reproduction. It is surrounded by the seed coat.

Endosperm   Food storage material in the seed. It is surrounded by the seed coat.

Entomologist   A person who specializes in the study of insects.

Environment   The sun of all external factors affecting an organism.

Enzyme   Protein that regulates cellular activity.

Epigeal germination   In dicots, germination in which the cotyledons rise above the soil surface.

Equilibrium   A state of natural balance. An ecosystem that is in equilibrium is self-perpetuating.

Essential elements   Elements essential for growth of a plant.

Exoskelton   A skeleton on the outside of a body.

Fanning mill   An air-screen seed cleaner in which screens and air flow separate weed seed and foreign material from seed.

Fat   Also called lipids; a group of organic compounds in the category of carbohydrates, but with a lower proportion of oxygen.

Fermentation   A form of anaerobic respiration that results in the production of alcohols and organic acids under some conditions. It is important in some industrial processes and in making silage.

Fertilizer   A material applied to the soil to provide one or more elements essential to growth of a crop.

Field capacity   Water held in a soil after it has been saturated and drained by natural (gravitational) forces.

Fine-textured soil   Soil with a relatively high percentage of silt and clay; includes all clay and loams. (See also Coarse-textured soil, Heavy soil, and Medium-textured soil.)

Food preservation   Chemical and physical processes used to retain the quality of food, and the methods of employing these processes.

Food processing   Chemical and physical methods used to preserve or protect food, to make using the food more convenient, or to create a variety of food materials from the same raw product.

Forage   Any plant material consumed by livestock. The most common forage crops are grasses and legumes.

Forbs   Any herbaceous (not woody) plants other than grasses or grasslike plants.

Fossil fuel   An energy source derived from the decay of organic material; crude oil is an example.

Freeze drying   Reducing the temperature of and removing the water from food material.

Frost-free period The time from the last killing frost of the spring until the first killing frost of the fall.

Fruit Mature ovary.

Fumigant A substance that forms a vapor that will destroy insects and pathogens. Fumigants are used most often in tightly closed structures.

Fungicide Chemicals used to control pathogens.

Fungus A plant that lacks chlorophyll and so must obtain all nourishment from other plants. These small plants cause rot, mold, and plant diseases. (Plural is fungi.)

Furfural A liquid solvent made from bran, corncobs, and similar plant materials.

Futures Contracts for buying and selling goods such as wheat, barley, corn and other products to be delivered in the future.

Futures markets A system of trading contracts of specific kinds and amounts of a particular commodity for delivery at a specific future time and place.

Gall A cancerous growth on the root or stem of a plant that interferes with cell division.

Gamete A specialized cell produced by the gametophyte (spore). They unite at fertilization to produce the zygote.

Geneticist A scientist who specializes in the science of heredity, such as plant breeders who develop new crop varieties or improve existing crops.

Germ The embryo of a cereal grain.

Germination A complex series of events during which the reserve food in a fruit or seed is respired. The embryo grows, develops, and ultimately becomes a self-sustaining plant.

Glycolysis The first step in respiration in which sugar is partially oxidized to yield a small amount of ATP for work in the cell.

Grafting Inserting buds or shoots from one plant into the roots or stems of another.

Grass Any plant of the Gramineae family.

Grazing pressure The amount of grazing on a range or pasture.

Gynophore A structure in peanut plants that connects the upper part of the plant with the underground fruits.

Hard pan A hard, subsurface layer of soil that may limit downward root growth and water drainage; may be caused by natural effects or by cultivations repeatedly at the same depth or by traffic over a field.

Harvester Any of various farm machines used to harvest farm crops.

Header A machine that cuts, gathers, and feeds grain into a threshing mechanism.

Heading back Pruning to remove only terminal portions of stems or branches, not entire stems or branches.

Heavy soil A soil with a relatively high proportion of clay and/or silt.

Hectare In the metric system, a measure of area equal to 10,000 square meters or about 2.5 acres.

Hedging The execution of sales and purchases in the futures market to offset each other. This marketing tool gives some protection to sellers and buyers of grain and other commodities against fluctuations in price.

Herbaceous cuttings Cuttings made from the stems of non-woody plants that are green and leaflike in appearance or texture.

Herbicide Chemicals used to kill plants,

usually weeds. Commonly called weed killers.

Humus[1] A thick paste, puree, made from crushed garbanzo beans (chick-peas) with lemon juice, garlic, and salt; a staple food of the Middle East.

[2]Decayed plant and animal material found in the soil.

Hybrid The product of cross pollination between genetically different plants.

Hypocotyl That portion of the stem from the top of the roots to the first node of the embryonic stem, or to the cotyledons.

Hypogeal germination In dicots, germination in which the cotyledons do not rise above the soil surface.

Increaser species The less palatable climax species on a range site that increases in amount when the range is overgrazed.

Indeterminate Having the axis or axes of plants not ending in a flower or bud, resulting in further elongation.

Indeterminate flowering A plant in which flowering is more or less continuous until climatic conditions become unfavorable.

Infiltration rate Rate at which a particular soil will absorb water.

Innoculation The placing of a culture containing bacteria (*Rhizobium*) on the seed of legumes prior to planting.

Insect Small animals which belong to the Anthropoda phylum. Their body is divided into three sections, head, thorax, and abdomen. Most adult insects have one or two pairs of functional wings.

Insecticide A chemical product designed to control insects.

Internode A portion of a plant stem between two nodes.

Invader species The weedy plants that move in on, or invade, a range site that has been badly abused.

Jointing The growth of an additional branch or leaf from a node.

Kilogram In the metric system, a measure of mass equal to 1000 g, or about 2.2 lb.

Kinetic energy Active energy, or energy currently doing work.

Larva A young insect that is entirely different from the adult. It is the feeding stage for those insects that go through complete metamorphosis or change. (Plural is larvae.)

Layering A propagation technique in which the stems develop roots while still attached to the parent plant.

Lecithin A substance used to treat oil used for food products and pharmaceuticals.

Legumes A family of plants that, in association with bacteria in the genus *Rhizobium*, participate in nitrogen fixation. Alfalfa, cloves, peas, beans, and soybeans are common legumes.

Light quality The relative proportions of light of different wavelengths.

Light quantity The amount of light energy received by a plant or leaf; commonly measured in foot candles, which measure brightness, not energy.

Light soil A soil with a relatively high proportion of sand.

Linter The short, fuzzlike fiber that is attached strongly to the cotton seed and is left after ginning.

Lysine One of the amino acids essential to good human nutrition.

Macronutrients   Essential minerals required in relatively large amounts by all higher plants.

Malnutrition   A condition arising from not receiving an adequate amount of nutrients, even if adequate calories are provided by the diet.

Market class   A classification of grain or other crop products, which may be further divided into subclasses, for the purpose of designating grade under the United States Grain Standards Act.

Marketing grade   Standards which have been established to describe the quality of a farm product for selling purposes.

Market spread   The difference between the price the producer receives and the price the consumer pays.

Maturity   A plant or plant part that is fully developed; no further growth occurs.

Medium-textured soil   Soil with a structure that is intermediate between fine-textured and coarse-textured; includes very fine sandly loam, silt loam, and silt. (See also Coarse-textured soil and Fine-textured soil.)

Meristem   Plant tissue usually made up of cells that are not fully differentiated and are capable of repeated mitotic divisions.

Metamorphosis   Change in shape from one stage to another in the life cycle of an insect.

Micronaire   A measure of fiber fineness that is dependent on fiber weight and diameter.

Micronutrients   Essential minerals required in relatively small amounts by all plants.

Microorganisms   Generally bacteria and fungi, but may also include nematodes and sometimes viruses.

Mineral component of soil   The portion of the soil derived from silt, sand, or clay. (See also Biotic component of soil.)

Minerals (in nutrition)   A variety of inorganic substances that are considered necessary for good health but that are not respired to yield energy. An example is calcium for strong bones and teeth.

Mitosis   Nuclear division in which chromosomes duplicate and divide to yield two nuclei that are identical to the original nucleus. It is the process by which new cells are formed for growth and tissue repair.

Monocotyledon (or monocot)   A plant that produces seed containing only one cotyledon.

Monogerm seed   Seed with only one embryo and from which only one plant will be produced.

Monophagous insects   A group of insects that restrict their feeding to a single species of plant.

Natural enemy   Environmental factors that help to control population levels.

Nematocide   Chemicals used to control nematodes.

Nematode   A small organism, such as a worm, that feeds on or in plants or animals. Nematodes live in moist earth, water, or decaying matter. They pierce the cells of plants and suck the juices.

Nitrogen fixation   The conversion of atmospheric nitrogen (N_2) into a stable, reduced, or oxidized form that can be assimilated by plants. Certain bacteria and algae are able to fix nitrogen biochemically.

Node   A swollen region of a stem to which

a leaf is attached and at which a bud is frequently located.

Nonrenewable natural resource A resource that is not replenished through natural means in a fairly short period of time, for example, a few years. Forests are considered to be renewable, the soil is not, due to the relatively long time required for soil formation.

No-till (or minimum-tillage) farming Farming practices that reduce the number of times a field must be cultivated to raise a crop. Usually a traditional seed bed is not formed, several operations are carried out at seeding, the crop is sown into a stubble or weedy mulch, and chemical control of weeds is very important.

Noxious Weed Act A Federal Act that authorizes the inspection of crop seed at port of entry and gives authority to conduct weed surveys.

Oilseed crop A crop that is grown for the oil that can be processed from its seed or fruits. Some crops are grown for many purposes including oil, for example, soybeans for the high protein seed meal and the seed oil and cotton for lint and oil.

Oligophagous insects A group of insects that restrict their feeding to a small group of similar plants.

Omnivorous insects A group of insects that consume a variety of organic food, both plant and animal.

Organic matter Material in the soil that includes plant and animal residue.

Ovary The enlarged base of the pistil in which female spores are produced and seeds are formed. The ovary matures into a fruit.

Overgrazing Allowing livestock to graze too long or intensely on a unit of land, which results in destruction of the plant community and soil erosion.

Oviparous insects A group of insects in which the female of the species lays eggs during the reproductive process.

Ovoviviparous insects A group of insects in which the female of the species retains the eggs in the body until hatched, at which time they give birth to active young.

Ovules The plant part that contains the embryo and the female germ cells.

Pappus The teeth, bristles, or awns that surround the achene of certain flowers of the Compositea family.

Parasite A living organism that must obtain part or all of its nutrients from other living organisms.

Parent material The rocks and minerals from which soil is formed by the various physical and chemical weathering processes.

Parthenogenesis The process by which a female insect's egg develops into a new insect without fertilization of the egg.

Pathogen Tiny microorganisms that live in and on host plants and cause plant diseases.

Pathogenic microorganism Tiny microorganism that lives in and on host plants and causes plant diseases.

Peg A stalklike structure formed on flowering peanut plants. A peg bends downward and forces the ovary into the soil, where it matures to form the pod, or unshelled peanut.

Perennial A plant that survives more than 2 years and frequently indefinitely. Perennial plants usually produce seeds each year; they then may become dormant before starting new vegetative growth and repeating the cycle.

Perishability The tendency of a food to become unfit for consumption or to lose nutritional value unless it is stored and/or packaged under special conditions.

Permanent wilting percentage Percentage of water remaining in a soil when a plant wilts and cannot recover unless water is added to the soil.

Photoperiod The relative length of light and dark periods as the season progresses.

Photoperiodic response The flowering response to the relative length of light and dark periods of a plant as the season (long summer days and short winter days) progresses.

Photosynthesis The complex series of chemical reactions involving light, as an energy source, and chlorophyll. During the process carbon dioxide (CO_2) and water (H_2O) are changed into a carbon compound rich in energy and oxygen is given off as a by-product.

Photothermal unit A measure of both heat and light energy.

Picker-sheller A machine that snaps the ears from the corn stalk, removes the husks, and shells the grain from the cob in one field operation.

Plant community Refers to a group of plants in a given environment that have similar living requirements.

Plant pathologist A scientist who studies plant diseases.

Plant population The number of plants per unit area.

Plant succession An orderly process of change in composition of a plant community until the vegetation stabilizes and is in balance with its environment.

Plume A large, long, conspicuous feather-like seedhead.

Pollination The transfer of pollen (male gametophytes) from the anther to the stigma of a flower.

Polyethylene A plastic material used to cover the soil, which aids germination, provides plant protection, and helps prevent weed growth.

Polyphagous insects A group of insects, such as the grasshopper, that eat a wide variety of plants.

Pome fruit A fruit that is composed of the pericarp (ovary wall) and stem or other nonreproductive tissue. Apples are the most common example.

Pore space The portion of a volume of soil not occupied by solids.

Potential energy A form of energy not currently doing work, but which can do work.

Primary noxious weed Usually includes a group of weeds that spread by sexual and asexual means and are very difficult to control.

Primary plant foods Nitrogen, phosphorus, and potassium, the three elements most frequently used as fertilizer.

Proboscis A highly specialized insect appendage used to extract nectar from plant flowers. When not being used, the proboscis coils up because of its natural elasticity.

Propagation The reproduction of plants by seed, cuttings, budding, grafting, or other means.

Protein Any of a group of complex organic compounds that contain special groups of molecules called amino acids. They are characterized by having hydrogen (H) and nitrogen (N) in a very specific arrangement.

Pruning General term to describe the removal by cutting of buds, stems, or entire branches.

Pubescent Covered with soft down.

Pulses Non-oil-producing cultivated legumes grown for their edible seeds.

Pupa The non-feeding stage in an insect's life cycle. The insect usually cannot move during this stage.

Pure live seed The percentage of a quantity of seed that is viable seed of the kind and variety (cultivar) designated.

Raceme An inflorescence with an unbranched central axis from which flowers are borne on short stalks.

Rangeland Land suitable for grazing by domestic lievestock. The vegetation consists mostly of native grasses, grasslike plants, forbs, and shrubs.

Range management The care and use of rangelands to get the highest continual production of forage and other products of the range resource (water, cover for wildlife, etc.).

Range site An area of land having a unique combination of soils, climate, and/or topography. Different sites have different vegetation and animal communities.

Raw sugar Sugar prior to the final steps in refinement.

Reel The large, paddle-wheel-like device at the front of a grain combine that rotates and pulls grain toward the cutter bar.

Registered insecticide One that has been approved, following extensive scientific research to prove its reliability, ability to selectively kill insects, and safety to humans when used at recommended rates on the appropriate crop.

Renewable natural resource See nonrenewable natural resource.

Respiration A complex series of chemical reactions in a cell during which food is broken down into carbon dioxide and water, if oxygen is available, and ATP is produced. (See also Aerobic respiration and Anaerobic respiration.)

Rhizobium Genus of bacteria that live symbiotically on the roots of legumes and fix nitrogen that is used by plants.

Rhizome A horizontal stem growing below the soil surface. (See also Stolon.)

Rick A large stack or pile of hay, cotton, corn, straw, or similar materials.

Ripeness of a fruit The physical and chemical changes in a fruit following maturity; these may include color development or drying, depending on the crop.

Root stock The trunk or root material to which buds, or scions, are transferred in grafting.

Rosette A cluster of spreading or radiating basal leaves.

Rotational grazing A management scheme in which livestock are systematically rotated from one pasture to another to maximize forage use and increase productivity.

Saturation The condition of a soil when all pores are filled to maximum capacity with water.

Scion The portion of a shoot or twig that is inserted into the "stock" of another plant when grafted.

Sclerotin A protein substance that helps the exoskelton of an insect to be tough, lightweight, and somewhat resistant to chemical substances.

Secondary noxious weed Weeds that are usually less difficult to control and somewhat less persistent than primary noxious weeds.

Seed An embryonic plant in a state of arrested development prior to germination.

Seed purity The percentage of a quantity of seed that is of the kind and variety (cultivar) listed on the label.

Seeding rate The amount of seed sown per acre; it may also include the spacing between rows of certain crops (e.g., vegetables) and within rows.

Sexual reproduction Reproduction of plant species that are propagated by true seeds.

Shrubs Woody plants that are relatively short in height.

Silage Moist forage preserved by partial fermentation in an airtight container.

Socioeconomic factors The complex of cultural traditions or religious beliefs and financial considerations that affect the culture of a country or region.

Sodium glutamate A chemical substance used by the food industry to bring out the flavor of certain cooked foods.

Soil The upper portion of the earth's crust that is capable of supporting plant life.

Soilage The term describes forage that is cut green, chopped, and fed directly to livestock.

Soil fertility The ability or potential of a soil to provide essential elements to plants.

Soil moisture-holding capacity The water held at field capacity in a foot of soil, generally expressed as inches of water per foot of depth of the soil. For example, clay has a greater holding capacity than sand due to grain pore space in a clay.

Soil ph The relative acidity or alkalinity of a soil, or the soil solution.

Soil structure The combination or arrangement of primary soil particles into secondary particles, normally larger in size.

Soil texture The relative proportion (or percentage) of sand, silt, and clay.

Soil triangle A geometric representation of soil texture that allows various textures to be characterized by the percentages of sand, silt, and clay.

Soil type A subdivision of a soil series that describes soil characteristics including texture of the surface horizon.

Speculation The practice of trading in futures contracts in which buyers take the risk of price change. Speculators underwrite the risk for the hedging process.

Spoilage Changes in foods or food products that render them unfit to eat or reduce their nutritional value. Spoilage is usually caused by microorganisms. (See also Contamination and Wastage.)

Spurs Stems formed from older wood that have short internodes and bear leaves, fruit, or both.

Square A cotton bloom just prior to opening.

Stand of crops The number of plants per unit of area that survive and grow; sometimes referred to as plant population.

Sterilization Using heat to destroy organisms that might cause food to spoil while it is in storage.

Stolon A horizontal stem growing above the soil surface. (See also Rhizome.)

Stomatal opening (or stoma) A small aperture or opening in the epidermis of leaves and stems. (Plural is stomata.)

Stone fruit A fruit with a single, hard, woody pit; also called drupe fruit. Examples include peaches and apricots.

Stubble-mulching Leaving crop residues in place on or partially incorporated into the soil subsequent to harvest and prior to and during preparation of the seedbed.

Stylet A spear-like mouthpart used by an insect to puncture a hole in plant tissue in order to remove plant fluid.

Subtropical crops Crops intermediate between tropical and temperate crops; they may be evergreen or deciduous, but do not tolerate temperatures below −4 to −5°C. Citrus is a good example.

Summer annual A weed that germinates in the spring, grows, sets seed, and dies before fall.

Summer fallow Uncropped land that is kept free of weeds throughout a growing season.

Symbiosis A mutually beneficial relationship between two living organisms of different species, living closely together.

Symbiotic nitrogen fixation The process by which legumes add extra nitrogen to the soil because of the unique ability of some bacteria.

Systemic Term applied to a substance that spreads throughout the entire body.

Temperate crops Crops that require or thrive in a relatively cool environment. The majority of crops produced in the United States are temperate crops.

Tendril Slender, coiling, modified leaves of climbing plants, such as peas, that support and attach the plant to surrounding surfaces.

Terminal elevator A crop-storage facility where large quantities of grain are delivered for further distribution and shipment.

Terminal transport The final steps in aerobic respiration in which some ATP is produced and the hydrogen removed from the "food" during oxidization is joined with oxygen to yield water.

Thresher A machine that separates grain from plants such as wheat and rye.

Threshing Removing the grain from seedheads, ears, or pods by a rubbing action while passing the material between a rapidly revolving cylinder and a stationary surface called a concave.

Tilth The physical conditions of a soil that make it suitable for the proper germination and growth of plants.

Transpiration The flow of water from the soil through a plant to the atmosphere.

Tropical crops Plants that are frequently evergreen and cannot stand freezing. Examples include bananas, pineapples, and coffee.

Tuber A specialized stem; the enlarged, fleshy tip of an underground stem. The potato is a tuber.

Undernutrition A condition arising from inadequate energy (calories) in the diet.

Variety See Cultivar.

Vector An agent that transmits a plant disease.

Vegetarian An animal that lives entirely on vegetative matter.

Vegetative growth The growth of the stems and leaves of a plant (as opposed to reproductive growth, which includes flowers, seeds, and fruits).

Vegetative reproduction See Asexual reproduction.

Vernalization Chilling plants or seeds to cause flowering or end dormancy. Winter wheat must be vernalized (exposed to several weeks or more of subfreezing temperatures) before plants will form floral parts.

Viable seed Seed that is alive, that is, able to germinate and grow.

Virus An infectious body too small to be seen without the aid of a powerful microscope. Viruses can multiply and act like living things when they are in living tissue.

Vitamins A variety of different organic substances necessary for life in animals but which are not respired to produce energy.

Viviparous insects A group of insects that reproduce in the same manner as humans and other mammals.

Wastage Loss of foodstuff or its nutritional value as a result of a variety of causes, many of which can be prevented or minimized. (See also Contamination and Spoilage.)

Water infiltration The maximum rate at which water will move into and through a soil.

Weather The current meteorological conditions, usually emphasizing temperature and precipitation.

Weathering The cumulative effects of precipitation (rain and snow), temperature, and their physical and chemical interactions that alter materials such as rock to yield a soil. Effects are usually measured over long periods of time.

Weed A plant out of place. Its location makes it more harmful than beneficial.

Windrow A row of unthreshed grain left in the field to dry before harvesting.

Winter annual A plant that germinates in the fall, overwinters, matures, sets seed, and dies in the spring or early summer.

Winter-hardiness The ability of plants to tolerate winter conditions.

Withdrawal time The time required by law to stop using a chemical prior to harvesting a crop. Different chemicals have different withdrawal times.

Zygote The immediate product of the union of egg and sperm. The zygote grows and develops into the miniature plant, or embryo, in a seed.

Index